SOUTH-WESTERN

GEOMETRY
FOR
DECISION
MAKING

JAMES E. ELANDER
North Central College
Naperville, Illinois

MG01AA
PUBLISHED BY
SOUTH-WESTERN PUBLISHING CO.
CINCINNATI, OH   DALLAS, TX   LIVERMORE, CA

Senior Acquisitions Editor: Eve Lewis
Production Editor: Joseph P. Powell III
Editorial Production Manager: Carol Sturzenberger
Designers: Barbara Libby and Darren K. Wright
Marketing Manager: Gregory Getter

Program Consultants:

Richard Sgroi
Assistant Professor
Mathematics Department
State University of New York
New Paltz, New York

Edgal Bradley
Lloyd High School
Erlanger, Kentucky

Editorial Advisers:

Mike Coit
Lisle High School
Lisle, Illinois

Tamilea Daniel
Educational Consultant
Mandeville, Louisiana

Tommy Eads
North Lamar High School
Paris, Texas

Tony Martinez
Leander High School
Leander, Texas

Cynthia Nahrgang
The Blake School
Hopkins, Minnesota

Emily Perlman
Educational Consultant
Houston, Texas

ISBN: 0-538-60293-7

Library of Congress Catalog Card Number: 90-61045

3 4 5 6 7 8 9 0 **AG** 0 9 8 7 6 5 4 3

Printed in the United States of America

CONTENTS

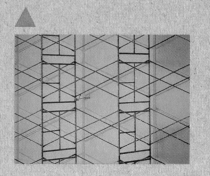

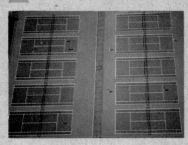

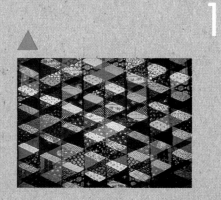

PREFACE

Geometry is the mathematical study of space. But your interpretation of space depends on your focus. Look at an object through different kinds of lenses and the image varies. Investigate space with different sets of postulates or premises and different worlds result. Euclidean geometry is just one approach. Others exist, but the Euclidean view seems more realistic in terms of everyday experience. Nevertheless, even ancient navigators had to bend a few of Euclid's rules when traveling long distances on the surface of a sphere they called Earth.

GEOMETRY

Whether you live in a large city or a remote, rural or underdeveloped area, you are surrounded by a vast number of examples or models of geometric concepts. Some are the results of human toil, others occur naturally. You can find geometry in the construction of complex highway systems and in the markings of a fragile butterfly. You need only look.

Geometry is all around you.

. . . In Theater

The Lyric Opera of Chicago premier of Gluck's timeless masterpiece, *Alceste*, starring Jessye Norman in the title role and Chris Merritt as Admète opened Lyric's 1990-91 season. Robert Wilson created this stunning new production.

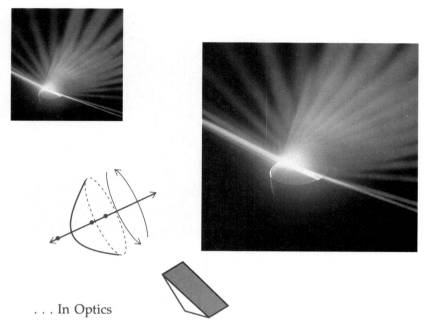

. . . In Optics

Lasers, devices that produce a very narrow beam of concentrated light, and fiber optics, a technology that uses specially developed bundles of transparent fibers to transmit light, have a wide range of applications including medicine and communications.

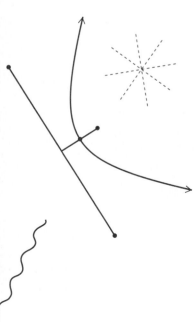

. . . In Chemistry

Mineralogists often use the geometric shape of crystals as one means of identification. Jewelers then cut some crystals geometrically to enhance their beauty. The abstraction below was created naturally. It depicts frost forming on a window pane.

. . . In Ecology

The experimental solar energy collector in New Mexico and wind turbines shown at sunset display a myriad of geometric shapes. Ecologists throughout the world are experimenting with a variety of techniques to produce energy in a way that avoids contamination of the environment. These endeavors and an educated public will preserve our world for future generations.

. . . In Architecture

This view of the East Wing of the National Gallery in Washington, D.C., shows how architects make use of geometric shapes to enhance the beauty and utility of their structures.

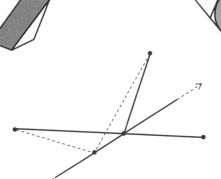

. . . In Nature

The mystery of the honeycomb still defies us. Bees produce these hexagonally shaped cells to store honey and reproduce their species.

. . . In Agriculture

Terracing was used by the early Babylonians to prevent soil erosion. The method is still practiced today in many areas of the world.

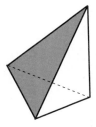

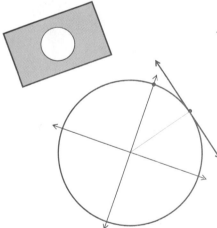

. . . In Sports

This photograph of the opening ceremony of the 1988 Summer Olympics in Seoul, Korea, illustrates how photographers position geometric shapes in a manner that creates a dramatic effect. But in sports many facets of geometry emerge. Measurement of distance and time is fundamental to Olympic events. The study of forces and vectors enable contestants to improve their techniques. Geometry is everywhere for all to see. You need only look.

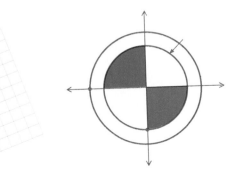

Letter to the Student

You are about to embark on a voyage through space. In your travels you will encounter objects so small that they have only one characteristic, position, and distances so large that to become meaningful they must be translated into time references, light years. The "navigational" tools you will need are a scientific calculator, a compass, a protractor, a ruler, and a notebook (log).

As a space traveler it would be wise to keep a log of the important facts you observe and any discoveries you might make. Each discovery can and should lay the groundwork for further discovery. Frequent review of your log will strengthen your hold on the material and make using it much easier.

When you encounter new objects or concepts in your travels, try to relate the new to the old. Relating new facts or relationships to familiar ideas will help you internalize the new and will make things easier to remember.

By the end of your journey you will have collected a vast amount of information, but it will be organized in a way that will stress relationships. A good understanding of these relationships will enable you to appreciate more fully the space around you, to discover additional relationships that exist in your world, and to use those relationships to solve real problems.

We hope that you will enjoy this trip through space and that you might find it sufficiently interesting to consider taking other excursions. Perhaps in the future you might consider trips to different kinds of space, where the few who dare visit are rewarded handsomely. They return home from alien lands with an appreciation for these new worlds, often with a greater understanding of their own world, but more importantly with a greater appreciation of the world around them.

James E. Elander
North Central College
Naperville, Illinois

James Elander, assistant professor of mathematics at North Central College, Naperville, Illinois, is the former mathematics department chair of Oak Park and River Forest High School, Oak Park, Illinois. He was named winner of the 1988 Burlington Northern Foundation Faculty Achievement Award for meritorious achievement in professional scholarship.

Mr. Elander served as chairman of the Illinois Section Mathematical Association of America Geometry Committee and served that organization's High School Lecture Program. He has helped organize programs for the National Council of Teachers of Mathematics, Illinois Council of Teachers of Mathematics, and the School Science and Mathematics Association.

WHAT IS GEOMETRY?

Mathematics is the gate and key to all sciences . . . he who is ignorant of it cannot know the other sciences or the things of this world.

Roger Bacon

Points, Lines, and Planes

1 2
3 4

5 6
7 8

After completing this section, you should understand
▲ the basic elements of geometry—points, lines, line segments, rays, and planes.

When you first begin to play a board game, you open the box, check the contents, and read the instructions. Then you discuss the rules, identify the players, and begin to play.

As a result of reading the instructions and beginning to play, you soon acquire the game's vocabulary and are able to communicate with all the other players. Communication is also very important in geometry. You need to learn its rules, definitions, and terminology.

To study geometry, you need to understand three basic terms—point, line, and plane. They are important because they are used to define other geometric terms.

PLANES, POINTS, AND LINES

A flat surface, such as a table top, suggests the idea of a **plane.** A table top is limited in size, but a geometric plane extends endlessly in all directions.

The flat surfaces pictured above suggest geometric planes. Unlike the flat surfaces in the photos, a plane has no boundaries.

2

The following drawings suggest how part of a plane, flat surface might look from different points of view.

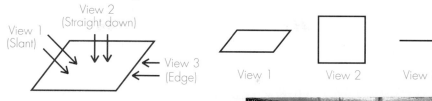

Planes are made up of points. The picture on a television screen consists of many small dots packed closely. You can think of each dot as representing a point. An individual point is only a location, but you can represent it by a dot.

In the diagram, *A* and *B* are two points in plane *P*. Passing through *A* and *B* is a line called **line *AB*.** The line is straight and continues endlessly.

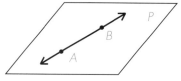

Like planes, lines are made up entirely of points. Lines and planes contain infinitely many points. Both of the figures below contain a infinite number of points. However, the figure to the right is not a line, because it is not straight.

This is a line.　　　　This is not a line.

DISCOVERY ACTIVITY

On a sheet of paper, draw a plane. On the plane, mark and label two points *S* and *T*, as shown.

1. How many lines can you draw through point *S*? through point *T*?
2. How many lines can you draw that contain *both* points?
3. On the plane, mark and label a third point *R*, as shown. How many lines can you draw that contain all three points?

You probably found that you were able to draw one and only one line through points *S* and *T*. Also, you probably discovered that no line can be drawn that passes through all three points.

In geometry, a statement that is assumed to be true in all cases is called a **postulate**.

POSTULATE 1-1-1:   **Two points determine exactly one line.**

According to Postulate 1-1-1, if *A* and *B* are any two points, there is one and only one line that goes through both. The written name for this line is $\overleftrightarrow{AB}$ or $\overleftrightarrow{BA}$, read "line *AB*" and "line *BA*", respectively.

LINE SEGMENTS, RAYS, AND ANGLES

Line segments and **rays** are parts of a line. This diagram shows the line segment connecting points *A* and *B*.

DEFINITION 1-1-1:   *A* and *B* are the **endpoints** of line segment *AB*. You can name the line segment in symbols by writing $\overline{AB}$ or $\overline{BA}$.

A LINE SEGMENT is a subset of a line, consisting of two points *A* **and *B* on the line and all points between *A* and *B*.**
How would you name this line segment?

This is ray *AB*. Point *A* is called the **endpoint** of ray *AB*.

Ray *AB* starts at point *A* and continues indefinitely in the direction of point *B*. The symbol for ray *AB* is $\overrightarrow{AB}$. When naming a ray, always name the endpoint first.

DEFINITION 1-1-2:   **A RAY is a subset of a line consisting of a point *A* on the line and all points of the line that lie to one side of point *A*.**

How would you name this ray?

Two rays that have the same endpoint form a special figure called an **angle.**

NOTEBOOK

DEFINITION 1-1-3:   **An ANGLE is a figure formed by two rays that have a common endpoint.**

The two rays are called the **sides** of the angle. Their common endpoint is called the **vertex** of the angle. The diagram shows the angle formed from $\overrightarrow{BA}$ and $\overrightarrow{BC}$. The angle

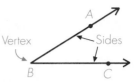

is called angle ABC or angle CBA—in symbols, $\angle ABC$ or $\angle CBA$. Notice that when you name an angle, you name the vertex in the middle. The other two letters name points on the two sides of the angle. If there is only one angle with a given vertex, you can simply write the angle symbol and the letter for the vertex.

Example: The angle in the diagram can be called $\angle LMN$, or $\angle NML$, or $\angle M$. The sides are $\overrightarrow{ML}$ and $\overrightarrow{MN}$. The vertex is point M.

How would you name this angle?

Sometimes a number is used to name an angle.

This is angle 1.   These are angles 2 through 5.

CLASS ACTIVITY

Name each of the following. Use symbols.
1. • P
2. A ———— B
3. ←——— F G
4. ←—•——•—→ M N
5. K ——•——•—— L
6. P ——•——•—— O
7. A / H / D
8. T R S
9. F G H

Draw and label each of the following.
10. line PQ
11. ray WR
12. angle BMT

13. line segment FG 14. angle TRG
15. ray ZX

Draw, label, and name each of the following.
16. $\overleftrightarrow{AB}$ containing a third point C
17. $\overrightarrow{XY}$ and $\overrightarrow{YX}$

18. $\overrightarrow{GF}$, point K not on $\overrightarrow{GF}$, and $\overrightarrow{KG}$
19. $\angle HIJ$ and $\overrightarrow{HJ}$

PROJECT 1-1-1 People in many careers use geometry daily in their work. Highway engineers use geometry when surveying. Choreographers use it to plan dance movements on a stage. Graphic artists, architects, package designers, and many others have frequent opportunities to put geometry to use on the job.

Choose a profession or career that interests you. Find someone who is working in that field and talk to them. Find out things such as the following:
- what a typical work day is like.
- what education and training are required.
- how geometry or other kinds of mathematics are used in the work.

Take notes during your discussion. Share your findings with the class.

HOME ACTIVITY

Name each of the following. Use symbols.

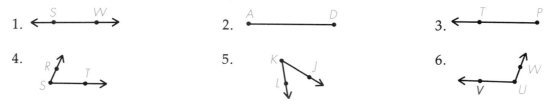

1. S W

2. A D

3. T P

4. R S T

5. K J L

6. W V U

Which of the geometric figures in this section do the following represent?

7. the surface of a still lake
8. a pencil
9. a beam of sunlight
10. clock hands showing 7:45

Use the diagram. Name the following.

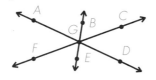

11. 2 rays

12. 2 lines

13. 2 segments

CRITICAL THINKING

14. Draw diagrams to show that two angles can have exactly one, two, three, or four points in common. Is it possible for two angles to have exactly five points in common?

SECTION 1.2

Measurement of Segments and Angles

After completing this section, you should understand
▲ how numbers are matched with points on a number line.
▲ how to measure line segments and angles.
▲ how to classify angles.

The number markings on a football field determine the exact placement of movements on the field between the two end zones. Football fields are marked in 1-yard, 5-yard, and 10-yard segments. The officials measure to see whether one team has advanced the ball far enough to gain a first down.

Some of the ideas that the football officials are using are also used in geometry.

THE NUMBER LINE

You may recall from algebra that the set of all positive and negative numbers, together with zero, make up the set of **real numbers.** You may also recall that the set of real numbers can be matched one-to-one with the points on a line to obtain a **number line.** The number matched with a point is called the **coordinate** of the point.

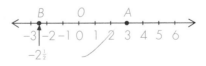

For the number line shown above, the coordinate of point *A* is 3 and the coordinate of point *B* is $-2\frac{1}{2}$. The point *O* whose coordinate is 0 is called the **origin.**

POSTULATE 1-2-1: **There is a one-to-one matching between the points on a line and the real numbers. The real number assigned to each point is its coordinate. The distance between two points is the positive difference of their coordinates. If A and B are two points with coordinates a and b such that $a > b$, then distance $AB = a - b$.**

Notice that $\overline{AB}$ (with a bar above the letters) means the line segment joining A and B. The symbol AB (no bar) means the distance between A and B, that is, the *length* of $\overline{AB}$.

Example: On the number line, -3 is the coordinate of point A, 4 is the coordinate of point B, and 8 is the coordinate of point C. The length of $\overline{AB}$ is $AB = 4 - (-3)$, or 7.

What is BC? What is AC?

MEASURING LINE SEGMENTS

When you use a ruler to measure a line segment, you are using the ruler as part of a number line. The number you assign will depend on which measurement you are using. The number you get for the length depends on whether the unit you are using is feet, inches, centimeters, meters, and so on.

Example: What is the length of $\overline{AB}$ in inches?

Place your inch ruler so that the mark that corresponds to 0 is at point A. Read the number at the other end of the segment.

The segment measures $3\frac{1}{8}$ inches.
What is its measure in centimeters?

CLASS ACTIVITY

Use the number line to help you answer the questions.

1. What are the coordinates of points A, B, C, and D?

2. What is the length of $\overline{AB}$? of $\overline{BD}$?

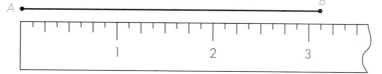

Use the number line below to help you answer the questions.

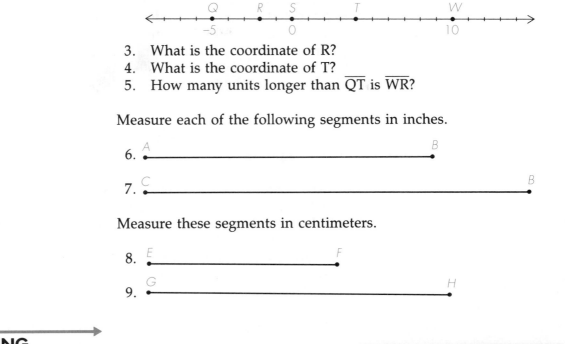

3. What is the coordinate of R?
4. What is the coordinate of T?
5. How many units longer than $\overline{QT}$ is $\overline{WR}$?

Measure each of the following segments in inches.

6. A ————————————————————————— B

7. C ———————————————————————————— B

Measure these segments in centimeters.

8. E ————————————————— F

9. G ——————————————————————— H

MEASURING ANGLES

Postulate 1-2-1 assures you that each line segment has one particular number as its length. The next postulate accomplishes something similar for angles.

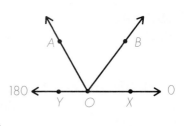

POSTULATE 1-2-2: **Let O be a point on $\overleftrightarrow{XY}$ such that X is on one side of O and Y is on the other side of O. Real numbers from 0 through 180 can be matched with $\overrightarrow{OX}$, $\overrightarrow{OY}$, and all the rays that lie on one side of $\overleftrightarrow{XY}$ so that each of the following is true:**

(1) 0 is the number matched with $\overrightarrow{OX}$.

(2) 180 is the number matched with $\overrightarrow{OY}$.

(3) If $\overrightarrow{OA}$ is matched with a and $\overrightarrow{OB}$ is matched with b and $a > b$, then the number matched with $\angle AOB$ is $a - b$.

The number matched with $\angle AOB$ is called the **measure** or the **degree measure** of $\angle AOB$. In symbols, you can write $m\angle AOB$. For the angle shown in the diagram for the postulate, $m\angle AOB = a - b$.

To measure angles, you use a **protractor**. The diagram shows a typical protractor. There is an outside scale and an inside scale, but both use numbers from 0 to 180. The unit of measure is shown with a small, raised °. $m\angle AOB = 115° - 50° = 65°$.

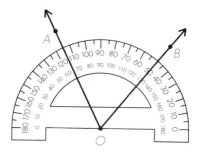

Follow these steps to measure an angle with a protractor:

1. Place the protractor over the angle so that the dot at the bottom center of the protractor is exactly over the vertex.
2. Make sure one side of the angle crosses the inside or outside scale at the 0-mark.
3. Follow around along that scale until you get to the place where the other side crosses the scale. The number for that place gives the degree measure of the angle.

Example:

1. Measure $\angle ABC$.

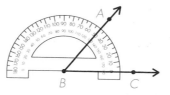

$\overrightarrow{BC}$ is at the 0-mark on the *outside* scale. So find the number where $\overrightarrow{BA}$ crosses the outside scale: $m\angle ABC = 50°$.

2. Measure $\angle DEF$.

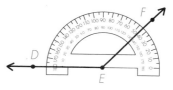

$\overrightarrow{ED}$ is at the 0-mark on the *inside* scale. So find the number where $\overrightarrow{EF}$ crosses the inside scale: $m\angle DEF = 135°$.

CLASSIFYING ANGLES

As you can see, angles can vary widely in size. Angles are classified according to the number of degrees they contain.

NOTEBOOK

DEFINITION 1-2-1: **If the number assigned to an angle is between 0° and 90°, then the angle is called an ACUTE angle.**

Example: You can trace $\angle ABC$ on a sheet of paper and check that $m\angle ABC = 35°$. Since 35° is between 0° and 90°, $\angle ABC$ is acute.

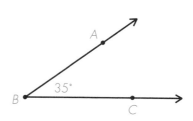

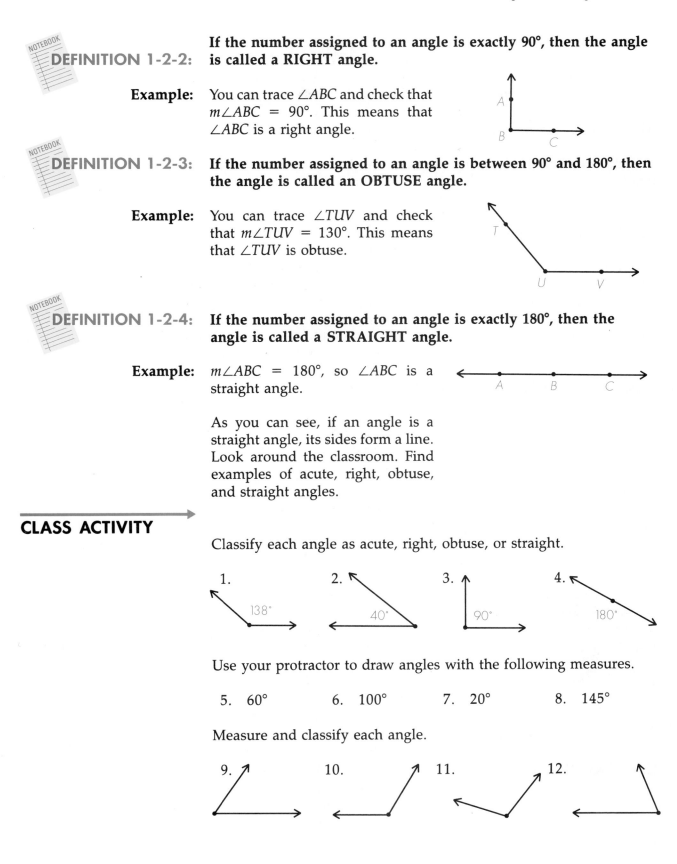

DEFINITION 1-2-2: If the number assigned to an angle is exactly 90°, then the angle is called a RIGHT angle.

Example: You can trace ∠ABC and check that m∠ABC = 90°. This means that ∠ABC is a right angle.

DEFINITION 1-2-3: If the number assigned to an angle is between 90° and 180°, then the angle is called an OBTUSE angle.

Example: You can trace ∠TUV and check that m∠TUV = 130°. This means that ∠TUV is obtuse.

DEFINITION 1-2-4: If the number assigned to an angle is exactly 180°, then the angle is called a STRAIGHT angle.

Example: m∠ABC = 180°, so ∠ABC is a straight angle.

As you can see, if an angle is a straight angle, its sides form a line. Look around the classroom. Find examples of acute, right, obtuse, and straight angles.

CLASS ACTIVITY

Classify each angle as acute, right, obtuse, or straight.

1. 138°   2. 40°   3. 90°   4. 180°

Use your protractor to draw angles with the following measures.

5. 60°   6. 100°   7. 20°   8. 145°

Measure and classify each angle.

9.   10.   11.   12.

HOME ACTIVITY

Use the number line to help you answer the questions.

1. What are the coordinates of points A, B, C, and D?
2. What is the length of $\overline{AB}$? of $\overline{BD}$?
3. What is the distance of each of the labeled points from the origin, zero?

Use the number line to find each of the following.

4. the coordinate of V.
5. the coordinate of P
6. the length of $\overline{VR}$
7. the length of $\overline{RP}$

Classify each angle as acute, right, obtuse, or straight.

8. 180°

9. 150°

10. 76°

Measure and classify each angle.

11.

12.

List all the angles in each figure. Then measure and classify them.

13.

14.

CRITICAL THINKING

Draw a figure consisting of three rays, $\overrightarrow{OA}$, $\overrightarrow{OB}$, and $\overrightarrow{OC}$, such that any pair of rays forms an angle that measures 120°.

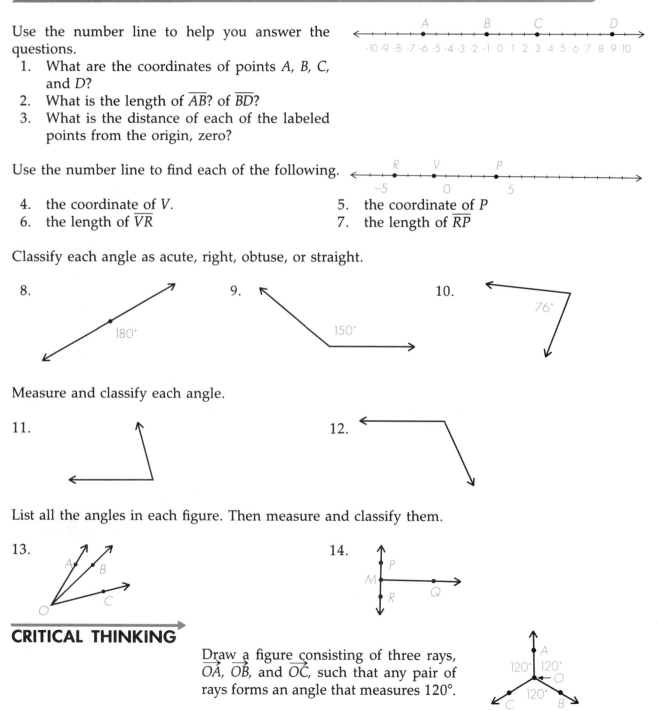

Midpoints of Segments and Angle Bisectors

After completing this section, you should understand
▲ how to find the midpoint of a line segment.
▲ how to find the bisector of an angle.
▲ the meaning of betweenness.

In a tug-of-war contest, a length of rope is marked in the middle. Players position themselves on opposite sides of this middle point and at equal distances from it.

The teams stand on opposite sides of a line on the ground. They pull the rope tight and situate the middle point straight above the line on the ground. Then, at the signal; the tugging begins.

MIDPOINT OF A LINE SEGMENT

In the tug–of–war contest, the rope is like a line segment. The middle point of the rope corresponds to the **midpoint** of the line segment.

DEFINITION 1-3-1: **The MIDPOINT of a line segment is the point which divides the segment into two equal segments.** C **is the midpoint of** $\overline{AB}$ **if** $AC = BC = \frac{1}{2}AB.$

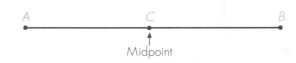

A   C   B

Midpoint

Since there is only one point that divides a given line segment equally, each line segment has one and only one midpoint.

Examples:

1. Suppose the length of $\overline{AB}$ is 4 in. and that C is the midpoint of $\overline{AB}$. Then $AC = BC = \frac{1}{2} \times 4 = 2$ in.

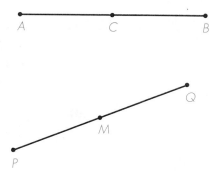

2. Suppose that $PQ = 13$ cm, $PM = 6.5$ cm, and $QM = 6.5$ cm. You can conclude that M is the midpoint of $\overline{PQ}$, since both PM and QM are equal to $\frac{1}{2}PQ$.

If point C is the midpoint of $\overline{AB}$ and $AB = 8.5$ cm, what is the length of $\overline{AC}$? of $\overline{CB}$?

CLASS ACTIVITY

Use your ruler to find the length of each segment in inches. Tell how far the midpoint of each segment is from the endpoints.

1. M ●────────────────────────● N

2. R ●─────────────────────────────────● S

Measure each segment in centimeters. Tell how far from the left-hand endpoint the midpoint of the segment will be.

3. J ●──────────────────────────────● K

4. O ●───────────────────● P

5. If point R is the midpoint of $\overline{QS}$, and $RS = 4$ in., how long is $\overline{QS}$? How long is $\overline{QR}$?

In the figure, $AB = 2$ cm, $BC = 1$ cm, and $AD = 6$ cm.

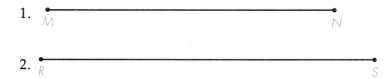

6. How long is $\overline{CD}$?

7. Which point is the midpoint of $\overline{AD}$?

8. Draw a line segment $\overline{XY}$ that is $3\frac{1}{2}$ in. long. Then mark the midpoint M of the segment you drew.

ANGLE BISECTORS

The midpoint of a line segment divides the line segment into two equal segments. An **angle bisector** divides an angle into two equal angles.

DEFINITION 1-3-2: An **ANGLE BISECTOR** of a given angle is a ray with the same vertex that separates the given angle into two angles of equal measure. $\overrightarrow{BD}$ is the angle bisector of $\angle ABC$ if $m\angle ABD = m\angle DBC = \frac{1}{2}m\angle ABC$.

Example: Suppose that $\angle BAC = 50°$ and that $\overrightarrow{AD}$ bisects $\angle BAC$. You can conclude that $m\angle BAD = m\angle DAC = \frac{1}{2} \times 50° = 25°$.

In the figure, $\angle RST$ has a measure of 110° and $\overrightarrow{SP}$ bisects $\angle RST$. What is the measure of $\angle PST$?

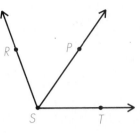

CLASS ACTIVITY

In each figure, $\overrightarrow{RS}$ is an angle bisector.

1.

$m\angle QRS = 60°$
$m\angle QRT = \underline{\qquad}$

2.

$m\angle PRS = 35°$
$m\angle SRQ = \underline{\qquad}$

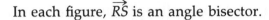

Use your protractor. Measure each angle and draw an angle of the same size. Classify each as acute, right, obtuse, or straight. With the aid of your protractor, draw the bisector of each angle. Give the measures of the two new angles you create.

3.

4.

5.

6.

7.

8.

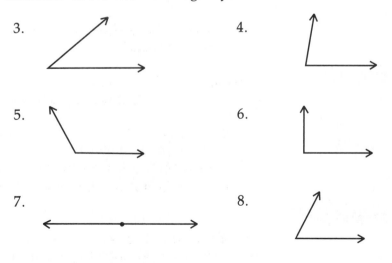

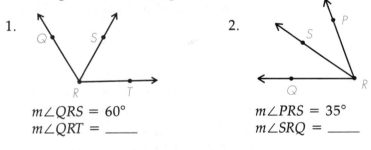

BETWEENNESS

Route 36 crosses the entire state of Kansas. Baileyville is between Fairview and Marysville along that route.

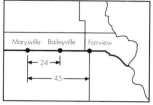

Route 36 goes in an east-west direction, so Baileyville is to the west of Fairview *and* to the east of Marysville. The distance from Fairview to Marysville is 45 miles. It is 24 miles between Baileyville and Marysville. How far is it from Baileyville to Fairview?

On the number line, if you know the coordinates of three points, it is easy to tell which point is between the other two.

DEFINITION 1-3-3: **On a number line, point C is BETWEEN points A and B if the coordinates of A, B, and C (a, b, and c) meet the condition that $a < c < b$ or $a > c > b$.**

If three points are on the same line and the length of each segment is known, you can tell which point is between the other two. Point L is between K and M if $KL + LM = KM$.

Examples: In the figure, the coordinates of P, Q, and R are $2\frac{1}{2}$, 5, and 6 respectively. Since $2\frac{1}{2} < 5 < 6$, point Q is between P and R.

CLASS ACTIVITY

Points P, J, and K are on a number line. Their coordinates are 2, 8, and 3, respectively.

1. Draw a figure to illustrate this.
2. Which point is between the other two?
3. Complete the following inequality relating the coordinates of the three points: $2 < \underline{\hspace{1cm}} < \underline{\hspace{1cm}}$

Find each of the following lengths.

4. $PK = \underline{\hspace{1cm}}$ 5. $KJ = \underline{\hspace{1cm}}$ 6. $PJ = \underline{\hspace{1cm}}$
7. The sum of the lengths of which two line segments equals the length of the third segment?

DISCOVERY ACTIVITY

You have learned under what conditions one point on a line is between two others. Under what conditions is a ray between two other rays?

The diagram shows $\overrightarrow{DA}$, $\overrightarrow{DB}$ and $\overrightarrow{DC}$.
Measure ∠BDA with your protractor.
Then measure ∠BDC and ∠CDA.

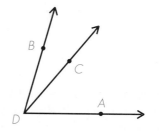

Is m∠BDC < m∠BDA? Is m∠CDA < m∠BDA?
What is the sum of the measures of ∠BDC and ∠CDA?

Would you say that $\overrightarrow{DC}$ is between rays $\overrightarrow{DB}$ and $\overrightarrow{DA}$? Explain why.

In the Discovery Activity, $\overrightarrow{DC}$ can be thought of as between $\overrightarrow{DA}$ and $\overrightarrow{DB}$. The rays have a common vertex, and the sum of the measures of two of the angles they form is equal to the measure of the third angle.

CLASS ACTIVITY

1. Use your protractor to help you make an accurate copy of ∠RST. Then draw a ray $\overrightarrow{SQ}$ between $\overrightarrow{SR}$ and $\overrightarrow{ST}$ such that m∠QST = 50°. What is m∠QSR?

   R
   90° T
   S

2. Use your protractor to help you make an accurate copy of ∠DEF. Then draw a ray $\overrightarrow{EG}$ between $\overrightarrow{EF}$ and $\overrightarrow{ED}$ such that m∠GED = 70°. What is m∠GEF?

   D
   85° F
   E

HOME ACTIVITY

In the figure, B is between A and D, and C is between B and D. Suppose AD = 12, AB = 3, and BC = 5.

A   B   C   D

1. CD = _____

2. Is C the midpoint of $\overline{AD}$?

In the figure, S is the midpoint of $\overline{QT}$, R is the midpoint of $\overline{QS}$, and QS = 9.

Q   R   S   T

3. QT = ____

4. QR = ____

Measure in inches and make an accurate copy of each segment. Mark and label the midpoint M and tell how far it is from the endpoints.

5. A ———————————— M ———————————— B

6. P ———————————— M ———————————— Q

Use your protractor. Measure each angle and draw an angle of the same size. Classify each as acute, right, obtuse, or straight. With the aid of your protractor, draw the bisector of each angle. Give the measures of the two new angles you create.

7.

A

B   C

8.

X

Y   Z

In the figure, $\overrightarrow{BD}$ bisects $\angle ABC$.

9. Name a right angle.
10. Name the bisector of $\angle BCD$.
11. What is the measure of $\angle CBD$?

Use your protractor to make an accurate copy of the given angle. Then draw a ray *between* the sides of the given angle to make an angle as indicated below the figure. Give the measure of the other new angle that is formed.

12.

F

45°   H

G

Draw $\overrightarrow{GK}$ to make
$m\angle KGH = 20°$.

13.

C   120°   A

P

Draw $\overrightarrow{PK}$ to make
$m\angle CPK = 100°$.

CRITICAL THINKING

Milt walks the same route to and from school each day. He always passes the fruit market, the post office, the park, the video store, and the cleaners, but not in that order. Going to school he passes the park before the video store, but after the post office. He passes the cleaners first. Going home, he passes the fruit market second. List the five places Milt passes as he walks to school. List them in the order in which he passes them.

SECTION 1.4

Supplements and Complements

After completing this section, you should understand
▲ the meaning of supplementary angles.
▲ the meaning of complementary angles.

Since angle measures are numbers, they can be added and subtracted. In this section you will learn about pairs of angles whose measures have sums of 90° or 180°.

DISCOVERY ACTIVITY

Draw $\overleftrightarrow{AB}$ on a sheet of paper.
Choose a point along the segment between A and B and label it C.
Next, draw and label $\overrightarrow{CF}$ where F is not on segment AB.
Using your protractor, measure $\angle ACF$ and $\angle BCF$.
Add the measures of the two angles. What do you notice?
Did the result surprise you? Explain.

You probably concluded that the two angles in the Discovery Activity have a sum of 180°, since $\angle ACB$ is a straight angle.

DEFINITION 1-4-1: **Two angles are SUPPLEMENTARY if their measures add to 180°.**

If two angles are supplementary, then each is called a **supplement** of the other.

Two angles do not have to have a common side in order to be supplementary.

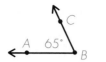

∠*ABC* and ∠*DEF* are supplementary because 65° + 115° = 180°. Knowing that two angles are supplementary is often helpful in solving problems.

Example: In the diagram, ∠*PQR* and ∠*RQS* are supplementary. Find the number of degrees in each.

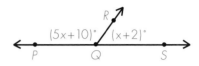

You can write an equation to express the fact that the angles are supplementary. Solve for *x* and then find the angle measures.

Use the given information. Remove parentheses. Combine the like terms. Subtract 12 from both sides. Divide both sides by 6.

$$(5x + 10) + (x + 2) = 180$$
$$5x + 10 + x + 2 = 180$$
$$6x + 12 = 180$$
$$6x = 168$$
$$x = 28$$

Substitute the value 28 for *x* in the expressions for the angle measures.
$(5x + 10)° = 150°$; $(x + 2)° = 30°$
$5x + 10 = 5 × 28 + 10 = 140 + 10 = 150$ and $x + 2 = 28 + 2 = 30$
The measures of the angles are 150° and 30°.

CLASS ACTIVITY

1. Identify three examples of supplementary angles in your classroom.
2. If $m∠A = 55°$, what is the supplement of ∠*A*?
3. What is the measure of a supplement of an angle of 145°? of an angle of 95°?
4. Complete the table to find the measure of a supplement of an angle having the given measure.

Angle Measure	Equation	Supplement
30°	30 + s = 180	150°
45°	_____	_____
80°	_____	_____
90°	_____	_____
135°	_____	_____
149°	_____	_____
$x°$	x + s = 180	$(180 - x)°$
$(x - 40)°$	_____	_____
$(2x + 20)°$	_____	_____

5. In the diagram, $\overleftrightarrow{CG}$ and $\overleftrightarrow{DF}$ intersect at E and m$\angle CED = 48°$. Find the measures of every other angle.

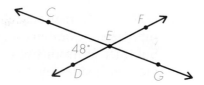

In Exercises 6 and 7, $\angle PIO$ and $\angle VEG$ are straight angles. Find the measures of the supplementary angles.

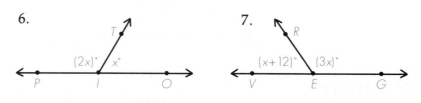

6.

7.

DISCOVERY ACTIVITY

Draw a right angle on a sheet of paper. Label it $\angle ACB$. From the vertex, C, draw a ray $\overrightarrow{CD}$ between $\overrightarrow{CA}$ and $\overrightarrow{CB}$. Use your protractor to measure $\angle ACD$ and $\angle DCB$. Add the two measures together. What do you notice?

In the Discovery Activity, you probably found that the sum of the measures of the two new angles is 90°.

NOTEBOOK

DEFINITION 1-4-2: **Two angles are COMPLEMENTARY if their measures add to 90°.**

If two angles are complementary, then each is called a **complement** of the other. Are these two angles complementary?

$\angle B$ and $\angle C$ are complementary, because $47° + 43° = 90°$.

Example: In the diagram, $\angle APB$ and $\angle BPC$ are complementary. Find the number of degrees in each angle.

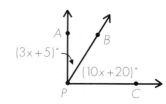

You can write an equation expressing the fact that the angles are complementary. Solve the equation and then find the angle measures.

Use the information. Remove parentheses. Combine like terms. Subtract 25 from both sides. Divide both sides by 13.

$$(3x + 5)° + (10x + 20)° = 90°$$
$$3x + 5 + 10x + 20 = 90$$
$$13x + 25 = 90$$
$$13x = 65$$
$$x = 5$$

Substitute the value 5 for x in the original expressions.
$$3x + 5 = 3 \times 5 + 5 = 15 + 5 = 20$$
$$10x + 20 = 10 \times 5 + 20 = 50 + 20 = 70$$
The measures of the angles are 20° and 70°.

CLASS ACTIVITY

1. If $m\angle A = 35°$, what is the measure of a complement of $\angle A$?
2. What is the measure of a complement of an angle of 25°? of an angle of 75°?
3. Complete the table to find the measure of a complement of an angle having the given measure.

Angle Measure	Equation	Complement
30°	$30 + c = 90$	60°
45°		
10°		
80°		
85°		
14°		
29°		
$x°$	$x + c = 90$	
$(x - 40)°$		
$(2x + 20)°$		

In Exercises 4 and 5, $\angle BAZ$ and $\angle SRV$ are right angles. Find the measures of the complementary angles.

4.

5.

In each of the following, assume that the two angles are complementary. Find the measures of both angles using the given information.

6. $m\angle T$ is twice $m\angle R$

7. $m\angle R$ is 10° more than $m\angle T$

8. $m\angle R$ is 40 less than $m\angle T$

9. $m\angle T$ is 4 times $m\angle R$

HOME ACTIVITY

In the figure at the right, $m\angle GOA = m\angle GOE = 90°$ and $\angle FOB$ is a straight angle. $\overrightarrow{OE}$ bisects $\angle FOC$. Name the angles described.

1. a pair of complementary angles
2. three acute angles
3. two pairs of supplementary angles
4. two obtuse angles
5. a straight angle other than $\angle FOB$
6. a right angle

Find the measure of an angle supplementary to an angle having the given measure.

7. 30°   8. 87°   9. 161.4°   10. $x°$

Find the measure of an angle complementary to an angle having the given measure.
11. 40°   12. 77°   13. 6.05°   14. $x°$

Tell which words you should use in the blanks—*acute, obtuse, right,* or *straight*—to make true statements.

15. A complement of an acute angle is _____.
16. A supplement of a right angle is _____.
17. A supplement of an acute angle is _____.
18. _____ angles do not have supplements.
19. _____ angles have acute supplements.
20. Any pair of _____ angles are supplementary.

In each of the following figures, the indicated pair of angles are either complementary or supplementary. Find the degree measure of each angle.

21.

22.

23.

24.

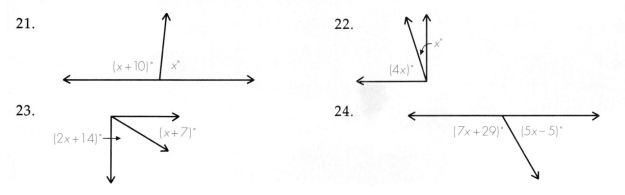

Find the value of x in each of the following. Assume that the lines that look like straight lines are straight lines.

25.

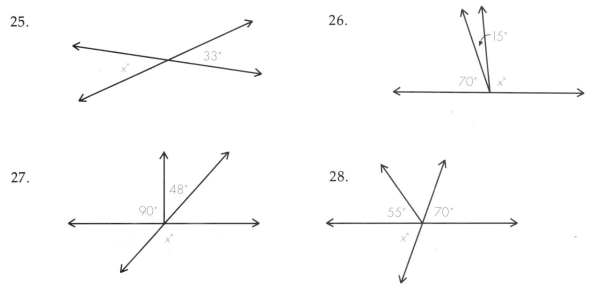

26.

27.

28.

Draw and label a diagram to show each of the following.

29. $\overrightarrow{CR}$ is between $\overrightarrow{CB}$ and $\overrightarrow{CA}$ in right angle ACB, and m$\angle ACR = 5$ m$\angle BCR$.

30. $\overrightarrow{BD}$ bisects $\angle ABC$ and $m\angle ABC = 90°$.

CRITICAL THINKING

The team equipment manager was counting the number of baseballs the team had. She noticed that if she counted the balls two at a time, there was 1 left over. If she counted them three at a time or four at a time there was also 1 left over. If she counted them five at a time, there were none left over. How many baseballs does the team have if they have between 75 and 100 baseballs?

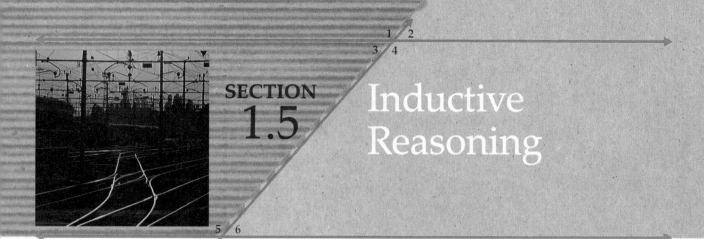

Inductive Reasoning

After completing this section, you should understand
▲ the meaning of inductive reasoning
▲ the pitfalls involved in inductive reasoning

Suppose it has rained everyday for the past week. On the basis of that information, is it reasonable to predict that it will rain tomorrow?

In your study of geometry, you will be discovering things that are true for all geometric figures of a certain kind. Also you will look for ways to justify your belief that the things you discover are *always* true.

In your attempts to discover things about geometric figures, you will often use what is known as **inductive reasoning.**

WHAT IS INDUCTIVE REASONING?

Inductive reasoning is the process of using a limited number of cases to arrive at a conclusion that will hopefully be true in *all* cases, including the cases you have not yet examined.

DISCOVERY ACTIVITY

How many squares of all sizes are there in the 5 × 5 square shown at the right?

Begin by looking for a pattern. Think of a systematic way to find all of the 1 × 1 squares, all of the 2 × 2 squares, all of the 3 × 3 squares, and so on.

What pattern did you notice?

You may have observed that a pattern involving sums of square numbers could be used to answer the question in the Discovery Activity. In a 5 × 5 square, there are twenty-five 1 × 1 squares, sixteen 2 × 2 squares, nine 3 × 3 squares, four 4 × 4 squares, and one 5 × 5 square, for a total of 55 squares of all sizes. Do you think this pattern would work if you had started with an 8 × 8 square? Here is another example where you can use inductive reasoning.

Example: If you mark one point, C, between A and B on $\overline{AB}$, 3 line segments are formed: $\overline{AC}$, $\overline{CB}$, and $\overline{AB}$.

If you mark two points, C and D, between A and B, you get 6 line segments: $\overline{AC}$, $\overline{AD}$, $\overline{AB}$, $\overline{CD}$, $\overline{CB}$, $\overline{DB}$.

Marking three points between A and B gives you 10 segments. Continue the pattern. How many segments do you predict will be created when you mark four points between A and B? five points?

Can you describe a pattern that would let you find the number of line segments formed for whatever number of points you wish to mark between A and B?

CLASS ACTIVITY

1. Triangular dot numbers are so named because that number of dots can be used to form a triangle with an equal number of dots on each side. Examine the following triangular dot numbers.

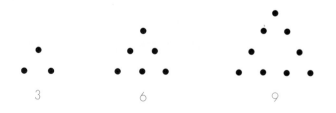

Predict what triangular dot number will have 10 dots on a side. Explain how you arrived at your prediction.

DISCOVERY ACTIVITY

Draw a circle and mark two points on it, as shown. Connect the points with a line segment and count the regions within the circle.

Draw another circle and mark three points, as shown. Connect the points and count the regions.

Draw another circle and mark four points. Again, connect the points and count the regions.

Draw a circle and mark five points as shown. Connect the points. First predict the number of regions. Then count the regions to check your prediction.

Now draw a circle and mark six points. Connect the points and predict the number of regions formed. Count the regions to check your prediction. What did you notice?

This Discovery Activity points out a weakness in inductive reasoning. Did you predict that when you connected six points on the circle you would have 32 regions? In fact, there are only 30. You have discovered that a pattern detected in a limited number of cases does not necessarily hold up in all cases.

Inductive reasoning may not always lead you to the right conclusion. Often this is because important factors have been overlooked. Here is an example from real life.

Example: Karl knows that the Tigers beat the Panthers in football and that the Panthers beat the Lions. The Tigers are playing the Lions in football today and he predicts the Tigers will win.

Although the Tigers may win this game, the reasoning is based on too few cases. Also, Karl is not thinking about some key things such as the line-up of players, the condition of the field, and so on.

CLASS ACTIVITY

Fill the blanks. Describe the pattern you found.

1. 1, 3, 7, 13, ____, ____, ____

2. 1, 3, 6, 10, 15, ____, ____, ____

3. 1, 3, 7, 15, 31, ____, ____, ____

Use a calculator for Exercises 4 and 5.

4. Divide 1 by 9, then 2 by 9. Predict the quotient of $5 \div 9$. Take rounding into account.

5. Divide 1 by 11, then 2 by 11. Predict the quotient of $4 \div 11$ and $9 \div 11$. Take rounding into account.

6. These are examples of square dot numbers.

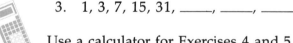

How many dots are on each side of the square dot number 28?

HOME ACTIVITY

Fill the blanks. Describe the pattern you found.

1. 2, 5, 8, 11, 14, _____, _____, _____

2. 1, 2, 6, 24, 120, _____,

3. 2, 3, 5, 9, 17, _____, _____, _____

4. $1, \frac{1}{2}, \frac{1}{4}, \frac{1}{8},$ _____, _____, _____

5. What fraction of the square region in the figure at the right is shaded? *Hint:* Discover the pattern of the shading.

Complete a table like the one below to show how many line segments are determined by n points that are evenly spaced around a circle.

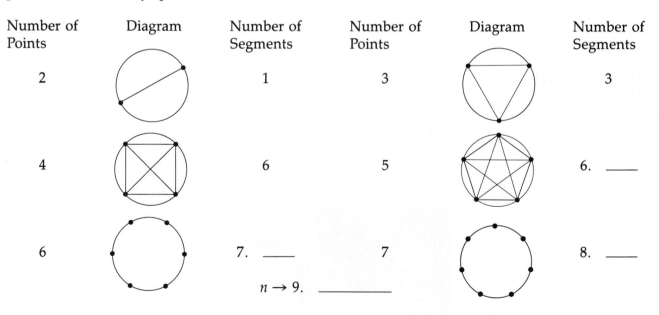

Number of Points	Diagram	Number of Segments	Number of Points	Diagram	Number of Segments
2		1	3		3
4		6	5		6. _____
6		7. _____	7		8. _____

$n \rightarrow$ 9. _____

CRITICAL THINKING

10. Hannah's bank has been charging her a monthly service charge of $3 plus $.10 for each check she writes. Now the bank is changing its policy to charge a $4 monthly fee and $.06 a check. The bank officer tells Hannah these changes will save her money. How many checks must Hannah write each month for this to be true?

LOGO and Geometric Drawing Tools

After completing this section, you should understand
▲ how to use some Logo commands to draw geometric figures.
▲ how to use a computer and geometric drawing tools to explore geometric concepts.

The computer language Logo was designed for use in education and artificial intelligence research. Today, Logo is used primarily in schools to introduce programming to students using turtle graphics. You can use turtle graphics to draw geometric figures by using relatively simple commands.

MOVING THE TURTLE

The commands used in turtle graphics, sometimes called primitives, manipulate a small triangular figure called a turtle. When you type the Logo command DRAW, the screen is cleared and the turtle is placed in its home position in the center of the screen, pointing up. The turtle shown in the first drawing on the next page is in home position.

To draw a picture using Logo, you use special commands that move the turtle around the screen. The turtle will leave a trail on the screen as it moves from one position to the next. After the turtle has been moved, it can be returned to its home position by typing the command HOME.

The command FORWARD or FD moves the turtle forward the number of spaces you specify. For example, FORWARD 6Ø or FD 6Ø tells the turtle to move 6Ø spaces forward. The command BACK or BK moves the turtle back the number of spaces you specify. For example, BACK 55 or BK 55 tells the turtle to move 55 spaces back. The command RIGHT or RT turns the turtle to the right the number of degrees you specify. For example, RIGHT 8Ø or RT 8Ø tells the turtle to turn to the right 80°. The command LEFT or LT turns the turtle to the left the number of degrees you specify. For example, LEFT 115 or LT 115 tells the turtle to turn to the left 115°.

Example: Show how the turtle moves when you type these Logo commands.

DRAW

The screen is cleared and the turtle is placed in its home position.

RT 9Ø

The turtle turns 90° to the right.

FD 12Ø

The turtle moves forward 120 spaces.

LT 45

The turtle turns 45° to the left.

BK 8Ø

The turtle moves back 80 spaces.

CLASS ACTIVITY

Type these Logo commands on a computer. Make a sketch of the figure drawn by the turtle.

1. DRAW	2. DRAW	3. DRAW
FD 5Ø	FD 8Ø	RT 9Ø
LT 9Ø	RT 9Ø	BK 25
FD 8Ø	FD 3Ø	LT 6Ø
LT 9Ø	RT 12Ø	FD 25
BK 3Ø	FD 3Ø	LT 6Ø
	LT 9Ø	FD 25
	BK 8Ø	

Write Logo commands that tell the turtle how to draw each figure. Use a computer to check your answers.

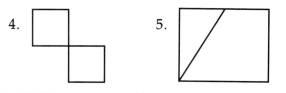

4.   5.   6. a 75° angle

DISCOVERY ACTIVITY

1. Type the commands from the example on page 31 on one line. What happens?

2. Type the commands from the example on page 31 as shown below. What happens?

   DRAW
   RT 9Ø FD 12Ø
   LT 45 BK 8Ø

3. What do you think would happen if you typed the commands from the example as shown below? Use a computer to check your answer.

   DRAW
   RT 9Ø
   FD 12Ø LT 45 BK 8Ø

4. Write a generalization that seems to be true.

You should have discovered that more than one command can be typed on one line.

MORE COMMANDS

Sometimes you may want to move the turtle without leaving a trail. The command PENUP or PU will accomplish this task. The command PENDOWN or PD tells the turtle to start leaving a trail again.

Example: Show how the turtle moves when you type these Logo commands.

DRAW
PU
LT 90
FD 50 RT 90
FD 50 RT 90
PD
FD 100 RT 90
FD 100 RT 90
FD 100 RT 90
FD 100 RT 90

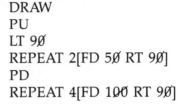

The commands in the
first six lines move
the turtle to this
position without
leaving a trail.

The result for the
remaining commands is a
square that is
centered on the
screen.

In order to draw the square in this last example, several groups of the same commands were typed over and over again. The REPEAT command can be used to avoid this.

DRAW
PU
LT 90
REPEAT 2[FD 50 RT 90]
PD
REPEAT 4[FD 100 RT 90]

In a REPEAT command, the number in front of the first bracket tells the turtle how many times to execute the command in the brackets. In the example above, the turtle will repeat the commands in the brackets in the fourth line 2 times and the commands in the brackets in the last line 4 times.

CLASS ACTIVITY

Write Logo commands that tell the turtle how to draw the figure described or shown. Use a computer to check your answers.

1. a horizontal segment and a vertical segment that intersects the horizontal segment at its midpoint

2. a horizontal segment and a nonvertical segment that intersects the horizontal segment at its midpoint

3.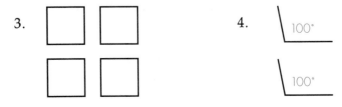

4. 100°

100°

PROCEDURES

In Logo, you can write procedures, or programs, that contain commands needed to complete a certain task. A name is assigned to the commands in the procedure. To execute the procedure, you simply type its name.

The commands in the example on page 33 can be assigned the name SQUARE as shown below.

```
TO SQUARE
PU
LT 9Ø
REPEAT 2[FD 5Ø RT 9Ø]
PD
REPEAT 4[FD 1ØØ RT 9Ø]
END
```

DISCOVERY ACTIVITY

1. Write a procedure named RECTANGLE to draw a rectangle that is centered on the screen. Use a computer to check your answer.
2. Write a procedure named TRIANGLE to draw a triangle that is centered on the screen. Use a computer to check your answer.
3. What do you think would happen if you typed the following commands? Use a computer to check your answer.

RECTANGLE   HOME   TRIANGLE

CLASS ACTIVITY

Show how the turtle moves when you type these Logo commands.

1. DRAW
   PU RT 9Ø BK 9Ø PD
   FD 3Ø LT 9Ø
   FD 5Ø LT 9Ø
   BK 8Ø

2. DRAW
   PU BK 8Ø RT 9Ø PD
   FD 1ØØ LT 135
   FD 7Ø LT 45
   FD 5Ø LT 9Ø
   FD 5Ø

Write Logo commands that tell the turtle how to draw each figure.

3. 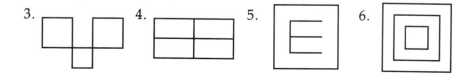 4. 5. 6.

GEOMETRIC DRAWING TOOLS

There are several geometric drawing tools that have been developed for the computer. These tools enable you to use the computer to explore some of the concepts you will study in this course.

To use one of these tools, simply follow the instructions that will be given to you on the screen. Be sure to record all important measurements so that you can quickly refer to them at a later time.

Example: Use a geometric drawing tool and follow the steps below to explore the concept of betweenness of points as presented in Definition 1-3-3.

(1) Draw a segment of any length. Label the endpoints A and B.

Suppose you tell the computer to draw a segment that has a length of 8. The segment will appear on the screen as shown below.

(2) Draw any point between points A and B. Label the point C.

The segment will now look something like this.

(3) Find the lengths of $\overline{AC}$, $\overline{CB}$, and $\overline{AB}$.

Tell the computer to measure $\overline{AC}$, $\overline{CB}$, and $\overline{AB}$. These measures will appear on the screen. Record the measures.

(4) Find the sum of AC and CB. Compare the sum with the length of $\overline{AB}$.

Suppose point C was drawn so that $AC = 5$ and $CB = 3$. Then the sum of AC and CB would be 8. In comparing this sum with the length of $\overline{AB}$, we see that they are the same. Record this result.

(5) Repeat steps 1–4 several times. Study the results and write a generalization that seems to be true.

After completing this step, you will see that when point C is between points A and B, the sum of the lengths of $\overline{AC}$ and $\overline{CB}$ is the same as the length of $\overline{AB}$.

CLASS ACTIVITY

Use a geometric drawing tool and follow the steps below.

1. Draw an acute angle.
2. Label the angle *ABC*, placing points *A* and *C* so that *BA* = *BC*.
3. Draw $\overline{AC}$.
4. Draw point *X* on $\overline{AC}$ so that *X* is the midpoint of $\overline{AC}$.
5. Draw $\overline{BX}$.
6. Measure ∠*ABX* and ∠*XBC*.
7. Compare the measures of ∠*ABX* and ∠*XBC*. What is $\overline{BX}$?

HOME ACTIVITY

Show how the turtle moves when you type these Logo commands.

1. DRAW
   REPEAT 6 [FD 2Ø RT 6Ø]

2. DRAW
   REPEAT 3[FD 5Ø RT 12Ø]
   RT 6Ø
   REPEAT 2[FD 5Ø RT 12Ø]

3. DRAW
   REPEAT 3[FD 50 RT 12Ø]
   LT 90
   REPEAT 3[FD 50 RT 90]

Write Logo commands that tell the turtle how to draw each figure.

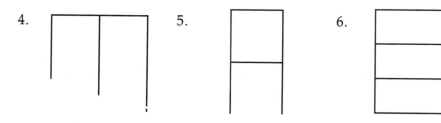

4.   5.   6.

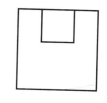

7.

Write procedures that tell the turtle how to draw each figure. Give each procedure the indicated name.

8. an acute angle; ACUTE

9. an obtuse angle; OBTUSE

10. a 130° angle and its bisector; ANGLEBISECTOR

CRITICAL THINKING

Write a paragraph explaining why it is important to give precise, detailed instructions to the computer when you want to use it to draw geometric figures. Would it be necessary to give similar instructions to a live human being who is going to draw a figure that you describe?

1.1 A **plane** is a flat surface that extends endlessly in all directions. Two points determine exactly one **line**. A **line segment** is part of a line and consists of two points A and B and all points between them. A **ray** consists of a point and all points to one side of that point. Two rays with a common endpoint form an **angle.**

In the diagram, plane P contains line CD ($\overleftrightarrow{CD}$), line segment AB ($\overline{AB}$), rays ED, EF, and EC ($\overrightarrow{ED}$, $\overrightarrow{EF}$, and $\overrightarrow{EC}$), and angles DEF, FEC, and DEC ($\angle DEF$, $\angle FEC$, and $\angle DEC$).

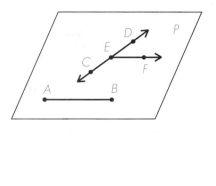

1.2 There is a one-to-one correspondence between points on a line and the real numbers. The distance between two points on a number line is the positive difference of their coordinates.

Acute angles have measures less than 90°, **right** angles measure exactly 90°, **obtuse** angles measure between 90° and 180°, and **straight** angles measure exactly 180°.

The diagram shows acute angles BCD and DCE, right angles BCE and BCA, obtuse angle ACD, and straight angle ACE.

1.3 The **midpoint** of a line segment is the point which divides the segment into two equal segments. The **bisector** of an angle is a ray with the same vertex that separates the angle into two angles, each of which has a measure equal to half of the original angle.

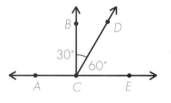

In the diagrams, C is the midpoint of $\overline{AB}$. $\overrightarrow{EF}$ bisects $\angle DEG$. $\overrightarrow{EF}$ is **between** $\overrightarrow{ED}$ and $\overrightarrow{EG}$.

1.4 Two angles are **supplementary** if their measures add to 180°. Two angles are **complementary** if their measures add to 90°.

In the diagrams, $m\angle ABD = 180°$ and $m\angle EFG = 90°$, $\angle ABC$ is supplementary to $\angle CBD$, and $\angle EFH$ and $\angle HFG$ are complementary.

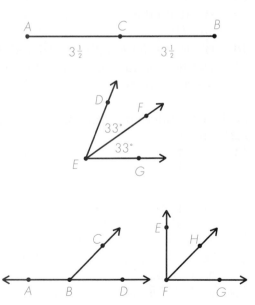

1.5 **Inductive reasoning** is the process of using a limited number of cases to arrive at a conclusion that you expect to be true in all cases.

1.6 **LOGO** is a computer language that can be used to investigate geometric properties.

Use symbols to name each of the following.

1.

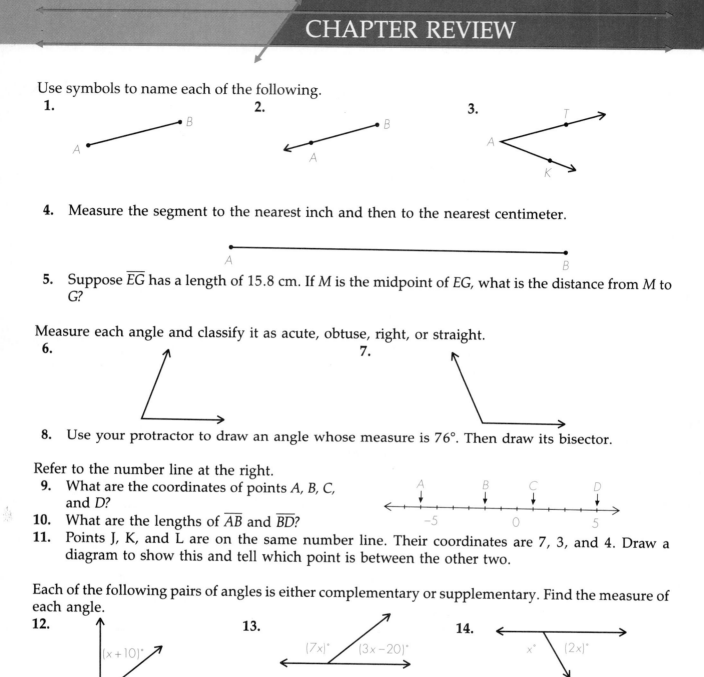

2.

3.

4. Measure the segment to the nearest inch and then to the nearest centimeter.

5. Suppose $\overline{EG}$ has a length of 15.8 cm. If M is the midpoint of EG, what is the distance from M to G?

Measure each angle and classify it as acute, obtuse, right, or straight.

6.

7.

8. Use your protractor to draw an angle whose measure is 76°. Then draw its bisector.

Refer to the number line at the right.

9. What are the coordinates of points A, B, C, and D?

10. What are the lengths of $\overline{AB}$ and $\overline{BD}$?

11. Points J, K, and L are on the same number line. Their coordinates are 7, 3, and 4. Draw a diagram to show this and tell which point is between the other two.

Each of the following pairs of angles is either complementary or supplementary. Find the measure of each angle.

12.

$(x+10)°$

$x°$

13.

$(7x)°$ $(3x-20)°$

14.

$x°$ $(2x)°$

15. Look for a pattern. Then fill in the blanks.
   a. 1, 3, 6, 10, 15, _____, _____, _____
   b. 2, 6, 18, 54, _____, _____, _____

16. Write LOGO commands for drawing a vertical line segment intersected at its midpoint by a horizontal line segment.

1. Complete the following table.

Angle	Complement	Supplement
41°	_____	_____
86°	_____	_____
5°	_____	_____
$(4x)°$	_____	_____
$(2x - 20)°$	_____	_____

Measure each angle and classify it as acute, obtuse, right, or straight.

2.

3.

4. What is the measure of each angle formed by the bisector of a 90° angle?

Tell what kind of figure each of the following is. Use symbols to name it.

5. *A*   *B*

6. *A*   *B*

7. *A*   *B*

8. *A*   *B*   *C*

On the number line at the right, O is the origin. P
and Q have coordinates -4 and 6, respectively,
and $PS = 2$.

9. What is the coordinate of S?
10. What is the length of $\overline{PQ}$? of $\overline{SQ}$?

Suppose B is between A and P. Point K is the midpoint of $\overline{BP}$. $AP = 8$ and $KP = 1\frac{1}{2}$.
11. Draw a diagram to show how A, B, K, and P are situated.
12. What is the measure of $\overline{BK}$? of $\overline{AB}$?

Look for a pattern, then fill in the blanks.
13. 1, 3, 7, 15, 31, _____, _____, _____
14. -8, -5, -2, _____, _____, _____
15. Write LOGO commands for drawing a 120° angle and its bisector.

ALGEBRA SKILLS

An algebraic expression consists of numbers and variables combined with operation signs ($+$, $-$, $\times$, $\div$, and so on). To evaluate an algebraic expression means to find its value given specific values for its variables.

To evaluate an expression, do the following things in this order.
1. Replace all variables by their values.
2. Simplify any expressions within parentheses.
3. Simplify any powers or roots.
4. Multiply or divide from left to right.
5. Add or subtract from left to right.

Example:
Evaluate $2x + y$ for $x = 10$ and $y = -2$.
$2(10) + (-2) = 20 - 2 = 18$
$2x + y = 18$

To solve an equation means to find the number or numbers that can replace the variable to make the equation true. Such a number is called a *solution* of the equation. To solve an equation, you isolate the variable in one member of the equation by using inverse operations. Your goal, in other words, is to get the variable alone on one side of the equal sign to find its value.

Example:
Solve for x.
$$2x + 6 = 34$$
$$2x + 6 - 6 = 34 - 6$$
$$2x = 28$$
$$2x \div 2 = 28 \div 2$$
$$x = 14$$
Check: $2(14) + 6 = 34$
$$28 + 6 = 34$$
$$34 = 34$$

Evaluate the following expressions for $x = 12$ and $y = -20$.

1. $x + y$
2. $x - y$
3. xy
4. $\frac{y}{5}$

5. $y - x$
6. y^2
7. $\frac{x}{y} - 1$
8. $x - (1 - y)$

Evaluate these expressions for $x = \frac{1}{2}$ and $y = \frac{2}{3}$.

9. $x + y$
10. $x - y$
11. $y - x$
12. $\frac{x}{y}$

13. xy
14. x^2
15. y^3
16. $2x + y$

Solve for x.

17. $4x - 7 = 53$
18. $2.5x + 8 = 23$
19. $\frac{x}{2} + 15 = 40$

20. $\frac{3}{4}x - 5\frac{1}{2} = 3\frac{1}{2}$
21. $12 - (2x - 5) = 9$
22. $3x = 8 - 5x$

23. Two angles are complementary. One is 5 times as great as the other. What is the measure of each angle?

24. Two angles are supplementary. The measure of one is 15° more than twice the measure of the other. What is the measure of the smaller angle?

2

INTERSECTING AND PARALLEL LINES

I value the discovery of a single even insignificant truth more highly than all the argumentations on the highest questions which fail to reach a conclusion.

Galileo Galilei

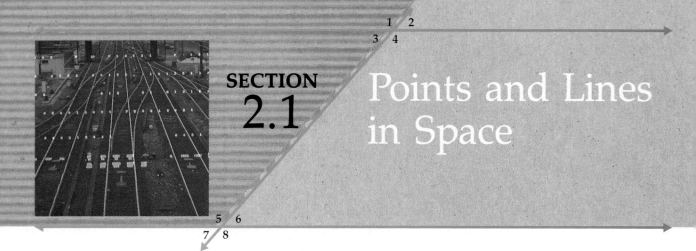

SECTION 2.1

Points and Lines in Space

After completing this section, you should understand
▲   collinear, noncollinear, coplanar, and noncoplanar points.
▲   intersecting, parallel, and skew lines.

In the diagrams below you see two groups of points. The group on the left is special because its points all lie on the same line. The group on the right is special because all the points are on the same plane (flat surface).

COLLINEAR AND COPLANAR POINTS

In geometry there are special terms for groups of points like those in the diagrams above.

DEFINITION 2-1-1:   **COLLINEAR POINTS are points on the same line.**

DEFINITION 2-1-2:   **COPLANAR POINTS are points on the same plane.**

Points that are not on the same line are *noncollinear*.
Points that are not on the same plane are called *noncoplanar*.

Examples: Refer to the diagram.

1.   *A, B,* and *D* are collinear.
2.   *C, D,* and *E* are coplanar.
3.   *C, D,* and *E* are noncollinear.
4.   *A, C, D,* and *E* are noncoplanar.

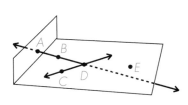

42

DISCOVERY ACTIVITY

You have learned that two points determine a line. You can experiment to find out how many points determine a plane.

Get a flat surface, such as a piece of cardboard, and some sharpened pencils. Using various numbers of pencils, try to balance the flat cardboard on the pencil points.

1. What was the least number of pencil points needed to balance the cardboard?

2. Were the pencil points collinear or noncollinear?

POSTULATE 2-1-1:   Three noncollinear points determine a plane.

CLASS ACTIVITY

Match the terms on the right with the descriptions on the left.

1.	a flat surface	a. noncoplanar points
2.	points on the same line	b. plane
3.	points on the same plane	c. collinear points
4.	points not on the same plane	d. noncollinear points
5.	points not on the same line	e. coplanar points

Tell whether each of the following statements is true or false.

6. Any three points are coplanar.
7. Any two points are collinear.
8. Two points can determine a plane.

Identify the following points as *collinear, coplanar,* or *noncoplanar.*

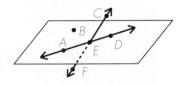

9.	*B, E,* and *C*	10.	*B, E,* and *D*
11.	*A, E,* and *D*	12.	*C, D,* and *F*
13.	*A, B, C,* and *D*	14.	*A, B, C,* and *F*

INTERSECTING LINES AND PARALLEL LINES

There are lines everywhere you look. Is the road shown at the left representative of a line or a line segment? Explain.

On a sheet of paper, draw several pairs of lines. Will every pair you can draw intersect?

For two lines in a plane there are only two possibilities: They will intersect or they will not.

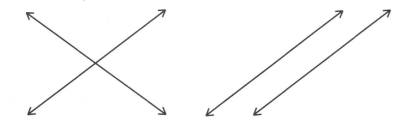

In Figure 1 the two lines intersect at one point. In Figure 2 the two lines will never intersect.

POSTULATE 2-1-2: **If two lines intersect, then they intersect at exactly one point.**

DEFINITION 2-1-3: **Two lines are PARALLEL if they are coplanar and do not intersect.**
The symbol ∥ means "is parallel to."

Look around your classroom. List three pairs of intersecting lines and three pairs of parallel lines.

Are all the lines represented in the photograph in the same plane?

DISCOVERY ACTIVITY

Place two pencils on your desk to illustrate a pair of intersecting lines. Then lift the top pencil straight up into the air. Do the two lines intersect? Are they parallel?

SKEW LINES

You have learned that in a plane, two lines will either intersect or be parallel. However, in 3-dimensional space it is possible to have two lines that are not parallel and do not intersect. The figure shows how this is possible. Lines of this type are called **skew lines.**

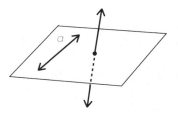

DEFINITION 2-1-4: **SKEW LINES are lines in three dimensions which do not intersect and are not parallel.**

CLASS ACTIVITY

Classify each pair of lines as parallel or intersecting. Assume the lines are in the same plane.

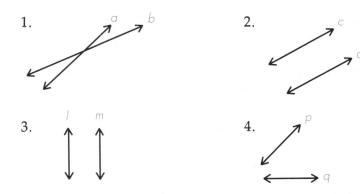

1.

2.

3.

4.

Use the figure at the right. Identify a pair of lines of each type.

5. intersecting lines
6. parallel lines
7. skew lines

Write true or false.

8. Intersecting lines are coplanar.
9. Skew lines are coplanar.
10. Parallel lines are noncoplanar.

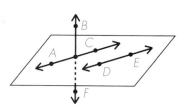

HOME ACTIVITY

Tell whether each set of points is collinear.

1. A, C, E   2. B, C, F   3. B, C, D   4. B, C

Use the drawing at the right. State whether each set of points is coplanar.

5. B, E, G
6. C, H, D
7. G, C, F
8. B, C, E, D

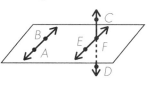

Use the drawing at the right. Identify the following.

9. parallel lines
10. skew lines
11. intersecting lines

 Write Logo procedures to tell the turtle to draw a figure representing a plane. Then tell the turtle to draw each pair of lines described below.

12. intersecting lines on the plane
13. parallel lines on the plane
14. a line on the plane and a line skew to that line

CRITICAL THINKING

15. Draw a figure that shows six points such that no three of the points are collinear. Draw all the lines possible through pairs of these points. How many of these lines are there?

SECTION 2.2

1 2
3 4

5 6
7 8

Corresponding Angles

After completing this section, you should understand
▲ parallel lines and transversals.
▲ corresponding angles.

You have learned that two lines are parallel if they are coplanar and do not intersect. Points in evenly spaced rows and columns form a **lattice**.

How many lines can be drawn through two or more points of this lattice that are parallel to the given line?

PARALLEL LINES AND TRANSVERSALS

Suppose $\overleftrightarrow{AB}$ and $\overleftrightarrow{CD}$ are parallel. If you draw a third line $\overleftrightarrow{EF}$ in the same plane, either it will be parallel to the other two, or it will intersect them.

Look at the examples below.

$\overleftrightarrow{EF}$, $\overleftrightarrow{AB}$, and $\overleftrightarrow{CD}$ are parallel.

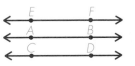

$\overleftrightarrow{EF}$ intersects $\overleftrightarrow{AB}$ and $\overleftrightarrow{CD}$.

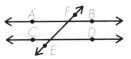

In the figure on the right, $\overleftrightarrow{EF}$ is called a **transversal** of $\overleftrightarrow{AB}$ and $\overleftrightarrow{CD}$.

DEFINITION 2-2-1: **A TRANSVERSAL is a line which intersects two or more coplanar lines at different points.**

Examples: 1. Line *t* is a transversal of lines *a* and *b*, because *t* intersects *a* and *b* in two different points.

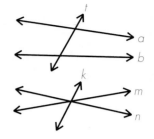

2. Line *k* is not a transversal of lines *m* and *n*, because it intersects them in only one point.

CLASS ACTIVITY

1. Which lines are parallel?

2. Name a pair of lines for which line *l* is a transversal.

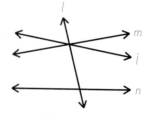

3. In which diagram is line *a not* a transversal of the other two lines?

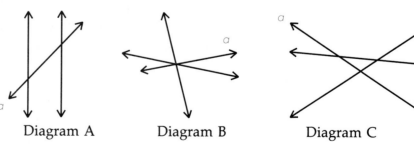

Diagram A   Diagram B   Diagram C

4. At the points where transversal *a* intersects lines *b* and *c*, how many angles are formed?

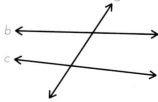

CORRESPONDING ANGLES

Take a sheet of paper and trace the diagram at the right, in which line t is a transversal. Notice that lines a and b are not parallel lines. Number the angles formed 1 through 8, as shown.

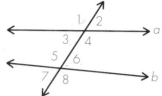

Notice that in your figure $\angle 1$ and $\angle 5$ are both on the left of the transversal and above the two lines cut by the transversal. You can say that $\angle 1$ **corresponds** to $\angle 5$.

Notice that $\angle 4$ and $\angle 8$ are both on the right of the transversal and below the two lines cut by the transversal. So $\angle 4$ **corresponds** to $\angle 8$.

What angle corresponds to $\angle 3$ in the figure? What angle corresponds to $\angle 2$?

DEFINITION 2-2-2: **CORRESPONDING ANGLES are angles which are in the same relative position with respect to the two lines cut by a transversal and the transversal.**

Use a protractor to find the degree measures of each of the eight angles in the diagram. Do any two corresponding angles have the same measure?

DISCOVERY ACTIVITY

Using both edges of a ruler, draw two parallel lines. Label the lines r and s.
Next, use your ruler to draw transversal t.
Then label all the angles formed by the transversal.
List all pairs of corresponding angles.
Finally, use a protractor to find the degree measure of each angle in the pair.
What do you notice? Compare your results with those of a classmate.

If you drew the lines correctly and measured carefully with your protractor, you probably discovered that the degree measures of the corresponding angles were the same.

POSTULATE 2-2-1: **If two parallel lines are intersected by a transversal, then the corresponding angles are equal in measure.**

You can use what you know about corresponding angles to solve problems.

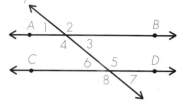

In the diagram at the right, $\overleftrightarrow{AB} \parallel \overleftrightarrow{CD}$. Line t is a transversal. What are the measures of angles 3 to 8 if $m\angle 1 = (x + 10)°$ and $m\angle 2 = (3x + 50)°$?

Since $\angle 1$ and $\angle 2$ are supplementary, their measures have a sum of 180°.

First find x.

$$(x + 10) + (3x + 50) = 180$$
$$4x + 60 = 180$$
$$4x = 120$$
$$x = 30$$

Then substitute.

$(x + 10)° = (30 + 10)°$ or 40°, so $m\angle 1 = 40°$.
$(3x + 50)° = (90 + 50)°$ or 140°, so $m\angle 2 = 140°$.

Using Postulate 2-2-1 and the meaning of supplementary angles:

$$m\angle 1 = m\angle 6 = m\angle 3 = m\angle 7 = 40°$$
$$m\angle 2 = m\angle 5 = m\angle 4 = m\angle 8 = 140°$$

CLASS ACTIVITY

In the diagram at the right, lines b and c are parallel. Identify each of the following.

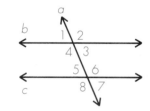

1. a transversal
2. two parallel lines
3. four pairs of corresponding angles

Refer to the diagram and complete the following.

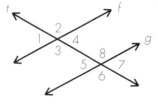

4. $\angle 1$ corresponds to $\angle$?
5. $\angle 2$ corresponds to $\angle$?
6. $\angle 3$ corresponds to $\angle$?
7. $\angle 4$ corresponds to $\angle$?

In the diagram, $p \parallel q$. Complete the following.

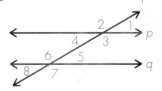

8. If $m\angle 1 = 30°$, then $m\angle 5 = $?
9. If $m\angle 2 = 150°$, then $m\angle 6 = $?
10. If $m\angle 3 = 150°$, then $m\angle 7 = $?
11. If $m\angle 4 = 30°$, then $m\angle 8 = $?

12. In the diagram, $j \parallel k$ and $m\angle 1 = 37°$. Find the measures of the other seven angles.

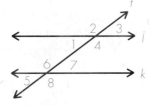

13. In the diagram, $r \parallel s$, $m\angle 1 = (2x - 10)°$, and $m\angle 2 = (x + 40)°$. Find the measures of all eight angles.

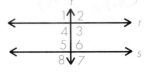

HOME ACTIVITY

In the diagram at the right, $a \parallel b$. Identify each of the following.

1. Two lines for which a is a transversal
2. Two lines for which l is a transversal
3. Two lines for which k is a transversal
4. Eight pairs of corresponding angles

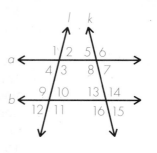

In the diagram, $r \parallel s$ and $m\angle 1 = 67°$. Complete the following.

5. $\angle 1$ corresponds to $\angle$?, which measures ?.
6. $\angle 2$ corresponds to $\angle$?, which measures ?.
7. $\angle 3$ corresponds to $\angle$?, which measures ?.
8. $\angle 4$ corresponds to $\angle$?, which measures ?.

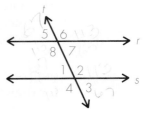

In the diagram, $l \parallel k$, $m\angle 1 = (4x - 15)°$, and $m\angle 2 = (15x + 5)°$.

9. $m\angle 1 = $?   10. $m\angle 2 = $?
11. $m\angle 3 = $?   12. $m\angle 4 = $?
13. $m\angle 5 = $?   14. $m\angle 6 = $?
15. $m\angle 7 = $?   16. $m\angle 8 = $?

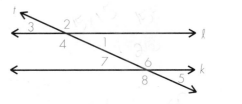

17. The diagram represents a street intersection. Calculate the measures of the indicated angles.

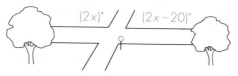

18. Write a Logo procedure to tell the turtle to draw two left-to-right parallel lines and a transversal so that the two supplementary angles formed by the top line and the transversal have measures of $(3x + 10)°$ and $(2x + 20)°$. Show how the turtle will move when you type in your procedure. Then label the angles and find the degree measure of all eight angles formed.

CRITICAL THINKING

You can use a compass and a straightedge to copy an angle. Here's how.

1. Draw a ray with endpoint P.
2. Construct an arc with its center at A. Then construct an arc with its center at P. Use the same compass opening.
3. Place your compass point at X. Adjust the opening so that you can draw an arc passing through Y. Then, using the same compass opening, construct an arc with its center at R.
4. Draw $\overrightarrow{PQ}$ with your straightedge. $m\angle YAX = m\angle QPR$.

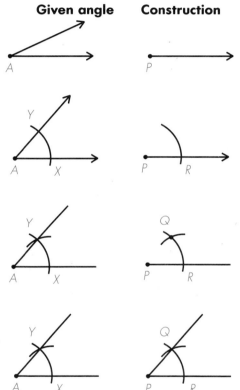

Suppose you have a line r and a point P not on r. Study this procedure for drawing a line s through P and parallel to r.

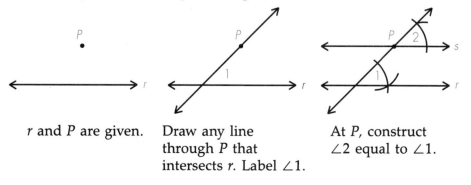

r and P are given.   Draw any line through P that intersects r. Label $\angle 1$.   At P, construct $\angle 2$ equal to $\angle 1$.

19. Explain why it is reasonable to think that $s \parallel r$.

SECTION 2.3

Vertical and Alternate Interior Angles

After completing this section, you should understand
▲ how deductive reasoning is used.
▲ the meaning of vertical angles.
▲ the meaning of alternate interior angles.

> I tell you, we've never had a tornado here. So you can stop worrying!

As you can see from the cartoon, drawing sweeping conclusions from limited information can be very risky. When you use limited information to arrive at a general conclusion, you are using **inductive** thinking. Everyone uses inductive thinking, but you have to be careful with it. In geometry you can use inductive thinking to arrive at good hunches about what is always true, but to **prove** that your hunches are correct, you need **deductive reasoning.**

DEDUCTIVE REASONING

To use deductive reasoning, you start with some given information. Then you present a series of statements, backed up by reasons, that lead logically, step-by-step to a final conclusion. The idea is to show that anyone who accepts the given information will have to accept the final conclusion.

Example: In the figure $\overrightarrow{XB}$ bisects $\angle AXC$ and $\overrightarrow{XC}$ bisects $\angle BXD$. Suppose you want to show that $m\angle 1 = m\angle 3$. $m\angle 1 = m\angle 2$, because of what *bisects* means. $m\angle 2 = m\angle 3$, for the same reason. So $m\angle 1 = m\angle 3$, by properties of algebra.

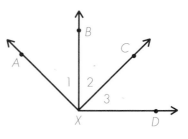

53

In geometry you will often encounter statements of the form "If . . . , then . . .", where certain statements are in place of the dots. Statements of this kind are called **conditional** statements. Within a conditional statement, the statement after "if" is called the **hypothesis.** The statement after "then" is called the **conclusion.**

conditional statement

If $m\angle A$ is between 0° and 90°, then $\angle A$ is acute.

hypothesis   conclusion

Suppose you know that the hypothesis is true. You can conclude the conclusion of the conditional statement must also be true.

True   → If (statement A), then (statement B).
True   → Statement A
You conclude → Statement B is true.

Example: Consider the figure shown here. Suppose a is parallel to b. This statement is true because it is a given. If two parallel lines are intersected by a transversal, then the corresponding angles are equal in measure.

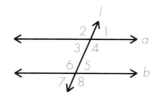

You know that $a \parallel b$ and l is a transversal. You can conclude that the corresponding angles are equal in measure.

CLASS ACTIVITY

Copy each statement. Underline the hypothesis and ring the conclusion.
1. If you live in Utah, then you live in the United States.
2. If you are in London, then you are in England.
3. If your pet is a terrier, then it is a dog.
4. If $x + 5 = 7$, then $x = 2$.

Suppose each pair of statements is true. What can you conclude?
5. If an angle is obtuse, then its measure is greater than 90°. $\angle PQR$ is obtuse.
6. If two angles are supplementary, then the sum of their measures is 180°. $\angle X$ and $\angle Y$ are supplementary.
7. If three points are noncollinear, then they determine a plane. Points L, M, and N are noncollinear.
8. If four points are collinear, then any three of them are collinear. Points A, B, C, and D are collinear.

VERTICAL ANGLES

Recall that two lines in a plane will either be parallel or intersect. When two lines intersect, they form four angles.

In the diagram at the right, the angles formed by the intersection of $\overleftrightarrow{PY}$ and $\overleftrightarrow{XZ}$ are labeled 1, 2, 3, and 4.

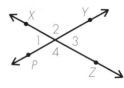

Notice that angles 1 and 3 are opposite or across from each other in the diagram. They are opposite or vertical angles. Angles 2 and 4 are another pair of vertical angles.

DEFINITION 2-3-1: **VERTICAL ANGLES are the opposite angles formed when two lines intersect.**

From the diagram, you can see that $\angle 1$ and $\angle 3$ look about equal in size. So do $\angle 2$ and $\angle 4$. By using deductive reasoning, you can show that they really are equal. First, notice that $\angle 1$ and $\angle 2$ are supplementary by the definition of supplementary angles (Definition 1-4-1)

$$m\angle 1 + m\angle 2 = 180.$$

But $\angle 3$ and $\angle 2$ are also supplementary, so that

$$m\angle 3 + m\angle 2 = 180.$$

Using the algebraic properties of equality, you can see that

$$m\angle 1 + m\angle 2 = m\angle 3 + m\angle 2$$

Now subtract $m\angle 2$ from both sides.

$$
\begin{array}{rcl}
m\angle 1 + m\angle 2 & = & m\angle 3 + m\angle 2 \\
- m\angle 2 & = & - m\angle 2 \\
\hline
m\angle 1 & = & m\angle 3
\end{array}
$$

You have come to the conclusion that vertical angles 1 and 3 are equal. You can use the very same kind of reasoning to show that angles 2 and 4 are equal.

The same reasoning would work for *any* intersecting lines and *any* vertical angles. When deductive reasoning has been used to show that a statement is true in all cases, the statement is called a **theorem.**

THEOREM 2-3-1: **If two lines intersect, then the vertical angles are equal.**
This theorem is called the *vertical angle theorem.*

Once you have proved a theorem, you can use it in any situation where it applies.

Example: $\overleftrightarrow{AB}$ and $\overleftrightarrow{CD}$ intersect to form $\angle APC$, having a measure of 75°. What is the measure of $\angle DPB$?

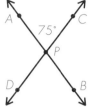

The vertical angle theorem tells you that $\angle APC$ and $\angle DPB$ have the same measure. Since $m\angle APC = 75°$, you can conclude that $m\angle DPB = 75°$.

CLASS ACTIVITY

Use a computer and a geometric drawing tool to complete Exercises 1-5.

1. Draw and label two intersecting lines.
2. Measure each angle formed.
3. Which angles have equal measures?
4. Repeat Exercises 1 through 3 several times.
5. Do your findings from Exercises 1 through 4 support Theorem 2-3-1?

Notice that $\angle 3$ and $\angle 5$ are on opposite or **alternate** sides of the transversal. They are also between the parallel lines. Such pairs of angles are called **alternate interior angles.** Which other pair of angles in the your diagram are alternate interior angles?

DISCOVERY ACTIVITY

Draw two parallel lines *m* and *n* intersected by transversal *t*. Label the angles formed using the numbers 1 through 8, as shown.

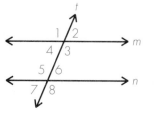

1. List all pairs of vertical angles.

   $\angle 1$ and $\angle 3$ $\angle 2$ and $\angle$?
   $\angle 5$ and $\angle$? $\angle 6$ and $\angle$?

2. Recall that corresponding angles are equal. List all of these pairs of equal angles.

   $\angle 1$ and $\angle 5$ $\angle 2$ and $\angle$?
   $\angle 3$ and $\angle$? $\angle 4$ and $\angle$?

3. Use your results from Exercises 1 and 2. What can you conclude about ∠3 and ∠5? about ∠4 and ∠6?

NOTEBOOK
DEFINITION 2-3-3: **ALTERNATE INTERIOR ANGLES are two angles formed by two parallel lines and a transversal such that they have different vertices, are between the parallel lines, and are on opposite sides of the transversal.**

NOTEBOOK
THEOREM 2-3-2: **If two parallel lines are intersected by a transversal, then the alternate interior angles are equal.**

You can use this theorem to help you solve problems.

Example: In the figure, t is a transversal of parallel lines a and b. $m∠1 = 110°$ and $m∠2 = 70°$. Find $m∠3$ and $m∠4$.

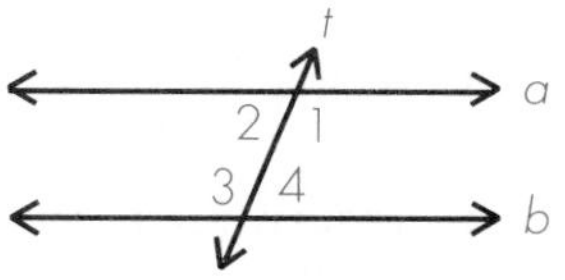

Use Theorem 2-3-2. You can conclude that $m∠1 = m∠3 = 110°$ and $m∠2 = m∠4 = 70°$.

CLASS ACTIVITY

In the diagram, $a \parallel b$ and t is a transversal.

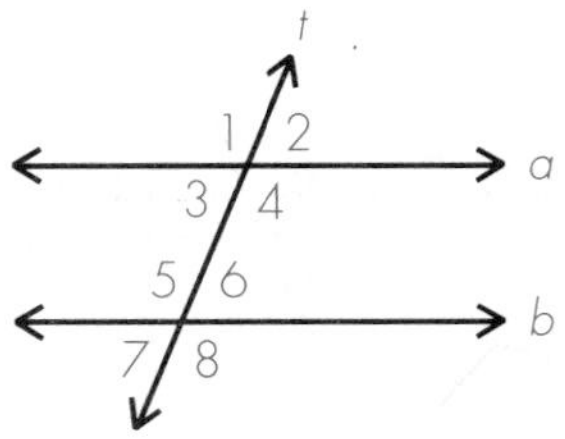

1. Identify a transversal
2. Identify a pair of parallel lines.
3. Identify four pairs of vertical angles.
4. Identify the corresponding angles.
5. Identify two pairs of alternate interior angles

Each diagram shows two parallel lines and a transversal. Find the measure of each numbered angle.

6.

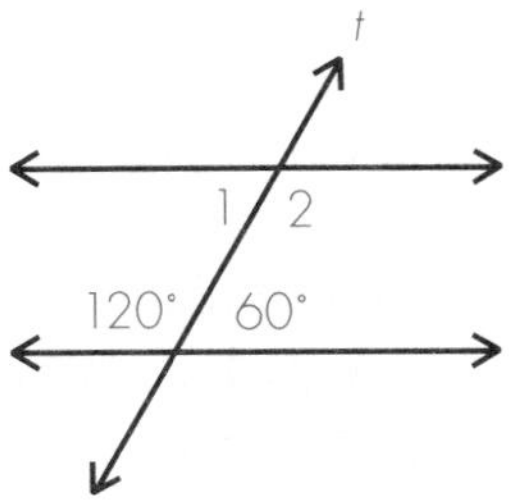

7.

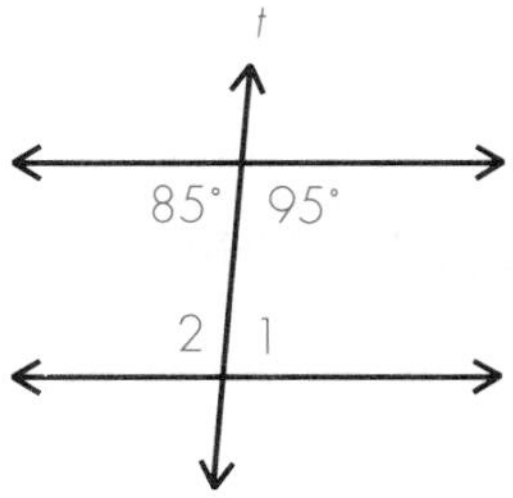

Use a computer and a geometric drawing tool to complete Exercises 8–12.

8. Draw and label two parallel lines and a transversal.
9. Measure each interior angle formed.
10. Which angles have equal measures?
11. Repeat Exercises 8–10 several times.
12. Do your findings from Exercises 8–11 support Theorem 2-3-2?

HOME ACTIVITY

Identify the hypothesis and conclusion for each statement.

1. If two angles are vertical angles, then they are equal.
2. If $n + 1$ is odd, then n is even.

Suppose each pair of statements is true. What can you conclude?
3. If a person is a taxi driver, then that person knows the city. Pete is a taxi driver.

4. If an angle is a straight angle, its sides form a straight line. $\angle ABC$ is a straight angle.

5. If a number is greater than 2, then its square is greater than 4. The number 2.001 is greater than 2.

In each figure t is a transversal of two parallel lines. Find the measure of each numbered angle.

6.

7.

8.

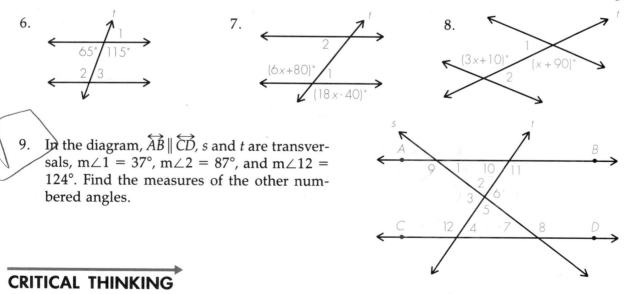

9. In the diagram, $\overleftrightarrow{AB} \parallel \overleftrightarrow{CD}$, s and t are transversals, $m\angle 1 = 37°$, $m\angle 2 = 87°$, and $m\angle 12 = 124°$. Find the measures of the other numbered angles.

CRITICAL THINKING

10. It is true that if two angles are vertical angles, then their measures are equal. If $\angle A$ and $\angle B$ are supplementary angles and vertical angles, what are their measures?

**SECTION
2.4**

1 2
3 4

Measurement and Estimation

5 6
7 8

After completing this section, you should understand
▲ the concept of error in measurement
▲ how to convert from one unit of measurement to another
▲ how to add and subtract measurements

Measurement is important in everyday life, and in fields such as mathematics, science, and economics. So far in this book, you have encountered two kinds of measures—length and angle measure.

ERROR IN MEASUREMENT

In everyday life you routinely put up with many inaccuracies and small errors. If you go to a cafeteria with a friend, you probably wouldn't ask the person serving rice to count the number of grains to be sure you and your friend get the same number. On the other hand, if you were cooking stew and used two teaspoons of salt instead of one, your might ruin the stew. If you are cutting wood for a book shelf, a hundredth of an inch more or less probably won't make much difference. But an automobile mechanic who ground a hundredth of an inch extra from the wall of a motor cylinder would probably be told to look for another job.

Absolutely exact measurements are never possible. The instruments in a high school science lab may permit more accurate measures of length than a tape measure in your house, but they wouldn't be good enough for a NASA engineer.

The way measures are reported tells you how accurate the person making the measurement was trying to be.

DISCOVERY ACTIVITY

Take your ruler and measure the segment below.

A _____ B

1. First, measure $\overline{AB}$ to the nearest inch.

2. Next, measure $\overline{AB}$ to the nearest $\frac{1}{2}$ inch.

3. Next, measure $\overline{AB}$ to the nearest $\frac{1}{4}$ inch.

4. Next, measure $\overline{AB}$ to the nearest $\frac{1}{8}$ inch.

5. Finally, if your ruler has markings of $\frac{1}{16}$ inch, measure AB to the nearest $\frac{1}{16}$ inch.

6. When you measured to the nearest $\frac{1}{8}$ inch, what is the greatest error you could have made?

7. If you measured to the nearest $\frac{1}{16}$ inch, what is the greatest amount of error likely?

When you report a length to the nearest unit or subunit, the *error* in measurement is half that unit or *subunit*. It tells you how much uncertainty there is in the measurement.

Example: To the nearest $\frac{1}{4}$ inch, $\overline{CD}$ measures $1\frac{3}{4}$ inch. The measurement is to the nearest $\frac{1}{4}$ inch, so the error is $\frac{1}{8}$ inch.

C _____
　　　　　　　　　 D

Similar ideas apply when you measure angles. Your protractor is marked off in degrees. When you use it to measure angles, the error is $\frac{1}{2}°$.

For more accurate measure of angles, degrees are divided into *minutes* and *seconds*.

　　　1 degree = 60 minutes. The symbol for minutes is '.
　　　1 minute = 60 seconds. The symbol for seconds is ".

Example: For a measurement of an angle to the nearest degree, the error is $\frac{1}{2}°$ or 30'. For a measurement to the nearest minute, the error is $\frac{1}{2}'$ or 30".

CLASS ACTIVITY

Measure each of these segments to the nearest inch and to the smallest subunit on your ruler. State the error for each measurement.

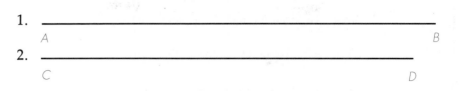

1.

 A   B

2.

 C   D

Measure each of these segments to the nearest centimeter and to the nearest millimeter. State the error for each measurement.

3. E   F

4.

 G   H

Measure each angle to the nearest 5 degrees and to the nearest degree. State the error for each measurement.

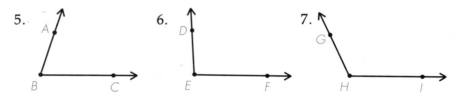

5. A

 B   C

6. D

 E   F

7. G

 H   I

CONVERTING FROM ONE UNIT OF MEASUREMENT TO ANOTHER

You can convert (change) from one unit of length to another in the customary or metric system. To do this you need to know how the units are related to one another.

The following equations show the relationships between some of the most common units of length in the customary and metric systems.

<u>Customary Measures</u>

1 foot (ft) = 12 inches (in.)
1 yard (yd) = 3 feet
1 mile (mi) = 5280 feet
1 mile = 1760 yards

<u>Metric Measures</u>

1 centimeter (cm) = 10 millimeters
1 meter (m) = 100 centimeters
1 kilometer (km) = 1000 meters

<u>Customary and Metric</u>
1 in. = 2.54 cm
1 yd = 0.91 m
1 mi = 1.6093 km

The prefixes *milli*, *centi*, and *kilo* mean one-thousandth, one-hundredth, and one thousand, respectively.

REPORT 2-4-1:   Prepare a report on the history of the metric system that includes the following information.

1. When and how the metric system was first established.

2. What kinds of quantities other than length are used in the metric system.

3. How the new International System (SI) units have refined the way basic metric units were originally defined.

Source:
The New Encyclopaedia Britannica, 15th edition, Encyclopaedia Britannica, Inc.,
Chicago: "The Metric System of Weights and Measures" from the Macropaedia article, "Measurement and Observation".

To convert from one unit to another, you multiply or divide.

Examples: 1.   A student's height is 70 in. What is her height in feet and inches? Divide to find the answer.

$$
\begin{array}{r}
5 \leftarrow \text{feet} \\
\text{Inches in 1 foot} \rightarrow 12\overline{)70} \\
\underline{60} \\
10 \leftarrow \text{inches}
\end{array}
$$

The student's height is 5 feet 10 inches.

2. How many meters is 124 centimeters?
100 cm = 1 m, and so 1 cm = $\frac{1}{100}$ m or 0.01 m. Multiply both sides of the second equation by 124. 124 cm = 1.24 m.

3. How many inches are equal to 124 centimeters? 1 in = 2.54 cm, and so 1 cm = $\frac{1}{2.54}$ in.
Multiply both sides of the second equation by 124.
124 cm = 124 × $\frac{1}{2.54}$ in.
Use your calculator and you should find that 124 cm = 48.8 in. (to the nearest tenth of an inch).

4. What is the total length of two boards that measure 5 ft 7 in. and 7 ft 8 in.? Add.

$$
\begin{array}{r}
5 \text{ ft} \quad 7 \text{ in.} \\
+ 7 \text{ ft} \quad 8 \text{ in.} \\
\hline
12 \text{ ft } 15 \text{ in.} = 12 \text{ ft} + 1 \text{ ft } 3 \text{ in.} = 13 \text{ ft } 3 \text{ in.}
\end{array}
$$

CLASS ACTIVITY

Use a calculator to help you answer the questions.

1. Tina is 58 inches tall. What is her height in feet and inches?
2. A car seat is 142 centimeters wide. What is its width in meters? In inches?
3. A basketball hoop is 10 feet above the floor. What is this height in inches? In meters?
4. A motel is 60 kilometers from Seattle. How far is it from Seattle in miles?

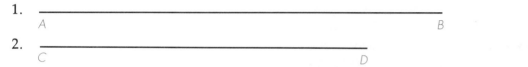

HOME ACTIVITY

Measure each of these segments to the nearest $\frac{1}{4}$ inch, the nearest $\frac{1}{8}$ inch, and the nearest centimeter. State the error for each measurement.

1. _____
   A   B

2. _____
   C   D

Measure each angle to the nearest degree. State the error for each measurement.

3.   4.   5.

Measure the sides of these figures to the nearest $\frac{1}{8}$ in. Then add to find the total distance around each figure.

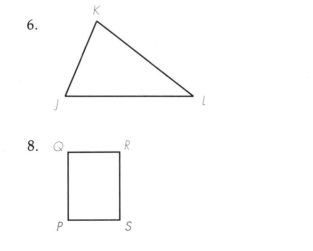

6.

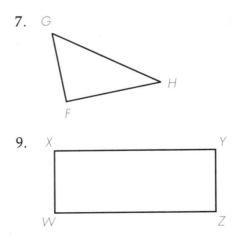

7.   G

8. Q   R   9.   X   Y

Find the sum of the angles of each figure.

10.
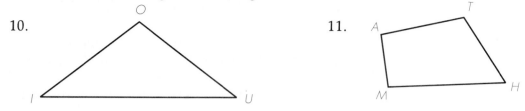

11.

12. The distance between two towns is 57 mi. What is the distance in kilometers?

13. In millimeters, how long is a segment that measures 60.9 cm?

14. A highway crew is going to pave a road that is 53,470 yd long. To the nearest tenth of a mile, how long is the road?

Add or subtract the measures.

15. 12 yd 1 ft
    +2 yd 2 ft

16. 12 yd 1 ft
    −7 yd 2 ft

17. 7 ft 10 in.
    +5 ft 6 in.

Recall that $1° = 60'$. Thus $73\frac{1}{2}° = 73° 30'$. Write each of the following measures in degrees and minutes.

18. $76\frac{1}{3}°$

19. $158\frac{3}{4}°$

20. $89.1°$

21. $118\frac{2}{3}°$

Add or subtract the angle measures.

22. 17° 25'
    +36° 30'

23. 114° 38'
    +26° 50'

24. 56° 40'
    −42° 50'

25. 147° 35'
    −80° 40'

CRITICAL THINKING

26. What is the measure of the angle formed by the hands of a clock when the time is 2:00? When it is 6:00?
Through how many degrees does the hour hand of a clock move in one hour?

Alternate Exterior Angles

After completing this section, you should understand
▲ alternate exterior angles.
▲ perpendicular lines.

Breakthrough designs using triangular braces have made it possible for architects to envision skyscrapers half a mile high. The Citicorp Building in Manhattan, for example, exerts less than half as much pressure on each square foot of ground below than the Empire State Building, which is only a few hundred feet taller.

In many modern buildings you can easily observe important relationships between the angles formed of parallel lines and a transversal.

ALTERNATE EXTERIOR ANGLES

In the diagram at the right, *a* and *b* are parallel lines intersected by transversal *t*. The angles formed are labeled 1 through 8.

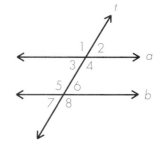

Look at angles 1 and 8. They are both *outside* the parallel lines. For this reason they are called **exterior** angles. Notice also that these two angles are on opposite or **alternate** sides of the transversal. Angles that are on opposite sides of the transversal and are outside the two parallel lines are called **alternate exterior angles.**

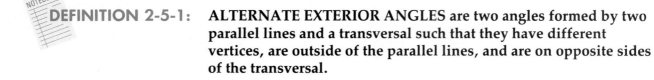

DEFINITION 2-5-1: **ALTERNATE EXTERIOR ANGLES are two angles formed by two parallel lines and a transversal such that they have different vertices, are outside of the parallel lines, and are on opposite sides of the transversal.**

Name another pair of angles in the diagram that are alternate exterior angles.

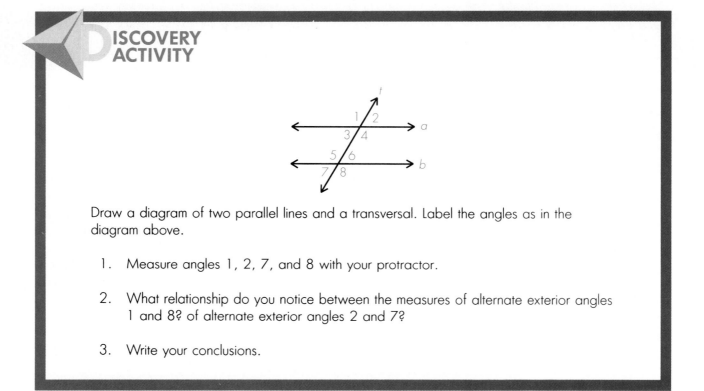

ISCOVERY ACTIVITY

Draw a diagram of two parallel lines and a transversal. Label the angles as in the diagram above.

1. Measure angles 1, 2, 7, and 8 with your protractor.

2. What relationship do you notice between the measures of alternate exterior angles 1 and 8? of alternate exterior angles 2 and 7?

3. Write your conclusions.

You may have concluded that alternate exterior angles formed by two parallel lines and a transversal are equal. By using known facts and deductive reasoning, you can prove that this is so.

Here, again, is the diagram.

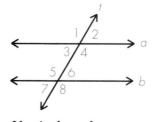

Step 1. $m\angle 1 = m\angle 4$ Reason: Vertical angles are equal.
Step 2. $m\angle 4 = m\angle 5$ Reason: Alternate interior angles are equal.
Step 3. $m\angle 5 = m\angle 8$ Reason: Vertical angles are equal.
Step 4. Since $m\angle 1 = m\angle 4$, $m\angle 4 = m\angle 5$, and $m\angle 5 = m\angle 8$, you can conclude that $m\angle 1 = m\angle 8$.

The *transitive property of equality*, which you learned when you studied algebra, states that if $a = b$ and $b = c$, then $a = c$. It is this property that allows the conclusion in Step 4. Similar reasoning will allow you to conclude that the other pair of alternate exterior angles are equal.

How do the results you have obtained through deductive reasoning compare with the results you obtained through measurement in the DISCOVERY ACTIVITY?

THEOREM 2-5-1: **If two parallel lines are intersected by a transversal, then the alternate exterior angles are equal.**

PERPENDICULAR LINES

If two lines intersect so that the angles formed are right angles, the two lines are said to be **perpendicular.**

In the diagram, line t is perpendicular to line a.

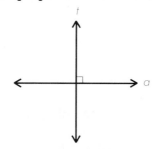

DEFINITION 2-5-2: **PERPENDICULAR LINES are lines that intersect to form right angles.**

The symbol $\perp$ means, "is perpendicular to." For the diagram above, you can write $a \perp t$ or $t \perp a$. The small ⌐ in the diagram, at the point where a and t intersect, is used to show that the angle indicated is a right angle.

Examples:

1. The lines in which the floor and walls of the room meet are perpendicular.

2. The horizontal and vertical sides of the picture frame shown here are perpendicular.

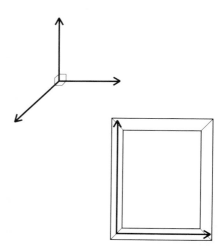

3. At 3 o'clock, the hands of this clock are perpendicular.

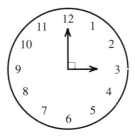

Look around you and see how many pairs of perpendicular lines you can see right where you are.

CLASS ACTIVITY

1. For an ordinary mailing envelope, how many pairs of parallel edges are there? How many perpendicular edges?

In the diagram, $a \parallel b$ and t is a transversal. For each of the following angles, name three angles equal to it.

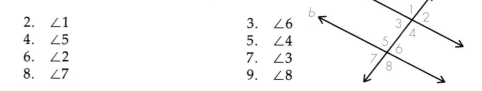

2. ∠1

4. ∠5

6. ∠2

8. ∠7

3. ∠6

5. ∠4

7. ∠3

9. ∠8

In the diagram, lines a and b are parallel and t is a transversal.

10. What is the measure of ∠1? of ∠2? and of ∠3?

11. What is the measure of each angle where t and b intersect?

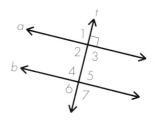

12. Is t perpendicular to b? Explain.

In the diagram, $j \parallel k$ and t is a transversal. Find the measures of angles 1 through 6.

13. $m\angle 1 =$

14. $m\angle 2 =$

15. $m\angle 3 =$

16. $m\angle 4 =$

17. $m\angle 5 =$

18. $m\angle 6 =$

In the diagram, $a \parallel b$ and $x \parallel y$. Name all the pairs of angles that fit the description.

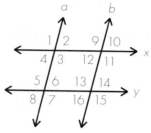

19. Alternate exterior angles with respect to a and b and transversal x.

20. Alternate exterior angles with respect to x and y and transversal b.

In the diagram, $p \parallel q$ and $f \parallel g$. Name all the angles that are equal to the given angle.

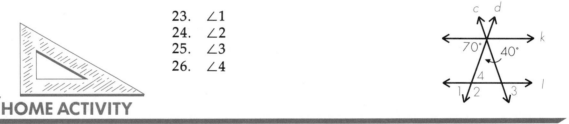

21. $\angle 1$

22. $\angle 16$

In the diagram, $k \parallel l$. Find the measures of the numbered angles.

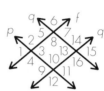

23. $\angle 1$
24. $\angle 2$
25. $\angle 3$
26. $\angle 4$

HOME ACTIVITY

In the diagram, t is a transversal of lines a and b. Name the following.

1. two pairs of alternate exterior angles.

2. four pairs of vertical angles

3. two pairs of alternate interior angles

4. four pairs of corresponding angles

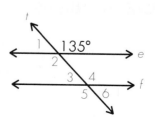

In the diagram, $e \parallel f$. Find the degree measure of each numbered angle.

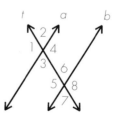

5. $\angle 1$ 6. $\angle 5$
7. $\angle 4$ 8. $\angle 2$
9. $\angle 3$ 10. $\angle 6$

In the diagram, $c \parallel d$. Find the measures of the numbered angles.

11. ∠1

12. ∠2

13. ∠3

14. ∠4

15. ∠5

16. ∠6

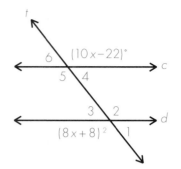

In the diagram, $p \parallel q$. Find the degree measure of each angle.

17. ∠1

18. ∠5

19. ∠4

20. ∠2

21. ∠3

22. ∠6

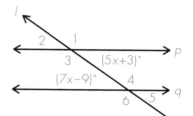

23. In the diagram, $c \parallel d$. Find the measure of ∠1.

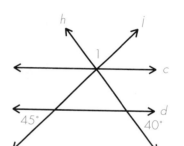

In the diagram, $s \parallel t$ and $v \parallel w$. Find the measures of the numbered angles.

24. ∠1

25. ∠2

26. ∠3

27. ∠4

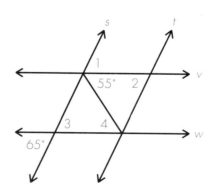

CRITICAL THINKING

28. In a diagram showing two parallel lines a and b and transversal t, Karen numbered the angles from 1 through 8. As labeled, ∠1 and ∠2 were vertical angles, ∠1 and ∠6 were alternate exterior angles, and ∠4 and ∠6 were supplementary angles. Name also another pair of angles that you can be sure are supplementary.

SECTION 2.6

Visualizing Three Dimensional Figures

After completing this section, you should understand
▲ how a three-dimensional figure appears when drawn on a two-dimensional surface.
▲ the relationship between a three-dimensional figure and its layout.

Highway systems, bridges and tunnels, buildings in apartment complexes and universities, and parks are planned by means of carefully conceived two-dimensional drawings. The ability to visualize a three-dimensional object and to represent it on a two-dimensional surface is an important technical skill for planners and designers of thousands of things around you.

DRAWINGS OF THREE-DIMENSIONAL FIGURES

All geometric figures (lines, rays, angles, and so on) consist of points. A line is one-dimensional, because a line has only length.
A plane is two-dimensional, because figures on a plane can have length and width.
Space is three-dimensional, because figures in space can have length, width, and depth.

Examples:

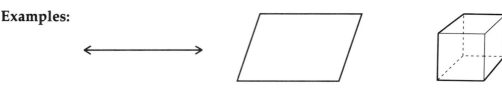

one-dimensional
or 1-D

two-dimensional
or 2-D

three-dimensional
or 3-D

Notice how the illusion of depth is created in the drawing of the three-dimensional figure. The sides of the figure that are in full view have edges drawn with solid lines. To indicate the part not in view, you make use of dotted lines.

The following drawing shows a plane, a line on the plane, and another line, not on the plane, that intersects the first line at point *A*.

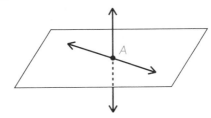

Again notice how a dotted line is used to create a three-dimensional effect.

The next drawing can be used to show three intersecting planes. The use of dotted lines is a little more complicated. Still, if you use your imagination, you can probably see how this drawing pictures something similar to three sheets of glass that intersect in the line *AB*.

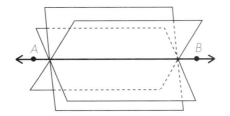

You can become good at drawing figures such as these. Practice picturing them and all their parts in your mind's eye.

DISCOVERY ACTIVITY

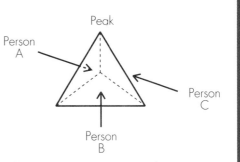

The drawing shows a three-dimensional figure called a pyramid. Imagine it as a big pyramid. It has three faces and a bottom. There are three people standing around it, looking at it from different points of view. If person B drew the figure from his point of view it would look much as it does in the drawing above.

1. Use solid lines and dotted lines to show the pyramid from person A's point of view.
2. Draw the pyramid from person C's point of view.
3. Imagine someone passing directly over the pyramid in a helicopter. Draw the pyramid from that point of view (looking down at the peak).

LAYOUTS OF THREE-DIMENSIONAL FIGURES

If you have ever eaten at a cafeteria you have probably had your food served in a cardboard pop-up container. Examine the flat **layout** and the container that it forms.

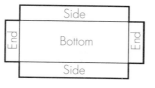

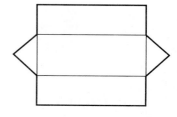

Layout   **Container**

How would you describe the closed figure that will be formed by this layout?

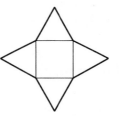

CLASS ACTIVITY

1. Draw the three-dimensional figure this layout would form.

2. Draw a layout for this round, can-shaped figure.

3. Draw the following views of a quarter placed on a tabletop.

   a. Looking directly down on it from above

   b. Looking at it from eye level with the table

   c. Looking at it from a standing position 3 feet from the table

HOME ACTIVITY

1. Draw plane P with line $\overleftrightarrow{BC}$ on it. Draw two perpendicular lines intersecting at B, one line on the plane, and the other not in the plane.

2. Fold a piece of paper in half. Open it. You now have two intersecting planes. What conclusion can you draw about the intersection of two planes?

 3. Write a Logo procedure to tell the turtle to draw a figure similar to the one shown at the right.

4. Draw the layout for this three-dimensional figure.

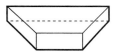

5. This drawing can be thought of as showing different figures. Describe as many as you can.

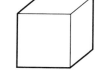

Draw closed, three-dimensional figures for each of these layouts.

6.

7.

8. Draw the three-dimensional view of the object with the following views:

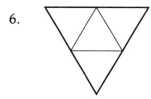

Top view   Side view   Front view

9. Draw the top view, end view, and front-side view of this building.

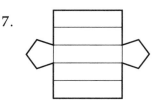

CRITICAL THINKING

10. If you were to fold each of these two layouts, which one would form a cube?

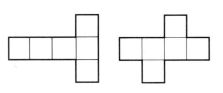

2.1 Collinear points are points on the same line. Noncollinear points are points not on the same line. Coplanar points are points on the same plane. Noncoplanar points are points that are not on the same plane. If two lines intersect, they intersect at exactly one point.

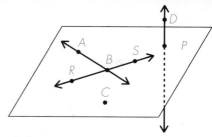

In the diagram, R, B, and S are collinear. Points R, S, and C are noncollinear. Points A, B, and C are coplanar. Points A, B, C, and D are noncoplanar. Lines $\overleftrightarrow{AB}$ and $\overleftrightarrow{RS}$ intersect at point B.

Two lines are parallel if they are coplanar and do not intersect. Skew lines are lines that do not intersect and are not parallel. In the diagram, lines a and b are parallel. Lines c and d are skew lines.

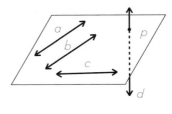

2.2 A transversal is a line that intersects two or more coplanar lines in different points. For two lines and a transversal, two angles in the same relative position are corresponding angles. If the two lines are parallel, the corresponding angles are equal. Line t is a transversal of parallel lines a and b. Since $\angle 1$ and $\angle 2$ are corresponding angles, they are equal.

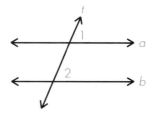

2.3 For two intersecting lines, opposite angles are vertical angles and are equal. For two parallel lines and a transversal, angles on opposite sides of the transversal and between the parallel lines are alternate interior angles. Alternate interior angles are equal. In the diagram, $p \parallel q$. $\angle 2$ and $\angle 3$ are vertical angles and $\angle 1$ and $\angle 2$ are alternate interior angles. $m\angle 1 = m\angle 2$ and $m\angle 2 = m\angle 3$.

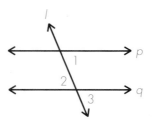

2.4 The error measurement is one-half the size of the unit or subunit to which you are measuring. A measurement to the nearest $\frac{1}{8}$ inch has an error of $\frac{1}{16}$ inch.

2.5 In the figure, $c \parallel d$. Lines s and t are transversals. $\angle 1$ and $\angle 6$ are on opposite sides of s and neither is between the parallel lines. Such angles are alternate exterior angles and are equal. Since lines t and c form $90°$ angles, they are perpendicular.

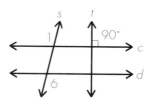

COMPUTER ACTIVITY

Once a Logo procedure has been defined, it can be used in other procedures. It is very convenient to be able to use a procedure within a procedure when the same figure needs to be drawn many times. In the example below, the procedure LINESEG is defined and then used in the procedure PARALLEL to draw parallel lines.

Example: Write a procedure named PARALLEL to tell the turtle to draw two parallel line segments.

First, write a procedure called LINESEG to draw one line segment.

TO LINESEG
RT 90
FD 60
END

Then use LINESEG in a procedure called PARALLEL to tell the turtle to draw two parallel line segments.

TO PARALLEL

LINESEG

The turtle draws a horizontal line segment 60 spaces long.

PU HOME FD 50 PD

The turtle moves back to its home position and then forward 50 spaces.

LINESEG

The turtle draws another horizontal line segment that is 60 spaces long and is parallel to the first line segment drawn.

END

The procedure PARALLEL can now be used in another procedure to draw two parallel line segments and a transversal.

Example: Write a procedure named TRANS to tell the turtle to draw two parallel line segments and a transversal.

TO TRANS

PARALLEL

The turtle draws two parallel line segments.

PU LT 90 FD 30 LT 135 PD

The turtle moves into position to draw the transversal.

FD 100

The turtle draws the transversal.

END

Modify the procedure LINESEG and the procedure PARALLEL to tell the turtle to draw each figure.

1. three parallel horizontal line segments

2. two parallel vertical line segments

3. two parallel line segments that are neither horizontal not vertical

Use the modified procedures from Exercises 1 through 3 and modify the procedure TRANS to tell the turtle to draw each figure.

4. three parallel horizontal line segments and a transversal

5. two parallel vertical line segments and a transversal

6. two parallel line segments that are neither horizontal nor vertical and a transversal

Modify any of the procedures from the examples and Exercises 1 through 6 and use them in procedures to tell the turtle to draw each of the following figures. The line segments that appear parallel are parallel.

7.   8.   9.   10.

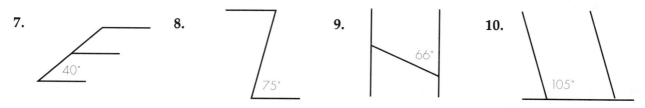

In the diagram, $\overleftrightarrow{AB}$ and $\overleftrightarrow{DC}$ do not intersect. Also, $m\angle FEA = 90°$. Identify the following.

1. intersecting lines
2. a transversal
3. perpendicular lines
4. parallel lines
5. skew lines
6. coplanar lines
7. three collinear points
8. three noncollinear points

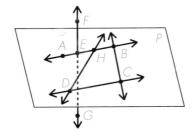

Use the diagram. Identify the following.

9. corresponding angles
10. alternate interior angles
11. alternate exterior angles
12. vertical angles

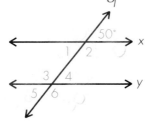

In the diagram, $x \parallel y$ give the degree measure of each angle.

13. $\angle 1$
14. $\angle 2$
15. $\angle 3$
16. $\angle 4$
17. $\angle 5$
18. $\angle 6$

Measure the segment as indicated. State the error in each measurement.

19. to the nearest $\frac{1}{2}$ inch

20. to the nearest $\frac{1}{8}$ inch

Measure each angle to the nearest degree. State the error in each measurement.

21.

22.

23. Draw a three-dimensional view of a shoebox with the top off.

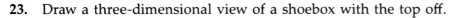

In the figure, $\overleftrightarrow{GH}$ and $\overleftrightarrow{EF}$ do not intersect. Identify the following:

1. three collinear points

2. three noncollinear points

3. four coplanar points

4. two parallel lines

5. two skew lines

6. three lines that intersect in one point

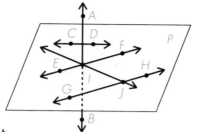

Use the diagram. Identify the following.

7. a pair of alternate interior angles
8. a pair of alternate exterior angles
9. a pair of corresponding angles
10. a transversal
11. a pair of vertical angles

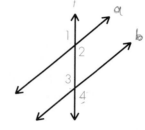

In the diagram, $p \parallel q$ and t is a transversal. Give the degree measures of each angle.

12. $\angle 1$
13. $\angle 2$
14. $\angle 3$
15. $\angle 4$
16. $\angle 5$
17. $\angle 6$

18. Draw the layout for this figure.

19. Draw the three-dimensional figure this layout would form.

20. What is the error in a measurement made to the nearest $\frac{1}{32}$ inch?

21. If a person studied hard for the test, then the person's score was high. Sarah studied hard for the test. What can you conclude about Sarah's score?

ALGEBRA SKILLS

To evaluate an algebraic expression is to find its value given specified values for its variables.

To evaluate an expression:
1. Replace the variables by their values.
2. Simplify any expressions within parentheses.
3. Simplify any powers or roots.
4. Do all multiplications and divisions, from left to right.
5. Do all additions and subtractions, from left to right.

Example:
Evaluate $3x + y$ for $x = 12$ and $y = -6$.
$$3x + y$$
$$3(12) + (-6)$$
$$36 - 6$$
$$30$$

To solve an equation means to find the number or numbers that can replace the variable to make the equation true. Such numbers are called solutions of the equation. Recall that solving equations is the process of isolating the variable in the equation in one member of the equation by using inverse operations.

Example:
Solve for x.
$$7x = 63 - 2x$$
$$7x + 2x = 63 - 2x + 2x$$
$$9x = 63$$
$$x = 7$$

Evaluate the following expressions for $x = 36$ and $y = -59$.

1. $x + y$
2. $x - y$
3. xy
4. $\frac{x}{12}$
5. $y - x$
6. x^2
7. $\frac{y - 1}{x}$
8. $2x - y$

Evaluate these expressions for $x = \frac{1}{4}$ and $y = -\frac{2}{3}$

9. $x + y$
10. $x - y$
11. xy
12. $\frac{x}{y}$
13. $y - x$
14. y^2
15. $\frac{y}{x}$
16. $\frac{x + y}{2}$

Solve for x.

17. $5x - 35 = 55$

18. $\frac{-3}{x - 4} = -\frac{3}{5}$

19. $\frac{4}{5} = \frac{2x}{20}$

20. $\frac{8}{3} = \frac{21}{3x}$

21. $10 - 2(3x - 5) = 80$

22. $12 - 5(2x - 20) = 3(2x - 8)$

23. Two angles are supplementary. One is 3 times as great as the other. What is the measure of the larger angle?

24. Angle 1 and angle 2 are corresponding angles. The measure of angle 1 is $(6x - 5)°$. The measure of angle 2 is $(x + 60)°$. What is the measure of each angle?

25. Angle 3 and angle 4 are alternate interior angles. The measure of angle 3 is $(2x - 15)°$ and the measure of angle 4 is $(85 - 3x)°$. What is the measure of each angle?

3

TRIANGLE
BASICS

My design . . . is not to explain the properties by [assuming them], but to prove them by reason and experiments.

Isaac Newton

The Sum of the Angles of a Triangle

After completing this section, you should understand
▲ the definition of a triangle.
▲ the sum of the measures of the angles of a triangle is 180°.

Recall the definitions of collinear points and noncollinear points.

Lines, line segments, and rays are determined by two points. According to Postulate 1-1-1, two points determine exactly one line. Given three points, the points can be collinear or noncollinear. If noncollinear, then three lines are determined instead of one line.

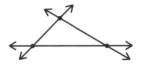

A closed figure is determined by the three lines.

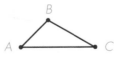

Observe the following about figure *ABC*.

Vertex	Side	Angle
A	$\overline{AB}$	∠A, or ∠BAC, or ∠CAB
B	$\overline{BC}$	∠B, or ∠ABC, or ∠CBA
C	$\overline{AC}$	∠C, or ∠ACB, or ∠BCA

The figure has three vertices, three sides, and three angles. The prefix "tri" means "three." Thus, the figures below are all triangles.

NOTEBOOK

DEFINITION 3-1-1 **A TRIANGLE is a figure consisting of three noncollinear points and their connecting line segments.**

A triangle is named by its vertices.

$\triangle ABC$

CLASS ACTIVITY

Tell whether each figure is a triangle. If the figure is not a triangle, explain why not.

1.

2.

3.

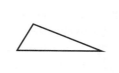

4.

THE SUM OF THE MEASURES OF ANGLES

Mathematicians have studied the properties of triangles for many years. About 300 B.C. a man named Euclid studied a property of triangles, the sum of the measures of the angles of the triangle.

DISCOVERY ACTIVITY

1. Mark three noncollinear points on a sheet of paper and label them A, B, and C. Connect the points to form a triangle.
2. Measure each angle, then find the sum of the angles.
3. Have one person record the sum for each student's triangle on the chalkboard. Using a calculator, find the average sum.
4. Draw a different triangle, and then find the sum of the angles for this triangle.
5. Write a generalization that seems to be true.

You probably found that the sum of the angles is close to 180°.

Euclid gave the following as a proof that the sum of the measures of the angles of a triangle is 180°.

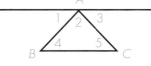

Draw a line through A parallel to $\overline{BC}$. m∠1 = m∠4 and m∠3 = m∠5. If parallel lines are intersected by a transversal, then the alternate interior angles are equal in measure. m∠1 + m∠2 + m∠3 = 180°, by definitions of straight angle and angle addition. Then m∠4 + m∠2 + m∠5 = 180° by substitution of equal numbers.

THEOREM 3-1-1 **If the figure is a triangle, then the sum of the measure of the angles is 180°.**

Euclid's proof for Theorem 3-1-1 depended on the fact that only one parallel line can be drawn through point A parallel to $\overline{BC}$. Euclid presented this as a postulate.

POSTULATE 3-1-1 **Through a point not on a line there is only one line parallel to the given line.**

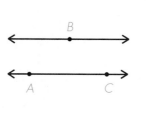

REPORT 3-1-1   The first geometry book was written by Euclid. His greatest contribution to geometry was the organization of theorems into a logical order. Use resources in the library to answer these questions about Euclid.

1. What is the name of his geometry book?
2. How many sections, or "books," did his text contain?
3. What is Euclid's famous fifth postulate?
4. Who was one of his teachers?
5. In which book can you find the Pythagorean Theorem?

Suggested sources:
Boyer, Carl B. *A History of Mathematics.* Princeton, NJ: Princeton University Press, 1985.

SOLVING FOR ANGLES

Example: Find the measure of each angle.

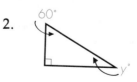

$$2x + (7x + 15) + (x - 5) = 180$$
$$10x + 10 = 180$$
$$10x = 170$$
$$x = 17$$

$\angle D$: $2(17) = 34°$
$\angle E$: $17 - 5 = 12°$
$\angle F$: $7(17) + 15 = 134°$

CLASS ACTIVITY

Find the measure of each unknown angle.

1.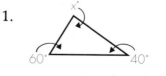

2.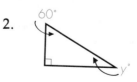

3.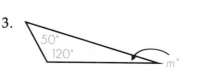

4.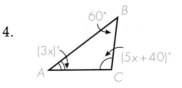

HOME ACTIVITY

For each triangle, a. name the triangle, using three letters; b. name the sides; c. name the vertices; d. name the angles (three-letter method).

1.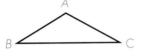

2.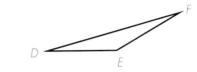

Use the information in each figure to find the measure of each unknown angle. Write the letter of the theorems or definition below that you used to solve the exercise.

A. If the figure is a triangle, then the sum of the measures of the angles is 180°.
B. If two parallel lines are intersected by a transversal, then the alternate interior angles are equal.
C. If two lines intersect, then the vertical angles are equal.
D. Two angles are supplementary if their measures add to 180°.

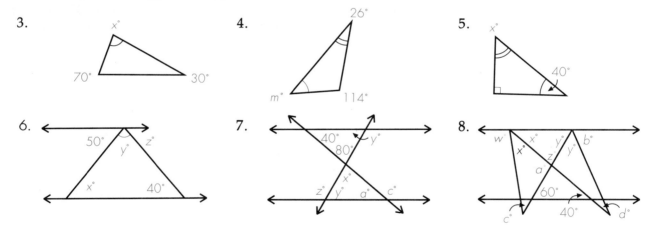

3.

4. 26°, $m°$, 114°

5.

6. 50°, $z°$, $y°$, $x°$, 40°

7. 40°, 80°, $y°$, $z°$, $x°$, $y°$, $a°$, $c°$

8.

9. Can a triangle have two obtuse angles? Explain your answer.

 Write Logo commands to draw each triangle described below. Use pencil and paper to find the measurement of the third angle.

10. $\triangle XYZ$ with m∠$X = 30°$ and m∠$Y = 90°$

11. $\triangle ABC$ with m∠$A = 60°$ and m∠$B = 60°$

CRITICAL THINKING

12. Randy and Jan predicted by inductive reasoning that the sum of any three consecutive counting numbers is divisible by three. Do you think they are right? Try a few cases and look for a pattern.

SECTION 3.2

Two Triangle Inequalities

5 6
7 8

After completing this section, you should understand
▲ indirect reasoning.
▲ the sum of the lengths of two sides of a triangle is greater than the length of the third side.
▲ the length of any side of a triangle is greater than the difference of the lengths of the other two sides.

In Section 3.1, the theorem about the sum of the measures of the angles of a triangle was presented. A proof given by Euclid was outlined to justify the theorem. Two more theorems about triangles appear in this section.

Jaime and his family live in a house with a corner lot. Jaime's dog is at *A*, and Jaime is at *B*. If Jaime calls his dog, do you think the dog will follow the sidewalk, or cut across the yard to Jaime? Why?

DISCOVERY ACTIVITY

On a sheet of paper, draw a large triangle. Measure the length of each side in inches or centimeters, and record the lengths.

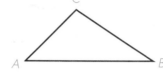

$AB =$ _____ $BC =$ _____ $AC =$ _____

C + BC. Choose the correct symbol. AC + BC (<, =, >) AB.

+ BC. Choose the correct symbol. AB + BC (<, =, >) AC.

AB. Choose the correct symbol. AC + AB (<, =, >) BC.

4. What is your conclusion in each case?

THE RELATIONSHIP BETWEEN THE SIDES OF A TRIANGLE

In the Discovery Activity, you may have found that the sum of the lengths of any two sides of a triangle is greater than the length of the third side.

This statement can be classified as a theorem if it can be justified. One method used to justify a theorem is indirect reasoning. Indirect reasoning involves listing all the possibilities for a situation, then showing that all but one of the possibilities are impossible.

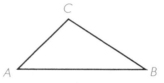

For the first part of the Discovery Activity, comparing $AC + BC$ to AB, the possibilities are

1. $AC + BC > AB.$ 2. $AC + BC = AB.$ 3. $AC + BC < AB.$

Case 2 says $AC + BC = AB$. If this were true, then by Definition 1-3-3, C is on $\overline{AB}$ between A and B. Therefore, all three points are collinear, and no triangle exists. This contradicts the given fact that figure ABC is a triangle. Therefore, Case 2 is impossible.

Case 3 says $AC + BC < AB$. This assumption implies that there is a shorter distance from A to B than the straight line path $\overline{AB}$. Therefore we must reject Case 3.

Case 1 which says $AC + BC > AB$ is the only possibility left.

This result can now be written as a theorem.

THEOREM 3-2-1 **The sum of the lengths of two sides of a triangle is greater than the length of the third side.**

Example: Use the figure to solve for all possible values for x.

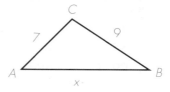

Use Theorem 3-2-1 to write three inequalities.

1. $AC + BC > AB$ 2. $AC + AB > BC$ 3. $AB + BC > AC$

Substitute the known values, and then solve the resulting inequalities for x.

1. $AC + BC > AB$ 2. $AC + AB > BC$ 3. $AB + BC > AC$
   $7 + 9 > x$ $7 + x > 9$ $x + 9 > 7$
   $16 > x$ $x > 2$ $x > -2$
   $AB < 16$ $AB > 2$ $AB > -2$

Since length must be positive, $AB > 2$ and $AB < 16$ are possible. Thus, AB is between 2 and 16. The range of possible values is $2 < x < 16$.

A SECOND TRIANGLE INEQUALITY

Another relationship can be developed from Theorem 3-2-1 and the example.

DISCOVERY **ACTIVITY**

1. Draw a triangle ABC on a sheet of paper. Let $\overline{AB}$ be the longest side of the triangle.

2. Draw a line ℓ. Measure $\overline{AB}$ and mark the length of $\overline{AB}$ on ℓ.

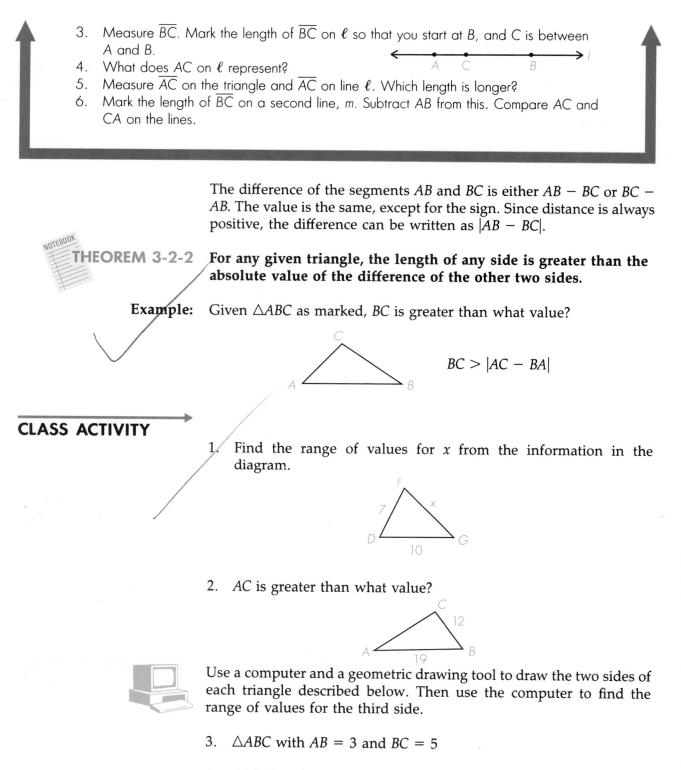

3. Measure $\overline{BC}$. Mark the length of $\overline{BC}$ on ℓ so that you start at B, and C is between A and B.
4. What does AC on ℓ represent? _____
5. Measure $\overline{AC}$ on the triangle and $\overline{AC}$ on line ℓ. Which length is longer?
6. Mark the length of $\overline{BC}$ on a second line, m. Subtract AB from this. Compare AC and CA on the lines.

The difference of the segments AB and BC is either $AB - BC$ or $BC - AB$. The value is the same, except for the sign. Since distance is always positive, the difference can be written as $|AB - BC|$.

THEOREM 3-2-2 **For any given triangle, the length of any side is greater than the absolute value of the difference of the other two sides.**

Example: Given $\triangle ABC$ as marked, BC is greater than what value?

$$BC > |AC - BA|$$

CLASS ACTIVITY

1. Find the range of values for x from the information in the diagram.

2. AC is greater than what value?

Use a computer and a geometric drawing tool to draw the two sides of each triangle described below. Then use the computer to find the range of values for the third side.

3. $\triangle ABC$ with $AB = 3$ and $BC = 5$

4. $\triangle MNP$ with $MN = 2$ and $NP = 6$

5. $\triangle RST$ with $ST = 15$ and $RT = 25$

HOME ACTIVITY

Could the following be three sides of a triangle?

1. 3, 4, 5

2. 4 cm, 5 cm, 7 cm

3. 10 ft, 20 ft, 30 ft

4. 2 m, 4 m, 6 m

5. 300, 500, 600

6. 1.5 ft, 27 in., 3.5 ft

The lengths of the sides of a triangle are given. Find the range of values of x for each.

7. 3, 5, x

8. 20 ft, 60 ft, x ft

9. 50 yd, 100 yd, x yd

10. 4.5 ft, 30 in., x ft

11. 60 km, 45 km, x km

12. Two sides of a triangular lot measure 70 feet and 90 feet. Between what two numbers is the measure of the third side?

13. BC is between what two values?

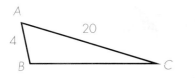

14. If Y, R, and E are three cities, what is the maximum number of miles you save by taking road $\overline{RE}$ instead of route $\overline{RY}$ and then $\overline{YE}$?

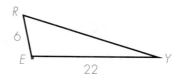

15. Find the range of values for HJ.

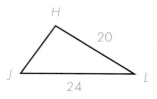

16. Use the diagram to complete the following:
_____ < SO < _____

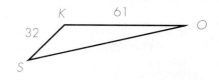

17. If the measurements are in kilometers, what is the range of values for VJ?

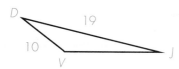

18. Is figure PRY a possible triangle? Justify your answer.

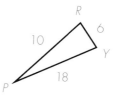

19. A city park designer was planning a new park that would be in the shape of a triangle. The designer informed the city council members that the park would measure 150′, 180′, and 340′. One of the council members directed the designer to go back and measure again. Why?

20. Are the measurements on the following drawing possible? If they are, determine the possible values for AK from both triangles.

$$6 < x < 30$$

21. Find the range of values for x that will make figure MNO a triangle.

$$5 < x < 15$$
$$6 < x < 15$$

$$x - 2 > 62$$
$$6x > 64 \quad \boxed{x > 16}$$
$$(3x - 6) - (x + 4) < 62$$
$$2x - 10 < 62$$
$$2x < 72 \quad \boxed{x < 36}$$

CRITICAL THINKING

22. Using the theorems in this section, explain why it is shorter for a dog to cut across the yard than to follow the sidewalk.

**SECTION
3.3**

Classification of Triangles

After completing this section, you should understand
▲ how triangles are classified by the length of their sides.

Thus far, three theorems that apply to all triangles have been presented. These theorems are:

The sum of the measures of the angles of a triangle equals 180°.

The sum of the lengths of any two sides of a triangle is greater than the length of the third side.

The length of any side of a triangle is greater than the absolute value of the difference of the lengths of the other two sides.

DISCOVERY ACTIVITY

Cut five strips of paper (no more than $\frac{1}{4}$-inch wide). Cut three of the strips to a length of 6 inches, one to the length of 5 inches, and one to the length of 4 inches. Using any three of the strips at a time, how many different shaped triangles can you form?

You should have found that you can make four different triangles. Certain triangles have special characteristics. One characteristic relates to the length of the sides of a triangle.

TYPES OF TRIANGLES

Triangles can be classified according to the lengths of their sides. Three possibilities exist for the lengths of the sides of a triangle.

1. All of the sides could be the same length.

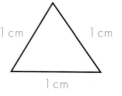

2. Two of the sides could be the same length.

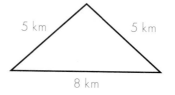

3. None of the sides are the same length.

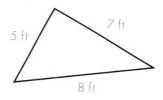

In the first type of triangle, all the sides are equal. This triangle is called equilateral.

DEFINITION 3-3-1 **An EQUILATERAL TRIANGLE is a triangle with all three sides having the same measure or length.**

In this drawing, a side note indicates that all three sides have equal measure. When notes are not included, you cannot assume from a drawing that lengths are equal. To indicate equal lengths, slash marks (/) will be included on the drawing.

Example:

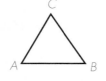

$$AB = BC = AC$$

In the second type of triangle, two sides are of equal length. This triangle is called isosceles. *Isosceles* is derived from a Greek word meaning two equal sides.

DEFINITION 3-3-2 **An ISOSCELES TRIANGLE is a triangle with two sides equal in measure or length.**

$\triangle MNO$ and $\triangle XYZ$ are isosceles triangles.

In an isosceles triangle, the two sides of equal length are called the LEGS of the triangle. The third side is called the BASE of the isosceles triangle.

The third type of triangle has sides that all have different lengths. The name given to this triangle is *scalene*—Greek for uneven.

DEFINITION 3-3-3 **A SCALENE TRIANGLE is a triangle with no equal sides.**

$$AB \neq BC \neq AC$$

Different numbers of slash marks on segments in a drawing indicate different lengths.

CLASS ACTIVITY

Classify each triangle.

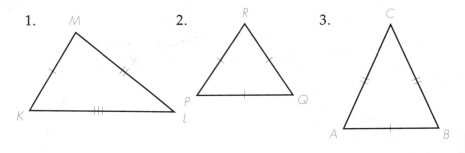

1.

2.

3.

Use figure *ABCD* to identify

4. a scalene triangle.
5. an isosceles triangle.
6. an equilateral triangle.
7. the total number of triangles shown.

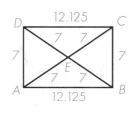

Choose the lengths of the sides for the type of triangles described. Then use a computer and a geometric drawing tool to draw each triangle. Compare your triangles with your classmates' triangles.

8. Isosceles triangle
9. Scalene triangle
10. Equilateral triangle

HOME ACTIVITY

Classify the triangles according to the markings.

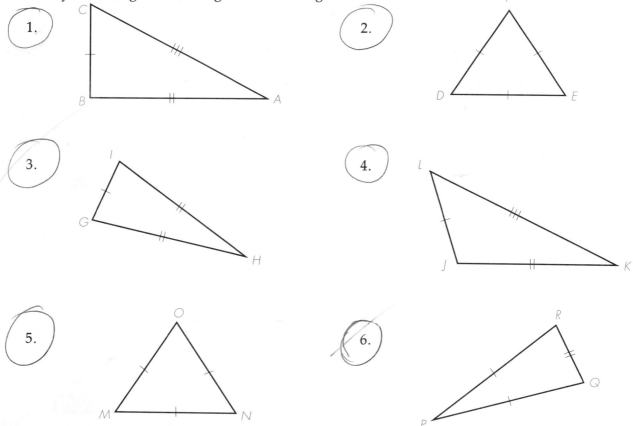

1.

2.

3.

4.

5.

6.

Classify each triangle based on figure *ABCD*.

7. △*ABC*

8. △*ADB*

9. △*ADO*

10. △*AOB*

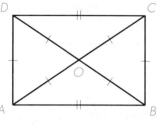

Classify each triangle based on figure *PQRT*.

11. △*PTR*

12. △*ROT*

13. △*QOR*

14. △*PTQ*

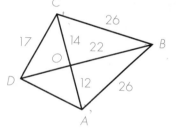

15. How many isosceles triangles can be found?

Classify each triangle based on figure *ABCD*.

16. △*ACB*

17. △*COB*

18. △*ABO*

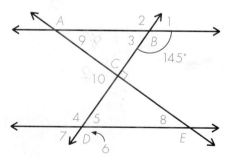

Lines *AB* and *DE* are parallel. Lines *AE* and *DB* are perpendicular.

19. Find the measures of angles 1–10.

20. Is m∠9 + m∠3 = m∠10?

21. What do you know about ∠*ACB* and ∠*DCE*?

22. What is the measure of ∠A?

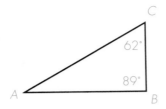

23. Is an equilateral triangle an isosceles triangle? Justify your answer.

24. Is an isosceles triangle an equilateral triangle? Justify your answer.

CRITICAL THINKING

25. The distances from the island of Sandwich to the island of Whitebread and the ports at Rye and Pumpernickel are correctly marked on the figure. The cargo boat *Robin* starts at the island of Whitebread and sails by the most direct route to Rye and Pumpernickel and then back to Whitebread. The skipper figures at the end of the trip that the *Robin* has traveled 25 kilometers altogether. Is this possible? Give a reason for your answer.

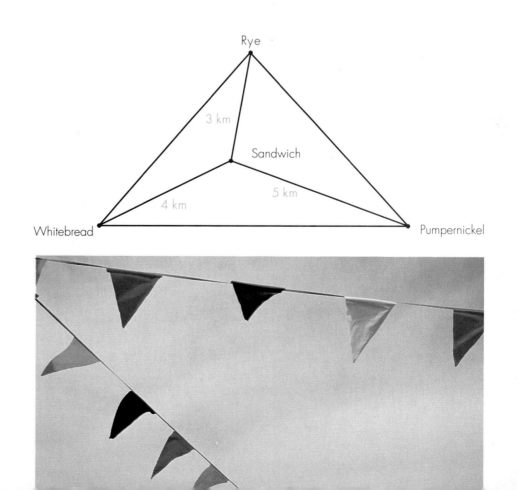

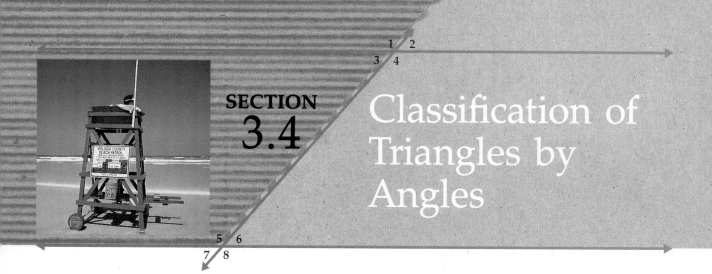

SECTION 3.4

Classification of Triangles by Angles

After completing this section, you should understand
▲ how to classify triangles by the measure of their angles.

Triangles were classified by the lengths of their sides in Section 3.3.

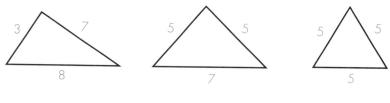

| Scalene | Isosceles | Equilateral |

CLASSIFYING TRIANGLES BY ANGLES

Triangles can also be classified according to the measure of their angles. Triangles with equal angles share special properties independent of the lengths of their sides.

Case 1: All the angles are equal in measure.

The arcs with slash marks indicate equal angles. The name given to this triangle is equiangular triangle. Since the sum of the measures of the angles of a triangle is 180°, each angle measures 60°.

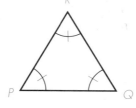

DEFINITION 3-4-1 **An EQUIANGULAR TRIANGLE is a triangle with the measure of each angle equal to 60°.**

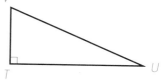

$$m\angle P = m\angle Q = m\angle R = 60°$$

Case 2: One of the angles is a right angle.

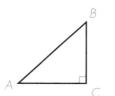

The symbol ∟ is used to indicate a right angle. $\triangle TUV$ is a right triangle, where $\angle T$ is the right angle. Recall that right angles measure 90°.

DEFINITION 3-4-2 **A RIGHT TRIANGLE is a triangle with a 90° angle.**

$$m\angle C = 90°$$

Case 3: All of the angles measure less than 90°. Recall that an angle measuring less than 90° is an acute angle. In this case the triangle is called an acute triangle.

DEFINITION 3-4-3 **An ACUTE TRIANGLE is a triangle with the measures of all three angles less than 90°.**

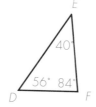

$\triangle DEF$ is an acute triangle.

Case 4: One angle measures greater than 90°. Recall that an angle measuring greater than 90° is an obtuse angle. In this case the triangle is called an obtuse triangle.

DEFINITION 3-4-4 **An OBTUSE TRIANGLE is a triangle with one angle measuring greater than 90°.**

∠A is obtuse, therefore △ABC is an obtuse triangle.

Example: Refer to figure *ABCFD*. Figure *ABCD* is a rectangle. Classify each triangle by its angles.

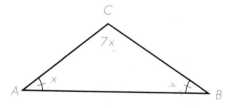

△ABC is right.   △DAO is acute.
△DOC is obtuse.   △DCF is obtuse.
There are nine triangles in all.

Example: A triangle has two equal angles. The third angle is seven times one of the equal angles. What is the measure of each angle?

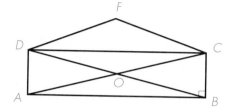

Solution:

Step 1: Draw a diagram.
Step 2: Label the angles.
   x is the measure of one of the equal angles.
   $7x$ is the measure of the largest angle.
Step 3: Write and solve an equation.
$$x + x + 7x = 180$$
$$9x = 180$$
$$x = 20$$
Step 4: Interpret the answer to the equation.
   $x = 20°$ the measure of each equal angle
   $7x = 140°$ the measure of the largest angle
Step 5: Check.
   $20° + 20° + 140° = 180°$

CLASS ACTIVITY

Solve for each unknown angle. Then classify each triangle according to its angles.

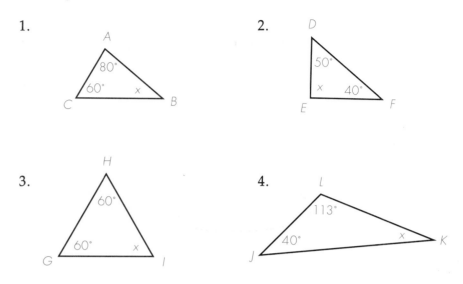

1.

2.

3.

4.

5. Solve for the measures of angles X, Y, and Z. Then classify the triangle according to its angles.

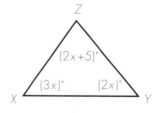

Use a computer and a geometric drawing tool to complete.

6. Draw an isosceles triangle. Measure each angle. Repeat this activity several times, using different lengths for the sides each time. How can an isosceles triangle be classified according to the measure of its angles?

7. Draw an equilateral triangle. Measure each angle. Repeat this activity several times, using different lengths for the sides each time. How can an equilateral triangle be classified according to the measure of its angles?

8. Draw a scalene triangle. Measure each angle. Repeat this activity several times, using different lengths for the sides each time. How can a scalene triangle be classified according to the measure of its angles?

HOME ACTIVITY

1. Solve for the measures of the unknown angles in △ABC, then classify the triangle according to its angles.

2. In the sketch of the ramp, what is the measure of the unknown angle? The ramp illustrates the case of what kind of triangle?

3. A city surveyor measured the angles of the triangular-shaped lot, but forgot to record the measure of ∠R. Find the measure of ∠R, and classify the triangle.

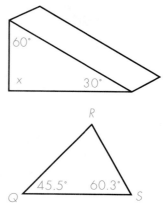

Refer to triangle *ABC*. Find the measures.

4. m∠ABC

5. m∠ABD

6. m∠A + m∠C

7. Compare the answers to Exercises 5 and 6.

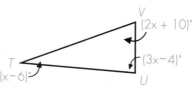

8. Refer to *TUV*. Solve for the measures of the angles of the triangle, then classify the triangle.

$$m\angle T = \underline{\qquad};$$
$$m\angle U = \underline{\qquad};$$
$$m\angle V = \underline{\qquad};$$

9. Refer to △*JTO*. Solve for the measures of the angles, then classify the triangle.

$$m\angle J = \underline{\qquad};$$
$$m\angle O = \underline{\qquad};$$
$$m\angle JTO = \underline{\qquad};$$

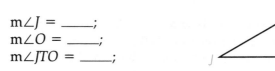

Refer to △ABC. $\overline{BE}$ and $\overline{CD}$ are angle bisectors.

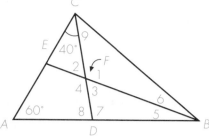

10. Find the measures of angles 1 to 9.

11. Use the three-letter method to name all the triangles in the figure.

12. Name one obtuse triangle.

CRITICAL THINKING

Answer true or false for Exercises 13–25.

13. A triangle can have three acute angles.

14. A triangle can have two acute angles.

15. A triangle can have one acute angle.

16. A triangle can have no acute angles.

17. A triangle can have no obtuse angles.

18. A triangle can have one obtuse angle.

19. A triangle can have two obtuse angles.

20. A triangle can have three obtuse angles.

21. A triangle can have no right angles.

22. A triangle can have one right angle.

23. A triangle can have two right angles.

24. A triangle can have an acute angle and a right angle.

25. A triangle can have an obtuse angle and a right angle.

26. Draw a right triangle that has sides or legs of length 10 cm and 3 cm. The length of the third side is between what two integers? What is the actual length of the third side? What is the name of the third side? [Hint: See Report 3-1-1, page 85.]

SECTION
3.5

Exterior Angles of a Triangle

After completing this section, you should understand
▲ the triangle exterior angle theorem.

The concepts in geometry have so far been divided into four different categories.

Category	Examples
1. Undefined term	point, line, plane
2. Defined term	segment, ray, angle, triangle
3. Postulate	If two parallel lines are intersected by a transversal, then the corresponding angles are equal in measure. Two points determine exactly one line.
4. Theorem	If the figure is a triangle, then the sum of the measures of the angles is 180°. If two lines intersect, then the vertical angles are equal in measure.

Recall that theorems must be justified.

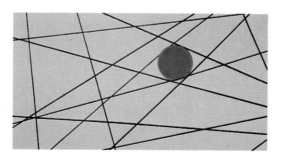

DISCOVERY ACTIVITY

1. On a sheet of paper, draw and label a triangle *ABC*. Extend segment $\overline{AB}$ past point *B*, and indicate point *D* on the extension.

2. Use a protractor to measure each of the following angles. Record the measurements.

   m∠A _____ m∠C _____ m∠A + m∠C _____ m∠CBD _____

3. Compare m∠A + m∠C to m∠CBD.

4. Compare your results in Part 3 with the results of your classmates. Were the results the same?

5. Draw a different triangle and repeat Steps 1–4. Were the results in Step 4 the same for everyone again?

The Discovery Activity shows another way to justify a theorem—using an inductive approach. With the inductive approach, several cases are examined, a pattern is established, and a conclusion is drawn from this pattern about all cases, not just those that were checked.

Some angles from the Discovery Activity need to be defined before the next theorem is introduced.

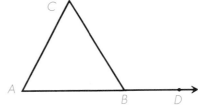

∠*CBD* is called an exterior angle of a triangle.

DEFINITION 3-5-1 **An EXTERIOR ANGLE of a triangle is the angle less than 180° in measure, formed by extending one side of the triangle.**

∠*A* and ∠*C* are the remote, or non-adjacent, interior angles for the exterior angle, ∠*CBD*.

You should have observed from the Discovery Activity that the measure of the exterior angle of a triangle seems to always be equal to the sum of the measures of the two remote interior angles. This fact is the CONCLUSION of the observation. The drawing that you started with showed the GIVEN facts.

Here is an outline of the justification of the conclusion.

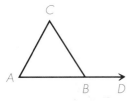

1. m∠A + m∠B + m∠C = 180°
    Reason: Sum of the measures of the angles of a triangle is 180°.
2. m∠ABD = 180°
    Reason: Definition of straight angle
3. m∠ABC + m∠CBD = 180°
    Reason: Definition of supplementary angles
4. m∠ABC + m∠CBD = m∠A + m∠B + m∠C
    Reason: Sums equal to 180° are equal to each other.
5. m∠ABC = m∠B
    Reason: Identity
6. m∠CBD = m∠A + m∠C
    Reason: Equal numbers (measurements) can be subtracted from both sides of an equality.

Step 6 shows the conclusion in the form of a statement. This outline serves as a proof of Theorem 3-5-1.

THE TWO-COLUMN PROOF

Proofs are often given with the statements shown in one column and the reasons shown in another column. This is called a two-column proof. Theorem 3-5-1 follows, with a two-column proof to justify it.

THEOREM 3-5-1   **The measure of an exterior angle of a triangle is equal to the sum of the measures of the two non-adjacent, or remote, interior angles.**

Given: △ABC with side $\overline{AB}$ extended to D.
Prove: m∠A + m∠C = m∠CBD

Statements	Reasons
1. m∠A + m∠B + m∠C = 180°	1. Theorem 3-1-1. Sum of the measures of the angles of a triangle = 180°.
2. Extend $\overline{AB}$ to D.	2. Two points determine a line.
3. ∠ABD is a straight angle.	3. $\overline{BD}$ drawn to extend $\overline{AB}$
4. m∠ABD = 180°	4. Definition of straight angle
5. m∠ABC + m∠CBD = 180°	5. Definition of supplementary angles
6. m∠ABC + m∠CBD = m∠A + m∠B + m∠C	6. Both sides equal to 180°.
7. m∠B = m∠ABC	7. Same angle
8. m∠B + m∠CBD = m∠A + m∠B + m∠C	8. Substitution
9. m∠CBD = m∠A + m∠C	9. Subtract m∠B from both sides.

CLASS ACTIVITY

What is the sum of m∠1 and m∠2 for each triangle?

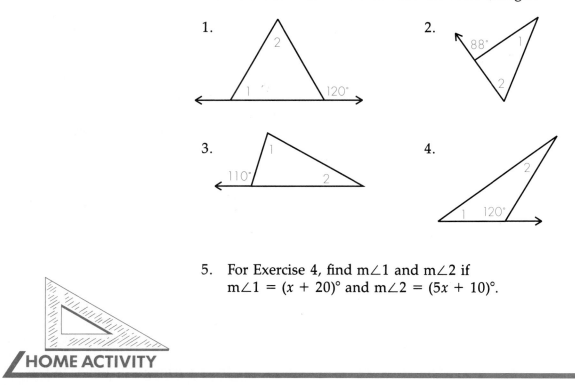

1.

2.

3.

4.

5. For Exercise 4, find m∠1 and m∠2 if
m∠1 = (x + 20)° and m∠2 = (5x + 10)°.

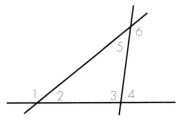

HOME ACTIVITY

1. Draw a triangle, △ABC. Extend the sides. Mark each exterior angle with an arc. How many exterior angles are there?

2. Using Theorem 3-5-1, write three equations involving the exterior angles marked in the figure.

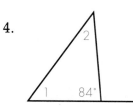

Find the sum of m∠1 and m∠2.

3.

4.

5.

6. Find m∠1 and m∠2.

$$m\angle 1 = (3x)°$$

$$m\angle 2 = (6x + 30)°$$

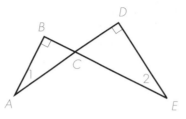

7. Find each angle measurement if $l_1 \parallel l_2$.

8. Show that m∠1 = m∠2. $\overline{BE} \perp \overline{AB}, \overline{DE} \perp \overline{AD}$

9. What is the sum of the measures of the six exterior angles of a triangle?

10. What kind of angles are an exterior angle and its adjacent interior angle of a triangle?

11. Can the measure of an exterior angle of a triangle ever equal the measure of its adjacent interior angle? Explain.

12. Can the measure of an exterior angle of a triangle ever equal the measure of one of the remote interior angles? Explain.

Refer to △ABC. $\overline{AC}$ represents a shortcut on a path.

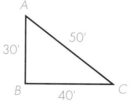

13. How many feet are saved by taking the shortcut?

14. If a student takes the shortcut twice a day, how many feet will she save in 5 days? in 20 days? in a school year (180 days)?

15. Convert the answer for the school year in Exercise 14 to miles.

Write Logo procedures to draw each triangle described below by moving the turtle forward and to the right each time.

16. Equiangular △MNP
17. △XYZ with m∠X = 40, m∠Y = 30, m∠Z = 110
18. △TUV with m∠T = 125, m∠U = 20, m∠V = 35
19. △CDE with m∠C = 62, m∠D = 90, m∠E = 28

Complete the two-column proof.

20. Given: $\overrightarrow{EA}$ bisects ∠DEF; m∠1 + m∠F = 90°

   Prove: ∠2 and ∠F are complementary.

Statements

a. $\overrightarrow{EA}$ bisects ∠DEF.
b. ∠1 = ∠2
c. m∠1 + m∠F = 90°
d. m∠2 + m∠F = 90°
e. ∠2 and ∠F are complementary.

Reasons

a.
b.
c.
d.
e.

CRITICAL THINKING

21. Roger, Bob, and Ruth are arguing over the number of points their bowling team scored during the season. They each made one statement. Only one person is telling the truth. Who is telling the truth?

Roger: The team scored at least 2800 points.
Bob: The total was not as great as Roger said.
Ruth: The team scored at least 1500 points.

22. A newspaper was discarded before everyone had a chance to read it. Four siblings, one of whom committed this act, made the following statements when questioned.

   Alan: Doris did it.
   Doris: Trina did it.
   Gary: I didn't do it.
   Trina: Doris lied when she said I did it.

a. If only one of these four statements is true, who was guilty?
b. If only one of these four statements is false, who was guilty?

Triangles on the *x-y* Plane

After completing this section, you should understand
▲ the *x-y* (Cartesian) coordinate plane.
▲ how the Cartesian coordinate plane is used in geometry.

In the early 1600's, a man named René Descartes (pronounced day cart) adapted plane geometry to a plane surface with a numbered grid. The numbered grid made locating points much easier for people.

DISCOVERY ACTIVITY

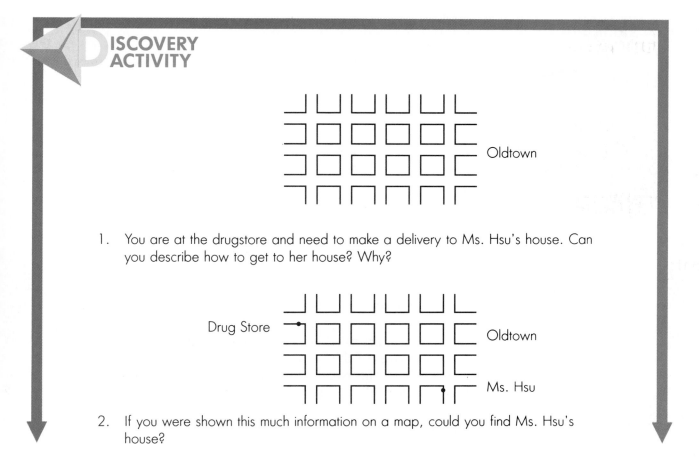

1. You are at the drugstore and need to make a delivery to Ms. Hsu's house. Can you describe how to get to her house? Why?

2. If you were shown this much information on a map, could you find Ms. Hsu's house?

3. What would make it easier for someone who didn't know the town to go from the drugstore to Ms. Hsu's house?

You should have found from the Discovery Activity that having labels such as street names and address numbers help people find specific points or locations.

THE CARTESIAN PLANE

Descartes' Cartesian plane consists of two number lines that are perpendicular to each other. The horizontal number line is the *x*-AXIS and the vertical number line is the *y*-AXIS. The point of intersection is the ORIGIN.

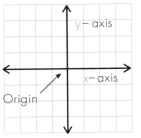

The positive numbers are to the right on the *x*-axis and up on the *y*-axis. Negative numbers are to the left on the *x*-axis and down on the *y*-axis. The arrows indicate that the axes extend indefinitely.

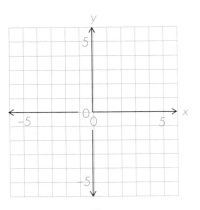

Two numbers are needed to locate a point. These numbers are called the coordinates of the point. The *x*-coordinate is given first, then the *y*-coordinate. The two coordinates form an ordered pair, a pair of numbers that follow a particular order.

DEFINITION 3-6-1   **A point on the *x-y* plane is an ORDERED PAIR of numbers, (*x,y*).**

Example: Locate the point $(3,-4)$.

Start at the origin. Move 3 positive units (to the right) on the x-axis and 4 negative units (down) on the y-axis.

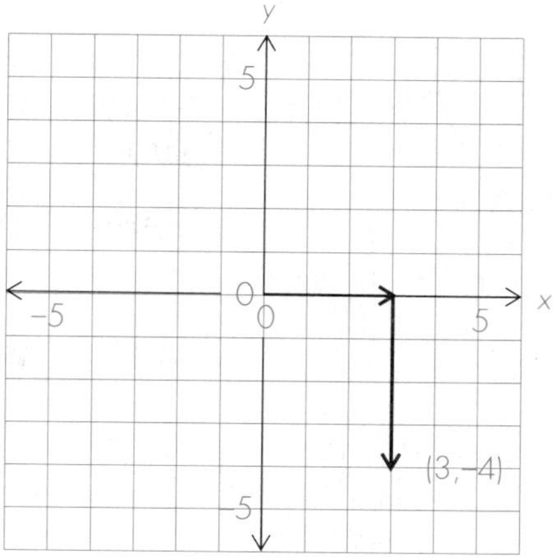

Example: On the x-y plane, locate the following points:

$A(2,4)$
$B(-5,1)$
$C(2,-5)$
$D(-1,-7)$
$E(0,0)$
$F(0,3)$
$G(-4,0)$
$H(0,-3)$
$I(7,0)$
$J(\frac{1}{2},\frac{1}{2})$

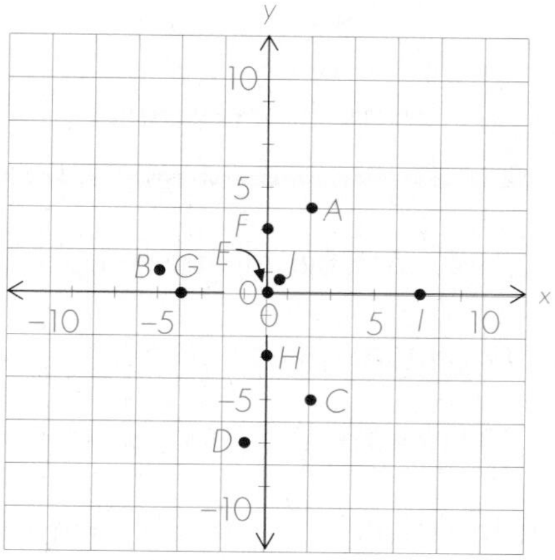

CLASS ACTIVITY

Draw the x-y plane and label the axes from -10 to 10. Locate the following points.

1. $A(2,6)$ 2. $B(3,5)$ 3. $C(4,0)$ 4. $D(3,-2)$

5. $E(-5,3)$ 6. $F(-6,-1)$ 7. $G(7,-4)$ 8. $H(-4,0)$

9. $I(-2,4)$ 10. $J(0,9)$ 11. $K(0,0)$ 12. $L(8,-4)$

13. $M(-4,-7)$ 14. $N(5,0)$ 15. $O(0,-3.5)$ 16. $P(-2.7,-7.3)$

The coordinates of the vertices of a triangle are given. Use a graphing calculator to graph each triangle. Then classify each triangle according to its sides and angles.

17. $A(0,5)$, $B(5,0)$, $C(0,0)$
18. $D(0,9)$, $E(7,0)$, $F(4,8)$
19. $G(-1,-1)$, $H(4,0)$, $I(-1,2)$
20. $J(5,-7)$, $K(1,-1)$, $L(8,-1)$

21. Give the coordinates of each point.

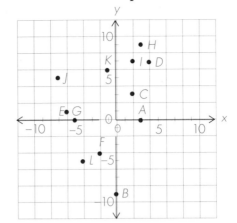

22. Plot the points (0,0), (5,5), and (10,0) on the *x-y* plane. Connect the three points, and describe the triangle formed.

HOME ACTIVITY

Graph each set of ordered pairs on a separate coordinate plane.

1. *A*(0,5), *B*(5,0), *C*(0,0)

2. *D*(0,9), *E*(7,0), *F*(4,8)

3. *G*(−1,−1), *H*(4,0), *I*(−1,4)

4. *J*(5,−7), *K*(1,−1), *L*(8,−1)

Connect the three points with line segments. Classify each triangle according to its sides and its angles.

5. Give the coordinates of each point.

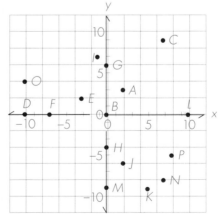

6. Draw the *x-y* plane and label the axes from −10 to 10. Locate the following points.

A(7,0)	*B*(1,2)	*C*(0,0)	*D*(−8,6)
E(4,6)	*F*(1,−10)	*G*(−2,−3)	*H*(−6,−6)
I(4.5,−6.5)	*J*(−4,4)	*K*(6,4)	*L*(2,−3)
$M(-\frac{2}{3},\frac{7}{3})$	*N*(−8.5,−3)	*O*(0,6)	*P*(−7,0)

Name the type of figure formed if each set of points is connected in order, left to right, as given. Use as many descriptive names as possible (i.e., use sides and angles to describe triangles).

7. (0,5), (0,0), (7,0), (0,5)   8. (0,0), (6,6), (7,7)

9. (−8,7), (0,0), (5,0), (−8,7)   10. (1,−5), (0,0), (−5,1), (1,−5)

11. All points that have an x-coordinate of 0 and a positive y-coordinate represent what geometric figure?

12. What is the minimum number of noncollinear points needed to determine a plane?

13. Is the point (3,4) the same as the point (4,3)? Explain.

14. Plot these points and identify the type of triangle outlined: (−3,0), (3,0), (0,5.2)

Give the relationship between each pair of lines and the coordinates for any point of intersection.

15. The line determined by points (0,3) and (0,−1), and the line determined by the points (2,4) and (2,−3)

16. The lines determined by points (3,3) and (−4,−4), and the line determined by the points (−3,3) and (2,−2)

17. The line determined by the points (−4,3) and (−1,−1), and the line determined by the points (3,−1) and (4,3)

18. The line determined by the points (0,1) and (2,2), and the line determined by the points (−6,−2) and (−2,0)

19. The figures in the coordinate plane at right share a number of properties. Give the coordinates of the sides of each figure, and describe two properties the figures all share.

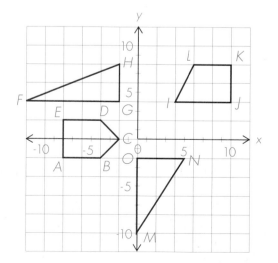

CRITICAL THINKING

20. Give 12 points such that when the coordinates of each point are reversed, the same point results.

21. Write an equation that relates the *x*-coordinate and *y*-coordinate of each point in Exercise 20. What geometric figure is formed when the points are connected?

A triangle is a rigid form. If you fasten three strips of cardboard together at their ends, the triangle that results holds its shape, even when it is moved. This property of triangles is the secret of the strength of many structures.

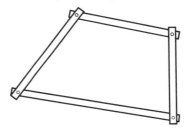

A quadrilateral, made with four strips fastened together, is flexible and can be changed into a variety of shapes.

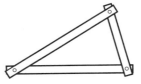

Decide whether the constructions shown below would be rigid or flexible.

22.   23.

24.   25.

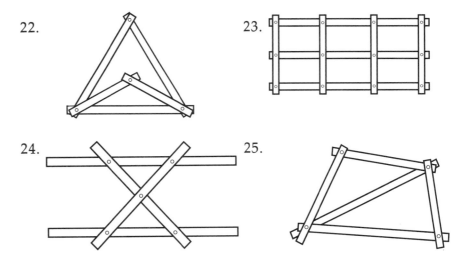

3.1 Figure *RST* is a triangle.

$$m\angle R + m\angle S + m\angle T = 180°$$

3.2 The lengths of the sides of any triangle have the following relationships:

$RS + ST > TR$ $\quad\quad\quad\quad\quad\quad$ $|RS - ST| < TR$
$ST + TR > RS$ $\quad\quad\quad\quad\quad\quad$ $|ST - TR| < RS$
$RS + TR > ST$ $\quad\quad\quad\quad\quad\quad$ $|RS - TR| < ST$

3.3 △*ABC* is an equilateral triangle; all three sides are equal in length.

△*DEF* is an isosceles triangle; two sides are equal in length.

△*GHI* is a scalene triangle; all three sides are of different length.

3.4 △*ABC* is an acute triangle; all three angles measure less than 90°. It is also equiangular.

△*DEF* is a right triangle; it has one 90° angle.

△*GHI* is an obtuse triangle; one angle measures more than 90°.

3.5 ∠*QST* is an exterior angle of △*QRS*. The measure of an exterior angle of a triangle is the sum of the remote interior angles.

$$m\angle R + m\angle Q = m\angle QST$$
$$70° + 30° = 100°$$

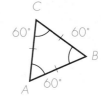

3.6 Point *F* on the *x-y* plane can be identified by an ordered pair of numbers, $(-3,4)$.

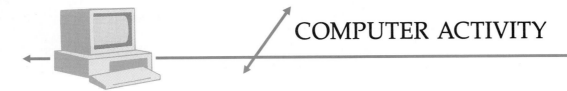
When using Logo, you can think of the computer screen as an *x-y* plane in which the turtle's home position is the origin (0,0). Every location on the screen can be represented by an ordered pair of numbers (*x,y*).

There are special Logo commands that can be used to move the turtle to any location on the screen's *x-y* plane. The command SETX moves the turtle horizontally to the *x*-coordinate that you specify. For example, SETX 10 tells the turtle to move horizontally to the *x*-coordinate 10. The command SETY moves the turtle vertically to the *y*-coordinate that you specify. For example, SETY 6 tells the turtle to move vertically to the *y*-coordinate 6. The command SETXY moves the turtle to the location with coordinates (*x,y*) that you specify. For example, SETXY 3 8 tells the turtle to move to the location with coordinates (3,8). If the *y*-coordinate in the SETXY command is negative, you must put parentheses around it.

Example: Write a procedure named XYPLANE that tells the turtle how to draw the *x*-axis and the *y*-axis with the origin at the turtle's home position.

```
TO XYPLANE

    SETX 139              These commands tell the
    HOME                 turtle to draw the x-axis.
    SETX −139
    HOME

    SETY 119
    HOME                 These commands tell the
    SETY −119            turtle to draw the y-axis.
    HOME

END
```

The screen will appear as shown below.

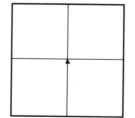

The procedure XYPLANE can now be used in other procedures to show the graph of ordered pairs on the *x-y* plane.

Show how the turtle moves when you type these Logo commands. Classify each triangle according to its sides and its angles.

1. XYPLANE	2. XYPLANE	3. XYPLANE	4. XYPLANE
PU	PU	PU	PU
SETXY 1 1	SETXY −1 2	SETXY −2 (−2)	SETXY 4 (−1)
PD	PD	PD	PD
SETXY 3 1	SETXY −2 4	SETXY 2 (−2)	SETXY 0 2
SETXY 2 4	SETXY −2 2	SETXY 2 2	SETXY −4 (−1)
SETXY 1 1	SETXY −1 2	SETXY −2 (−2)	SETXY 4 (−1)

Use the procedure XYPLANE in procedures that tell the turtle to draw each figure. Give each procedure the indicated name.

5.

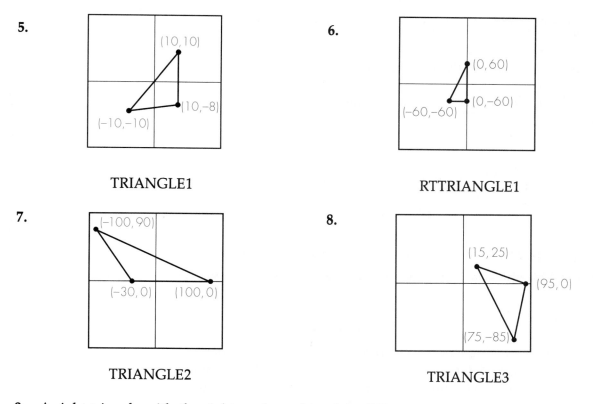

TRIANGLE1

6.

RTTRIANGLE1

7.

TRIANGLE2

8.

TRIANGLE3

9. A right triangle with the right angle at the origin; RTTRIANGLE2

10. An isosceles triangle with the midpoint of the base at the origin; ISOSCELES

Match the types of triangles on the left with the figures on the right. Each term may describe more than one triangle. List as many as possible. Triangles may be used more than once.

1. Equilateral

2. Isosceles

3. Scalene

4. Equiangular

5. Right

6. Acute

7. Obtuse

8. If a figure is a triangle, then the sum of the measures of the angles is _____.

Refer to $\triangle ACD$. Find the measure of each angle.

9. $\angle 1$

10. $\angle 2$

11. $\angle 3$

12. $\angle 1 + \angle 3$

13. $\angle 5$

14. $\angle 4$

Refer to $\triangle RAT$. If $RT = 37$, $RB = 29$, and $BA = 20$, then find the limits for the line segment.

15. ____ $< TB <$ ____

16. ____ $< TA <$ ____

17. Use indirect reasoning to reach the proper conclusion. A guidance counselor questioned three students, trying to determine which one had damaged a car and which one had damaged a bike. Only one of the following statements is true. Who damaged the car, and who damaged the bike?

   Alice: Carl did not damage the car.   Carl: Bev did damage the bike.   Bev: Carl is lying.

Plot the following points on the x-y plane and identify the type of triangle that is formed when the points are connected.

18. $(0,-5)$, $(-5,0)$, $(0,0)$

19. $(5,7)$, $(1,1)$, $(-8,1)$

Write the letter of each triangle that fits the description. There may be more than one triangle that fits, and triangles may be used more than once.

1. Acute triangle

2. Scalene triangle

3. Equilateral triangle

4. Obtuse triangle

5. Right triangle

6. Isosceles triangle

7. Equiangular triangle

a.

b.

c.

d.

e.

f.

g.

h.

8. The length of the unknown side in the triangle is between what two numbers?

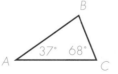

9. What is the measure of ∠B?

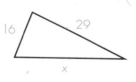

10. Given the following triangle and the information indicated, find the measure of angles 1–5.

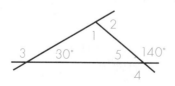

11. Plot and connect the following points. Classify the type of triangle formed. $A(0,5)$, $B(-5,5)$, $C(-5,0)$

ALGEBRA SKILLS

Example:

Evaluate $x + 3y$ using $x = 1.6$ and

$y = 2.8$.

$1.6 + 3(2.8)$

$1.6 + 8.4$

10

Example:

Solve for x.

$3x - 5 = 7x + 9$

$3x - 14 = 7x$

$-14 = 4x$

$x = -3.5$

In Exercises 1–8, evaluate each expression using $x = 2.1$ and $y = -5.2$.

1. $x + y$
2. $x - y$
3. xy
4. $x/7$

5. $y - x$
6. x^2
7. $\frac{y - 1}{2}$
8. $|y|$

In Exercises 9–16, evaluate each expression using $x = \frac{1}{5}$ and $y = -\frac{2}{3}$.

9. $x + y$
10. $x - y$
11. xy
12. $\frac{x}{y}$

13. $y - x$
14. y^2
15. $\frac{y}{x}$
16. $|y|$

In Exercises 17–22, solve each equation for x.

17. $\frac{2}{x} = \frac{54}{27}$
18. $\frac{-3}{x - 4} = \frac{45}{27}$
19. $\frac{5}{4} = \frac{2x}{20}$
20. $\frac{3}{8} = \frac{21}{2x}$

21. $20 - 2(3x - 5) = 90 - 2x$
22. $12 - 5(2x - 20) = -3(2x + 8)$

Solve each problem.

23. If the ratio of two angles is $\frac{2}{3}$ and the angles are complementary, then what is the measure of the larger angle?

24. If m$\angle A$ in a given triangle is twice as great as m$\angle B$, and m$\angle B$ is three times as great as m$\angle C$, then what is the measure of each angle?

4

SIMILAR TRIANGLES

It is in mathematics we ought to learn the general method, always followed by the human mind in its positive researches.

Auguste Comte

SECTION 4.1

Mappings and Correspondences

After completing this section, you should understand
▲   how to map figures.
▲   how to write correspondences.

In order to solve problems found in different professions, it is often useful to be able to map triangles (and other geometric figures). Mapping can help you find corresponding parts of two triangles.

MAPPINGS

Maps of cities or states are used to help people locate key points, such as streets or borders. Maps are drawn so that you can compare the map to the actual area where you are, and see how to get to the key point. Each feature of the area is keyed to one point on the map.

If a state were viewed from far above the earth, the cities would look like the dots representing them on the map. They would appear to be the same distances apart, even though one area encompasses square kilometers and the other square centimeters. The correspondence between the earth and the map would not depend on the actual size—they appear to have the same shape.

This idea of fitting two look-alike figures on top of each other and matching corresponding point to corresponding point can be applied to triangles or other geometric figures.

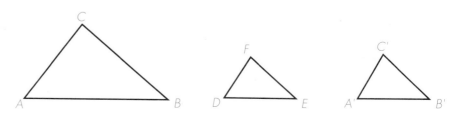

Two methods may be used to label triangles that are being mapped. The first method uses new letters to label the second triangle.

$\triangle ABC$ is mapped to $\triangle DEF$.

The second method uses the same letters as the first triangle, but with a mark following the letter. A', read A prime.

$\triangle ABC$ is mapped to $\triangle A'B'C'$.

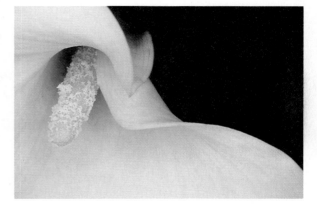

![WRITING arrow]

WRITING
CORRESPONDENCES

When mapping triangles, you must be given enough information to make sure that the shapes of the triangles will be the same.

Example: In the following triangles, $\angle A \cong \angle D$ and $\angle B \cong \angle E$.

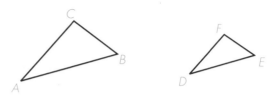

Nine correspondences can be written.

Vertices	Sides	Angles
A corresponds to D.	$\overline{AB}$ corresponds to $\overline{DE}$.	$\angle A$ corresponds to $\angle D$.
B corresponds to E.	$\overline{BC}$ corresponds to $\overline{EF}$.	$\angle B$ corresponds to $\angle E$.
C corresponds to F.	$\overline{CA}$ corresponds to $\overline{FD}$.	$\angle C$ corresponds to $\angle F$.

To abbreviate the correspondences, a double arrow is used.

$A \leftrightarrow D$ Read: A corresponds to D.
$\overline{BC} \leftrightarrow \overline{EF}$ Read: $\overline{BC}$ corresponds to $\overline{EF}$.
$\angle C \leftrightarrow \angle F$ Read: $\angle C$ corresponds to $\angle F$.

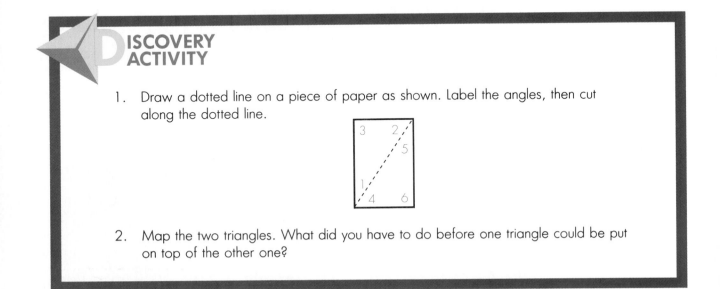

DISCOVERY ACTIVITY

1. Draw a dotted line on a piece of paper as shown. Label the angles, then cut along the dotted line.

2. Map the two triangles. What did you have to do before one triangle could be put on top of the other one?

You may have discovered that sometimes you have to turn one of the triangles before a mapping can be done.

CLASS ACTIVITY

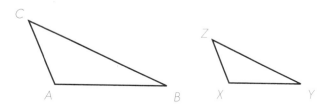

1. Trace the triangles on a sheet of paper so that one is fitted on top of the other (mapped).

2. Write the vertices that correspond to each other.

3. Write the angles that correspond to each other.

4. Write the sides that correspond to each other.

5. What condition is necessary in order to map two figures?

HOME ACTIVITY

Refer to triangles *PQR* and *ABC*.

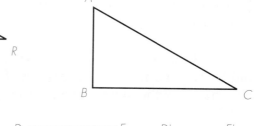

1. Map the two triangles.

2. Write the vertices, sides, and angles that correspond to each other.

Refer to triangles *DEF* and *D'E'F'*.

3. Map the two triangles.

4. Write the vertices, sides, and angles that correspond to each other.

Refer to triangles *ABC* and *SVT*.

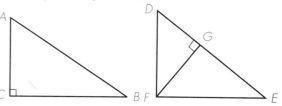

5. Map the two triangles.

6. Write the correspondences.

Use the XYPLANE procedure from page 118 and the command SETXY to write a procedure to draw the triangles whose coordinates are given. Then map the triangles and write correspondences.

7. △*ABC* where *A*(10,10), *B*(50,10), *C*(10,40)
   △*DEF* where *D*(0,−15), *E*(20,−15), *F*(0,0)

8. △*MNP* where *M*(0,0), *N*(30,0), *P*(15,20)
   △*XYZ* where *X*(−3,−4), *Y*(3,−4), *Z*(0,0)

9. Given these correspondences, draw the two figures after they are mapped: $A \leftrightarrow B$, $C \leftrightarrow F$, $D \leftrightarrow X$.

10. Given these correspondences, draw the two figures after they are mapped: $A \leftrightarrow C$, $B \leftrightarrow X$, $E \leftrightarrow Y$, $D \leftrightarrow Z$.

Refer to triangles *ABC*, *DFG*, and *EFG*. You may want to trace all three and cut the two smaller ones apart.

11. Draw a picture of the three triangles mapped.
12. How many correspondences can be written?
13. Write the correspondences for the sides.

Two isosceles triangles are shown, with $AC = BC$, and $DF = EF$. $\angle C \cong \angle F$.

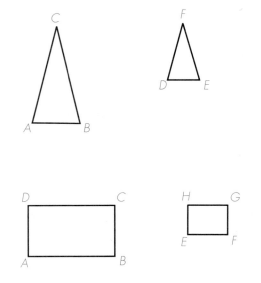

14. Map the two triangles and write the angle correspondences.

15. Maria said there was another way to map the two triangles. Explain. Write the angle correspondences.

16. Wayne said that the rectangles on the right could be mapped four ways. Show the four ways, and write the vertex correspondences for each.

17. Name pairs of objects in your classroom that could be mapped, and explain their correspondences.

18. Name one pair of objects in your home that could be mapped.

19. Draw two triangles that can be mapped by sliding one on top of the other.

Refer to triangles ABC and DEF.

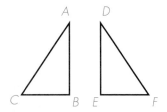

20. Try to map the two right triangles by sliding one on top of the other. Trace one triangle and cut it out.

21. Map the two triangles if you are permitted to turn and flip one triangle.

CRITICAL THINKING

22. On the physical education field, the coach noticed that a number of dogs had joined his P.E. class. He counted 70 legs and 21 heads. How many dogs and how many students were on the field?

SECTION 4.2

Ratios and Proportions

After completing this section, you should understand
▲ ratios and proportions.
▲ how to solve problems involving ratios and proportions.

You have probably encountered ratios and proportions many times each day, when you talk about sports statistics with your friends or look at a recipe in a cookbook.

RATIOS

In algebra, you learned about rational numbers, numbers that can be written in the form $\frac{integer}{integer}$, with the denominator not equal to 0. The word *ratio* is found in *rational*. A ratio is a comparison of two numbers by division.

DEFINITION 4-2-1   **A RATIO is one number divided by another number, or a fraction N/D where D is not zero.**

Example: Some ratios: $\frac{2}{3}$, 34/62, 3:4, and 5 to 7
Not a ratio: $\frac{5}{0}$

WRITING A RATIO IN LOWEST TERMS

In the example above, $\frac{2}{3}$, 3:4, and 5 to 7 are written in lowest terms. The ratio $\frac{34}{62}$ is not in lowest terms.

Example: Write $\frac{34}{62}$ in lowest terms.

$$\frac{34}{62} = \frac{2(17)}{2(31)} = \frac{17}{31}$$

Example: Write 0.5 in lowest terms.

$$0.5 = \frac{5}{10} = \frac{5(1)}{5(2)} = \frac{1}{2}$$

129

Example: Write $\frac{2}{3}/\frac{3}{4}$ in lowest terms.

$$\frac{\frac{2}{3}}{\frac{3}{4}} = \frac{2}{3} \div \frac{3}{4} = \frac{2}{3} \cdot \frac{4}{3} = \frac{8}{9}$$

PROPORTIONS

You know that an equation is a statement that says two expressions are equal.

$$x = 560 \qquad\qquad x + 350 = 468 \qquad\qquad |x - 27| = 68$$

A proportion is a special type of equation. Here are some examples:

$$\frac{1}{2} = \frac{6}{12} \qquad\qquad \frac{3}{4} = \frac{x}{5} \qquad\qquad 1{:}2 = 7{:}14 \qquad\qquad 15{:}x = 30{:}6$$

$$\frac{15}{25} = \frac{60}{y} \qquad\qquad \frac{x}{0.5} = \frac{11}{10} \qquad\qquad \frac{(7-x)}{12} = \frac{20}{36}$$

DEFINITION 4-2-2 **A PROPORTION is an equation consisting of two equal ratios.**

Notice that proportions are written using a fraction bar or a colon (:). Each equation consists of two equal fractions. When one of the ratios contains a variable, you can solve for the value of the variable that makes the statement true.

Read $\frac{3}{4} = \frac{x}{5}$ as 3 is to 4 as x is to 5.

Example: Solve the proportion.

$$\frac{4}{5} = \frac{x}{15} \qquad \frac{4}{5} \cdot 15 = \frac{x}{15} \cdot 15$$
$$12 = x$$

CLASS ACTIVITY

1. Which of the following are ratios?

   a. $\frac{2}{3}$ b. $\frac{5}{7}$ c. 4:9   d. 0.8   e. $\frac{6}{0}$ f. $\frac{0}{6}$

2. Which of the following are proportions?

   a. $\frac{2}{3} = \frac{4}{6}$ b. $\frac{x}{3} = \frac{4}{6}$ c. $\frac{2}{3} = \frac{5}{6}$ d. $0.8 = \frac{x}{10}$

In Exercises 3–7, write each ratio in lowest terms.

3. $\frac{16}{64}$ 4. $\frac{128}{64}$ 5. 0.25

6. $\frac{256}{1024}$ 7. $\frac{\frac{5}{7}}{\frac{10}{14}}$

8. A volleyball team has 5 coaches and 75 players. What is the ratio of players to coaches?

Solve each proportion using your calculator.

9. $\frac{2}{x} = \frac{10}{12}$

10. $\frac{5x}{16} = \frac{72}{160}$

11. $\frac{37}{78} = \frac{7x}{81}$

SOLVING PROBLEMS WITH PROPORTIONS

You can use the rules you've already learned for solving equations to solve proportions. Remember that when you perform an operation on one side of the equal sign, you must perform the same operation on the other side of the equal sign. Recall that you cannot divide by zero.

Example: If 3 footballs cost $110, what will 5 footballs cost at the same price per football?

3 footballs are to 5 footballs as $110 is to $x.
$\frac{3}{5} = \frac{110}{x}$ (Multiply by 5x.)
$5x\left(\frac{3}{5}\right) = 5x\left(\frac{110}{x}\right)$
$3x = 550$
$x = 183.3\overline{3}$

Rounded to the nearest penny, 5 footballs will cost $183.33.

Notice that the proportion used to solve the problem could also have been set up as $\frac{3}{110} = \frac{5}{x}$. The number of footballs appears in the numerators, with the appropriate cost in the denominators.

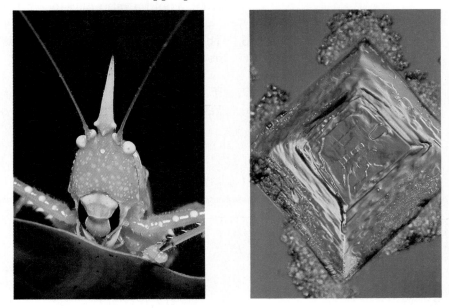

Example: A map has a scale that shows 1 inch on the map equals 4.5 miles. Two cities are 2.3 inches apart on the map. How far apart are the cities in miles?

$$\frac{1 \text{ inch}}{2.3 \text{ inches}} = \frac{4.5 \text{ miles}}{x \text{ miles}}$$
$$\frac{1}{2.3} = \frac{4.5}{x} \quad \text{(Multiply by 2.3x)}$$
$$x = 2.3(4.5) = 10.35 \text{ miles}$$

Since the problem gave measurements to the nearest tenth, round 10.35 miles to 10.4 miles.

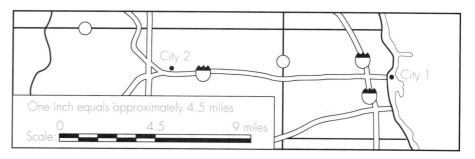

One inch equals approximately 4.5 miles

Scale: 0   4.5   9 miles

CLASS ACTIVITY

Write each ratio in the form N/D in lowest terms.

1. $\frac{16}{28}$

2. $\frac{\frac{5}{5}}{6}$

3. $34:48$

4. $\frac{0.2}{\frac{1}{5}}$

5. If the ratio of boys to girls is 2 to 3 (2 boys for every 3 girls), then what would a ratio of 5 to 7 mean?

6. The ratio of teachers to students is 1:16. How many teachers are needed for 560 students?

7. If a car gets 35 miles per gallon of gas (35 to 1), then how many gallons of gas will be consumed on a trip of 1435 miles?

HOME ACTIVITY

Write each ratio in the form N/D in lowest terms.

1. $\frac{2}{4}$

2. $\frac{10}{16}$

3. $\frac{20}{32}$

4. $\frac{39}{65}$

5. $\frac{2730}{3094}$

6. $\frac{\frac{4}{7}}{\frac{20}{35}}$

Refer to the drawing below. Write each ratio in the form N/D, in lowest terms.

```
┌─────────────────────────┐
│ ○ □ □ △ △ △ ○ □         │
│ □ △ △ □ ○ △ □ △         │
│ △ ○ △ □ △ △ □ △         │
└─────────────────────────┘
```

7. Triangles to squares

8. Circles to triangles

9. Circles to all figures

10. Squares to triangles

11. Triangles to circles

12. Squares to all figures

13. Circles to squares

14. Squares to circles

15. Triangles to all figures

16. Circles and squares to triangles

17. Squares and triangles to all figures

18. Write a sentence to explain what Exercise 17 means.

Solve each proportion for x using your calculator.

19. $\frac{3}{8} = \frac{x}{12}$

20. $\frac{9}{4} = \frac{2x}{12}$

21. $\frac{15}{x} = \frac{3}{4}$

22. $\frac{(2x + 4)}{15} = \frac{40}{30}$

23. $\frac{40}{(3x - 2)} = \frac{5}{21}$

24. $\frac{(x + 3)}{6} = \frac{3}{4}$

25. $\frac{3x}{6} = \frac{(4x + 1)}{5}$

26. $\frac{(2x - 3)}{8} = \frac{126}{(-6)}$

CRITICAL THINKING

27. Which is a better buy, an 18-oz economy-size box of cereal for $2.56, or a 13-oz box of the same cereal for $1.82?

28. What is the sum of the first 50 even counting numbers? Look at the cases below, and continue the table until you see a pattern.

Case	Numbers	Sum
1	2	2
2	2 + 4	6
3	2 + 4 + 6	12

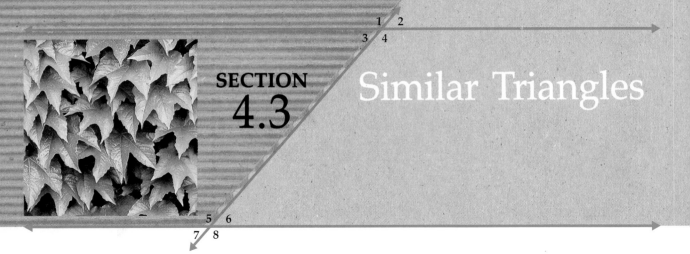

Similar Triangles

After completing this section, you should understand
▲ what similar triangles are.
▲ how to solve problems involving similar triangles.

So far in this chapter, you have studied mappings, correspondences, ratios, and proportions. These ideas are now all tied together in work with triangles.

DISCOVERY ACTIVITY

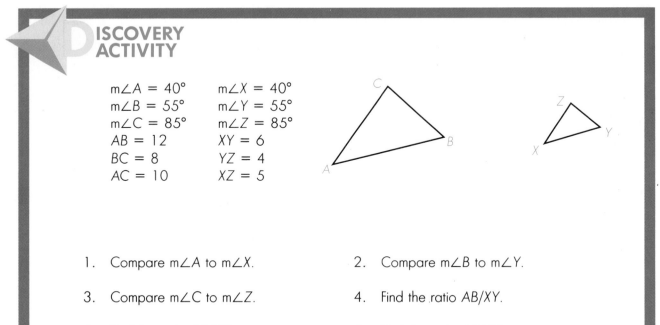

$m\angle A = 40°$	$m\angle X = 40°$
$m\angle B = 55°$	$m\angle Y = 55°$
$m\angle C = 85°$	$m\angle Z = 85°$
$AB = 12$	$XY = 6$
$BC = 8$	$YZ = 4$
$AC = 10$	$XZ = 5$

1. Compare $m\angle A$ to $m\angle X$.

2. Compare $m\angle B$ to $m\angle Y$.

3. Compare $m\angle C$ to $m\angle Z$.

4. Find the ratio AB/XY.

5. Find the ratio BC/YZ.

6. Find the ratio AC/XZ.

7. Compare the ratios in questions 4–6.

8. Can you map $\triangle ABC$ and $\triangle XYZ$?

SIMILAR TRIANGLES

In the Discovery Activity, you may have found that the corresponding angles were all equal in measure, and that the ratios of the corresponding sides were all equal. The two triangles are similar triangles.

DEFINITION 4-3-1 **Two triangles are SIMILAR if**

a. the corresponding sides have the same ratio;

Abbreviated SSS
(side-side-side)

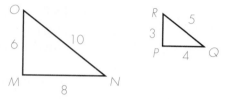

$\frac{6}{3} = \frac{10}{5} = \frac{8}{4} = \frac{2}{1}$

or

b. two angles of one triangle are equal in measure to two angles of another triangle;

Abbreviated AA
(angle-angle)

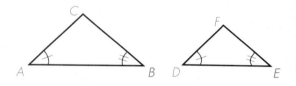

or

c. one angle of a triangle is equal in measure to an angle of the other triangle and the corresponding sides that include the equal angles are in the same ratio.

Abbreviated SAS
(side-angle-side)

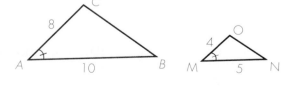

$m\angle A = m\angle M$
$\frac{10}{5} = \frac{8}{4} = \frac{2}{1}$

The symbol $\sim$ is read *is similar to*. The drawing in part c of the definition could be labeled

$$\triangle ABC \sim \triangle MNO.$$

It is important to list the vertices of each triangle in the correct order. Corresponding vertices must match when writing similarities.

Definition 4-3-1 has three parts separated by the word *or*. This means that only one of the three parts must be satisfied in order to have similar triangles. You do not need to show that all three parts are true, only one. Two, or even three, of the parts may be true, but only one is necessary.

Remember that a valid definition is true when reversed. From Definition 4-3-1, if two triangles are similar, then the corresponding angles are equal in measure and the corresponding sides are in equal ratio. Reversing the definition,

if the corresponding sides are in equal ratio (SSS),

or

if two angles of one triangle are equal in measure to two angles of another triangle (AA),

or

if one angle of a triangle is equal in measure to an angle of the other triangle and the corresponding sides that include the equal angles are in the same ratio (SAS),

then

the two triangles are similar.

Example:

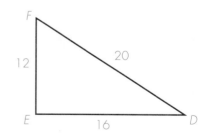

Find the ratio for each pair of corresponding sides.

$$\frac{4}{16} = \frac{1}{4}; \frac{3}{12} = \frac{1}{4}; \frac{5}{20} = \frac{1}{4}$$

Are the triangles similar?

$\triangle ABC \sim \triangle EDF$

The side correspondences are $\overline{AB} \leftrightarrow \overline{ED}$, $\overline{BC} \leftrightarrow \overline{DF}$, and $\overline{AC} \leftrightarrow \overline{EF}$.

Example:

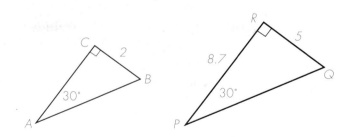

Two of the angles of $\triangle ABC$ are equal in measure to two of the angles of $\triangle PQR$. One pair of corresponding angles are 30°, and one pair are right angles and therefore of equal measure.

Therefore, $\triangle ABC \sim \triangle PQR$, by AA.

The angle correspondences are $\angle A \leftrightarrow \angle P$, $\angle B \leftrightarrow \angle Q$, and $\angle C \leftrightarrow \angle R$.

Find AC.

$RQ/CB = \frac{5}{2}$, therefore $PR/AC = \frac{5}{2}$.

$$\frac{8.7}{AC} = \frac{5}{2}$$

$$AC = 3.5$$

CLASS ACTIVITY

Refer to triangles DEF and ABC.

1. Why are the triangles similar?

2. Map the triangles with the correct correspondences.

3. Write the angle correspondences.

4. Write the side correspondences.

5. Solve for EF and ED.

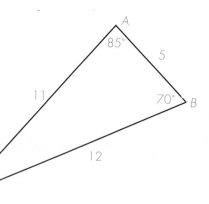

Refer to triangles *GBM* and *RKJ*.

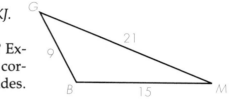

6. Are the triangles similar? Explain. If yes, identify the corresponding angles and sides.

Use a computer and a geometric drawing tool to complete the exercises.

7. Draw any △*ABC*.

8. Draw $\overline{DE}$ such that *D* is on $\overline{AC}$ and *E* is on $\overline{BC}$ and $\overline{DE} \parallel \overline{AB}$.

9. Measure each side in each triangle.

10. Is △*ABC* ~ △*DEC*? If so, why?

11. If they are similar, map the triangles.

12. If they are similar, write the ratio of the corresponding sides.

HOME ACTIVITY

The two triangles are similar. Write the correspondences of the angles and of the sides.

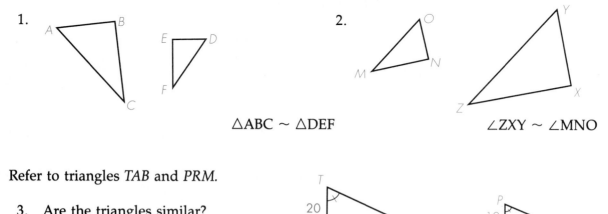

1.

△ABC ~ △DEF

2.

∠ZXY ~ ∠MNO

Refer to triangles *TAB* and *PRM*.

3. Are the triangles similar?

4. Why or why not?

Refer to triangles *OHM* and *ALP*.

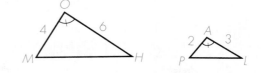
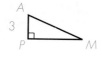

5. Why are these triangles similar?

6. Find the ratio of the sides.

Refer to triangles *MAP* and *ABC*.

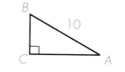

7. Are the triangles similar?

8. Why or why not?

Refer to triangles *ADC* and *PTQ*.

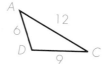

9. Are the triangles similar?

10. Why or why not?

In the figure at the right $l_1 \parallel l_2$; $\overline{AC}$ and $\overline{CB}$ are transversals.

11. $\triangle ABC \sim \triangle DEC$. Why?

12. Map $\triangle ABC$ and $\triangle DEC$.

13. Write the ratio of the corresponding sides of $\triangle ABC$ and $\triangle DEC$, and solve for x.

CRITICAL THINKING

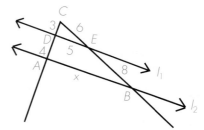

14. *AG* is a flagpole that casts a shadow to *H*. I have only a tape measure, a pencil, and a piece of paper. Explain how I can figure out the height of the flagpole. (Hint: I know my own height.)

SECTION 4.4

Similar Figures and Scale Drawings

After completing this section, you should understand
▲ scale drawings.
▲ applications of scale factors.

In Section 4.3, you studied the conditions for similar triangles and wrote correspondences for angles and sides of the similar figures. The work you have done will now be expanded to applications in both two dimensions and three dimensions.

SIMILAR FIGURES

As with similar triangles, similar figures have equal corresponding angles and corresponding sides that are in equal ratio.

Examples: Two-dimensional

Three-dimensional

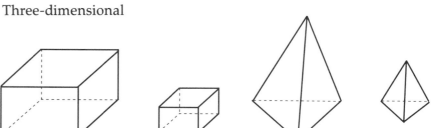

NOTEBOOK
DEFINITION 4-4-1 **If two figures are SIMILAR, then**

a. the measures of the corresponding angles are equal

and

b. the corresponding sides are in equal ratio.

Note the word *and* in the definition. Both conditions must exist for figures to be similar.

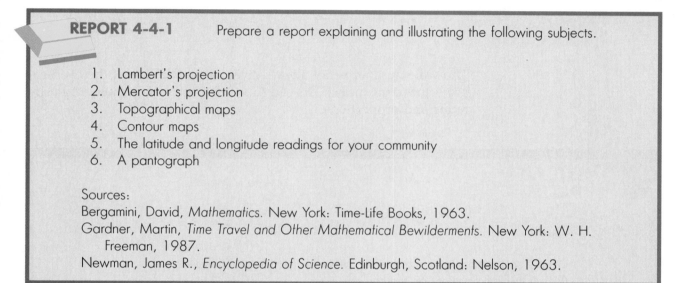

REPORT 4-4-1 Prepare a report explaining and illustrating the following subjects.

1. Lambert's projection
2. Mercator's projection
3. Topographical maps
4. Contour maps
5. The latitude and longitude readings for your community
6. A pantograph

Sources:
Bergamini, David, *Mathematics*. New York: Time-Life Books, 1963.
Gardner, Martin, *Time Travel and Other Mathematical Bewilderments*. New York: W. H. Freeman, 1987.
Newman, James R., *Encyclopedia of Science*. Edinburgh, Scotland: Nelson, 1963.

SCALE DRAWINGS

One use of similar figures is in scale drawings. Scale drawings can reduce the size of a large object in its representation, or increase the size of a small object. You see an example of scale drawings when a teacher uses the overhead projector or when you attend a movie. What you see on the screen is a projection of the picture on the transparency or the film.

The picture on the screen is larger than the one on the film. This enlargement factor, or multiplier, is the same as the ratio number you worked with in Section 4.3. A ratio number also appears on maps and is identified as the scale or scale factor.

Scale:
```
        0     5    10    15    20    25    30    35
```

Scale: one inch = approximately
13.7 miles or 22.5 kilometers

The scale number is not always given, as in the case of the overhead projector or the movie. You can solve for these scale numbers by using ratios and proportions.

DISCOVERY ACTIVITY

1. On a transparency, draw a line segment one centimeter in length.
2. Project the line segment on the screen.
3. Measure the segment on the screen.
4. What is the scale factor?
5. Draw a scalene triangle on the transparency. Measure and record the lengths of the sides in centimeters.
6. Project the triangle on the screen.
7. Predict the length of the sides of the triangle on the screen, using the scale factor from question 4.
8. Measure the sides of the triangle on the screen and compare the lengths with the prediction.

You may have predicted the correct lengths of the sides of the triangle. Scale factors can be determined by careful measurement, then used to predict unknown lengths.

APPLICATION USING SCALE DRAWING

Many people use scale drawings in their work, including drafters, architects, and toy manufacturers.

The directions for a model car may state that the scale is 1:18. This means the model will be similar to the actual car, with 1 inch on the model car representing 18 inches on the actual car. Scales can also be written as 1 inch to x feet, such as 1 inch to 1.5 feet.

Example: A model car was built to the scale 1:18. If the model is 14.5 inches long, what is the length of the actual car?

$$\frac{1}{18} = \frac{14.5}{x} \quad \text{(Multiply both sides by } 18x.\text{)}$$
$$x = 18(14.5)$$
$$x = 261 \text{ inches, or } 21.75 \text{ feet}$$
The length of the actual car is 21.75 feet.

Example: The actual car is 5.4 feet wide across the inside. What is the width of the model, if the scale is 1 inch to 1.5 feet?

$$\frac{1}{1.5} = \frac{x}{5.4} \qquad \text{(Multiply both sides by } 1.5(5.4).\text{)}$$
$$5.4 = 1.5x \qquad \text{(Divide both sides by 1.5.)}$$
$$x = 3.6 \text{ inches}$$
The inside width of the model is 3.6 inches.

CLASS ACTIVITY

On a road map, the distance from Houston to San Francisco measured 11.5 inches. The scale used was 1 inch to 150 miles.

1. How far apart are the two cities?

2. A chart on the map listed the mileage between the two cities as 1953 miles. Explain why the two distances might be different.

3. The shadow of your hand on your desk is a projection. Name some other examples.

Use a computer, a geometric drawing tool, and $\triangle WXY$ to complete the exercises.

4. Draw a similar triangle with sides twice as long (scale factor of 2).

5. Draw a similar triangle with a scale factor of $\frac{1}{2}$.

HOME ACTIVITY

1. Find the scale for the projection illustrated.

2. A line segment on an overhead transparency is 2.5″ in length. The same segment on the screen is 9.5″. What is the scale?

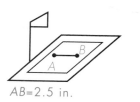

$AB = 2.5$ in.

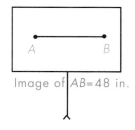

Image of $AB = 48$ in.

3. A model airplane kit indicates the scale is 1 inch to 4 feet. If the wingspan is 34 feet, what is the wingspan on the model?

The scale for a map of Texas is 1 inch to 63 miles.

Scale: 0 ... 50 ... 100 ... 150

4. If the longest east-west segment on the map is 14.9 inches, what is the distance in miles?

5. If the north-south distance is 710 miles, how many inches is this on the map? Round to the nearest tenth.

A model of a steam locomotive was made to the scale of 1:120.

6. Explain what 1:120 means.

7. If the model is 8.4 centimeters long, what is the length of the actual steam locomotive in meters?

8. A chuck wagon used by the early settlers measured 10.5 feet in length. If the scale for the model is 1 to 16, then how many inches long is the model?

9. A battleship is 860 feet long. What is the length of the model if the scale is 1 inch to 53.75 feet?

10. Copy and complete the following table. Use your calculator to solve the proportions.

Inches	1	2	3	4	5	7.5	10
Centimeters	2.54	5.08	___	___	___	___	___

11. What is the scale for converting inches to centimeters?

Determine the scale for the indicated conversions.

12. Feet to yards

13. Feet to miles

14. Inches to feet

15. Yards to meters

16. Centimeters to meters

17. Yards to feet

18. Miles to feet

19. Feet to inches

20. Meters to yards

21. Meters to centimeters

△ABC has sides of length 5 cm, 3 cm, and 4 cm.

22. Draw a similar triangle with sides twice as long (scale factor of 2).

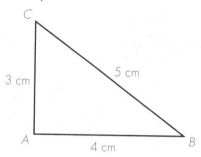

23. Draw a similar triangle with a scale factor of $\frac{1}{2}$.

CRITICAL THINKING

24. Julie found a shortcut for reducing 16/64 to lowest terms. She crossed out the 6's.
$$\frac{1\cancel{6}}{\cancel{6}4} = \frac{1}{4}$$

Lucy didn't think that this would always work. Give one other example where Julie's method will work, and one where it won't work.

Similar Triangles and Conclusions

After completing this section, you should understand
▲ how to justify conclusions involving similar triangles.
▲ how to solve for missing lengths, using similar triangles.

Previously, you learned how to tell if triangles were similar. You practiced mappings, wrote correspondences, solved proportions, and used scale drawings. In this section, two theorems will be introduced. These theorems, along with definitions, will be used to show similarity of triangles.

JUSTIFYING CONCLUSIONS

Often you are presented with a situation where you must analyze given information. You then make a guess, called a conjecture, that you think is correct. You have reached a conclusion. Conjectures are formed many different ways. You can form conjectures by observing patterns, by looking at examples, or by looking at all the possibilities. Sometimes conjectures are given to you. In geometry, once you have reached a conclusion, you know that the conclusion must be justified.

DISCOVERY ACTIVITY

1. Draw a large scalene triangle on a sheet of paper. Label the triangle *ABC*.

2. Using your ruler, find the midpoint of $\overline{AC}$. Label this point *D*. Do the same for $\overline{BC}$, labeling this point *E*.

3. Measure $\overline{DE}$ and $\overline{AB}$.   4.   Find the ratio of *DE* to *AB*.

5. On the chalkboard, record the measurements and ratios of each student's triangle. Make a conjecture that appears to be true.

In the Discovery Activity, you may have concluded that $DE = \frac{1}{2}(AB)$, or that $DE:AB = 1:2$. This conclusion now needs to be justified (proved).

A proof is started by listing the given information, drawing a diagram, and stating the conclusion that needs to be justified.

Given: $\triangle ABC$ with midpoints D and E

Prove: $DE = \frac{1}{2}(AB)$

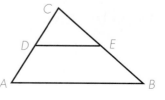

Statements	Reasons
1. $\triangle ABC$ with midpoints D and E	1. Given
2. $CD = \frac{1}{2}(CA)$ and $CE = \frac{1}{2}(CB)$	2. Definition of midpoint
3. $m\angle C = m\angle C$	3. Identity
4. $CD/CA = CE/CB = \frac{1}{2}$	4. Algebra
5. $\triangle ABC \sim \triangle DEC$	5. SAS
6. $DE/AB = \frac{1}{2}$	6. Corresponding sides of similar triangles are in equal ratio.
7. $DE = \frac{1}{2}(AB)$	7. Multiplication

 THEOREM 4-5-1 **If a segment connects the midpoints of two sides of a triangle, then the length of the segment is equal to $\frac{1}{2}$ the length of the third side.**

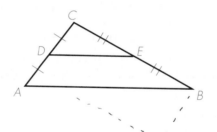

$DE = \frac{1}{2}(AB)$

Example: Given: △ABC, with AB = 12, BC = 15, AC = 21. D is the midpoint of $\overline{AC}$, E is the midpoint of $\overline{BC}$, and F is the midpoint of $\overline{AB}$. How long are segments $\overline{DE}$, $\overline{EF}$, and $\overline{DF}$?

$DE = \frac{1}{2}(AB) = \frac{1}{2}(12) = 6$
$EF = \frac{1}{2}(AC) = \frac{1}{2}(21) = 10.5$
$DF = \frac{1}{2}(BC) = \frac{1}{2}(15) = 7.5$

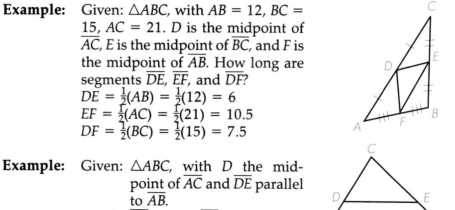

Example: Given: △ABC, with D the midpoint of $\overline{AC}$ and $\overline{DE}$ parallel to $\overline{AB}$.

Justify: $\overline{DE}$ bisects $\overline{CB}$.

We are given △ABC with D the midpoint of $\overline{AC}$, and $\overline{DE}$ parallel to $\overline{AB}$. Since $\overline{DE}$ is parallel to $\overline{AB}$, m∠A = m∠CDE. △ABC ~ △DEC by AA (m∠C = m∠C). Then the ratios of corresponding sides are equal. D is the midpoint of $\overline{AC}$, so $CD/AC = \frac{1}{2}$. Then $CE/CB = \frac{1}{2}$. Therefore, $CE = \frac{1}{2}(CB)$. So E is the midpoint of $\overline{CB}$. This means that $\overline{DE}$ bisects $\overline{CB}$.

DISCOVERY **ACTIVITY**

1. On a sheet of paper, draw △ABC. Draw a line parallel to $\overline{AB}$, intersecting $\overline{AC}$ at D and $\overline{BC}$ at E.

2. Measure the lengths of $\overline{CD}$, $\overline{DA}$, $\overline{CE}$, and $\overline{EB}$.

3. What is the ratio of CD to DA? of CE to EB?

4. Are the two ratios equal?

5. Compare your results to those of your classmates. Did everyone have equal ratios?

6. Write a conclusion that appears to be true.

THEOREM 4-5-2 **If a line is parallel to one side of a triangle and intersects the other sides at any points except the vertex, then the line divides the sides proportionally.**

Example: Given △ABC with $\overline{DE} \parallel \overline{AB}$. What is the length of $\overline{BE}$?

Let $BE = x$.
Using Theorem 4-5-2, the proportion is

$\frac{5}{4} = \frac{9}{x}$.
$5x = 36$
$x = 7.2$

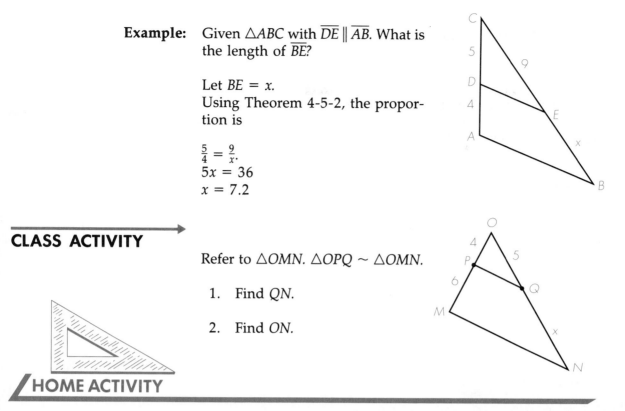

CLASS ACTIVITY

Refer to △OMN. △OPQ ~ △OMN.

1. Find QN.

2. Find ON.

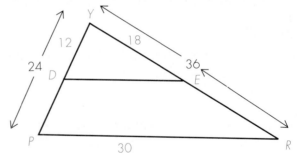

HOME ACTIVITY

Refer to △PRY.

1. Are △PRY and △DEY similar? If so, why?

2. Solve for DP and RE.

Refer to △ABC. AB = 8 and BC = 6.

3. Are the triangles similar? If so, why?

4. Solve for DE and AD.

Refer to △WXY. △WXY ~ △WAT.

5. Find WX.

6. Find YT.

7. Find XA.

8. Find AT.

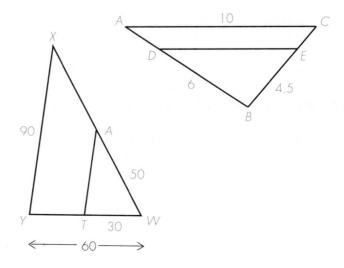

Refer to △ABC. AB = 38, BC = 54, AC = 24, CD = 16, CE = 36.

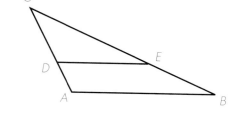

9. Are the triangles similar? If so, why?

10. Solve for DE and BE.

Refer to △ABC.

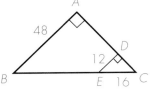

11. Are the triangles similar? If so, why?

12. Solve for BC.

Refer to △NOP. $\overleftrightarrow{QR} \parallel \overline{NO}$.

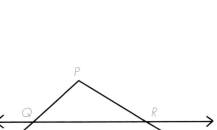

13. Is △NOP ~ △QRP? If so, why?

14. Write an if-then statement for Exercise 13.

In the figure to the right, $\overleftrightarrow{AB} \parallel \overleftrightarrow{CD}$ with transversals $\overleftrightarrow{AD}$ and $\overleftrightarrow{CB}$ intersecting at O.

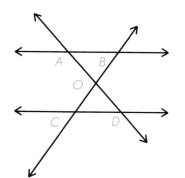

15. Are the triangles similar? If so, why?

Refer to △ABC with ∠C a right angle. $\overline{CD}$ is perpendicular to $\overline{AB}$, m∠A = 60°.

16. Find m∠1, m∠2, m∠3, and m∠4.

17. How many triangles are there? List them.

18. △ABC ~ △CBD ~ △ACD. Why?

19. Map the similar triangles.

20. Is it true that $(CD)^2 = (AD)(DB)$?

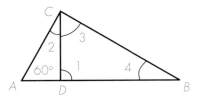

21. Write a Logo procedure that tells the turtle to draw $\triangle XYZ$ with $XY = 70$, $YZ = 60$, and $XZ = 90$. Then tell the turtle to draw $\overline{MN} \parallel$ $\overline{YZ}$ where M is the midpoint of $\overline{XY}$ and N is the midpoint of $\overline{XZ}$. Is $\triangle XYZ \sim \triangle MXN$? If so, why?

Refer to $\triangle ABC$. $\triangle ABC \sim \triangle DEC$.

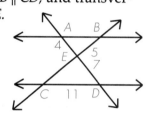

22. Find CB.

23. Find AD.

24. Find EB.

25. Find DE.

In the drawing at the right, $\overleftrightarrow{AB} \parallel \overleftrightarrow{CD}$, and transversals $\overleftrightarrow{AD}$ and $\overleftrightarrow{BC}$ intersect at E.

26. Find EC

27. Find AB.

28. In the drawing at the right, $\overleftrightarrow{AB} \parallel \overleftrightarrow{CD} \parallel \overleftrightarrow{EF}$. Find the unknowns w, x, y, and z.

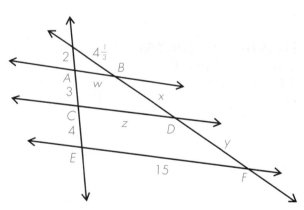

CRITICAL THINKING

This was once called the Golden Rule in arithmetic:

Multiply the last number by the second, and divide the product by the first number.

For example, if 100 notebooks cost $40, how much money should 15 cost?

Following the rule, $15 \cdot 40 = 600$
$$\frac{600}{100} = \$6$$

29. Explain why this method works. Hint: Use a proportion.

Solving Problems involving Similar Triangles

After completing this section, you should understand
▲ how to solve problems involving similar triangles.

The last section dealt with justification of similar triangles from a theoretical viewpoint. This section will deal with applications of similar triangles. It extends the theories to situations in life where they can be useful.

SOLVING PROBLEMS USING SIMILAR TRIANGLES

Some of these problems are taken from situations that may be found in your community. A couple of examples are given to show you how the people in your community might make use of similar triangles.

Example: A surveyor was hired to determine the width of a pond. She took some measurements, then made a drawing.

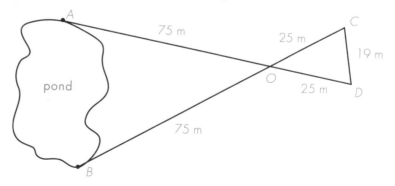

$$\frac{CO}{BO} = \frac{25}{75} = \frac{1}{3} \qquad \frac{DO}{AO} = \frac{25}{75} = \frac{1}{3} \qquad \text{Two ratios are equal.}$$

The angles included between the sides are equal.

$$m\angle COD = m\angle BOA$$
$$\triangle COD \sim \triangle BOA \text{ by SAS.}$$

Map the triangles so that the corresponding parts are easier to see.

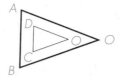

Therefore, $\frac{CD}{BA} = \frac{1}{3}$.

$\frac{19}{BA} = \frac{1}{3}$

$BA = 57$

The width of the pond is 57 meters.

Example: Some hikers noticed that a lookout tower cast a shadow that was 50 feet long. One of the hikers, who was 6 feet tall, cast a shadow that was 8 feet long. What is the height of the tower?

Both the person and the tower can be assumed to be perpendicular to the ground, thus forming right angles.

The sun beaming down forms angles that are equal, as long as the person and the object being measured are close together.

Therefore, similar triangles can be used to solve this problem.

Use the proportion $\frac{x}{6} = \frac{50}{8}$.

$$8x = 300$$

$$x = 37.5$$

The tower is 37.5 feet high.

Example: A surveyor wants to measure the width of a stream. She puts a stake down at *A*, directly across from a lone tree at *B*. She measures and marks 10 meters and then 2 meters along the bank. She moves away from the bank until her second stake (*C*) and the tree line up. This point she finds is 9 meters from the bank. She now has enough information to find the distance across the stream.

$$\triangle ABC \sim \triangle DEC \text{ (By AA)}$$
$$\frac{AB}{DE} = \frac{AC}{DC}$$
$$\frac{AB}{9} = \frac{10}{2}$$
$$AB = 45 \text{ m}$$

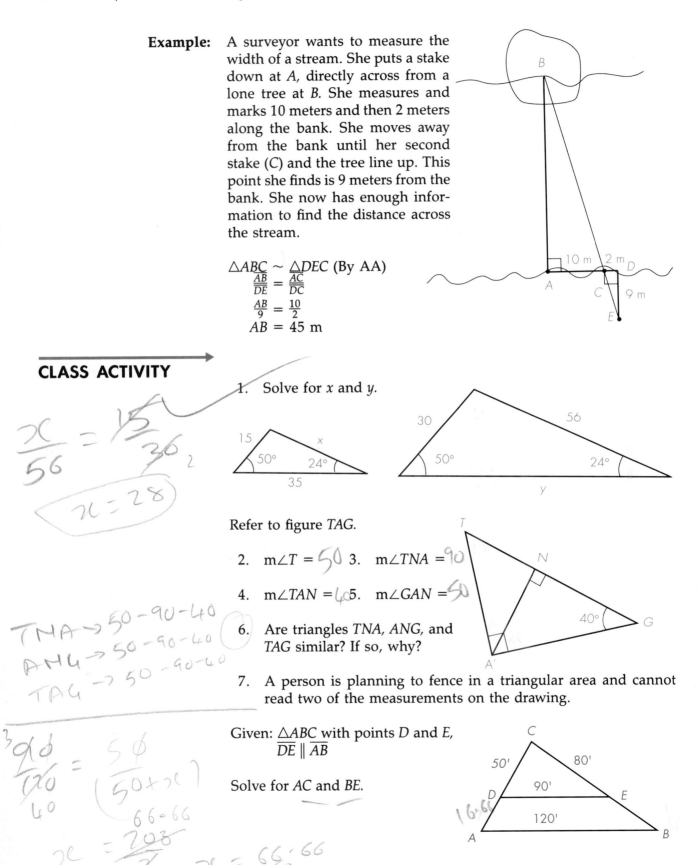

CLASS ACTIVITY

1. Solve for *x* and *y*.

$$\frac{x}{56} = \frac{15}{30}$$

$$x = 28$$

Refer to figure *TAG*.

2. m∠*T* = 50   3. m∠*TNA* = 90

4. m∠*TAN* = 65   5. m∠*GAN* = 50

TNA → 50 – 90 – 40
ANG → 50 – 90 – 40
TAG → 50 – 90 – 40

6. Are triangles *TNA*, *ANG*, and *TAG* similar? If so, why?

7. A person is planning to fence in a triangular area and cannot read two of the measurements on the drawing.

Given: $\triangle ABC$ with points *D* and *E*, $\overline{DE} \parallel \overline{AB}$

Solve for *AC* and *BE*.

$$\frac{90}{x0} = \frac{50}{(50+x)}$$
$$40$$
$$66 - 66$$
$$x = \frac{203}{3}$$

$$x = \frac{66 \cdot 66}{50.00}$$
$$16 \cdot 66$$

HOME ACTIVITY

1. Triangle *ABC* is projected on the screen by the overhead projector. The original triangle has sides 3, 4, and 5 inches long. The shortest side of the projected triangle is 14 inches long. What are the lengths of the other two sides of the projected triangle?

2. A surveyor needs to calculate the length of a pond. He drew the following diagram and made the measurements as indicated. What is the length of the pond?

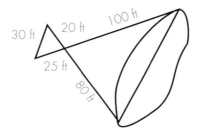

3. Photography makes use of similar figures. Solve for *x* and *y*, the dimensions of the final enlargement.

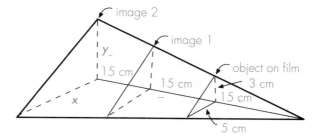

4. On a state map, city A is 5.5 inches from city Z. The scale is 1 inch to 22 miles. To the nearest mile, how far are the two cities from each other?

5. Solve for x and y.

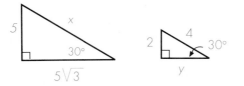

6. Find the width of the river in the figure.

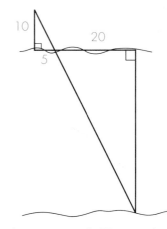

7. A student drew △ABC on a field, with $\overline{DE} \parallel \overline{AB}$. CA equals 25 m and CD equals 10 m. The student then had only enough time to measure CE (7 m) before it began to rain. What is the length of $\overline{CB}$?

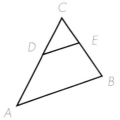

8. A group of students noticed the shadow of the flagpole was 47 feet long at the same time a person 6 feet tall cast a shadow 7 feet long. What is the height of the flagpole to the nearest foot?

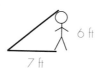

Refer to △FGH. m∠H = 90°, $\overline{HD} \perp \overline{FG}$, and m∠F = 35°.

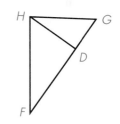

9. m∠G =   10. m∠FDH =
11. m∠FHD =   12. m∠GHD =
13. Are FGH, HGD, and FHD similar? If so, why?

Refer to △*ABC*.

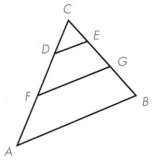

14. Write a Logo procedure to tell the turtle to draw the figure at the right, given that $\overline{DE} \parallel \overline{FG} \parallel \overline{AB}$ with $CE = 7$, $CD = 10$, $DF = 12$, $FA = 13$, and $CB = 24.5$.

15. Are the triangles in figure *ABC* similar? If so, why?

Refer to △*JKL*.

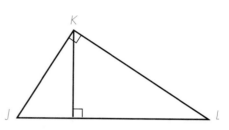

16. How many triangles are there?

17. Are any triangles similar? If so, why?

Refer to △*ABC*. A student was asked to measure the sides of this triangle, but the tape measure wasn't long enough. The only measurements that could be taken are indicated below.

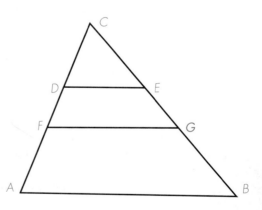

Given: $\overline{DE} \parallel \overline{FG} \parallel \overline{AB}$

  $CD = 18$ ft
  $DF = 12$ ft
  $EG = 15$ ft
  $AF = 19$ ft
  $DE = 20$ ft

Find the length of each segment.

18. *CE*

19. *BG*

20. *AC*

21. *CB*

22. *FG*

23. *AB*

CRITICAL THINKING

When a beam of light strikes a mirror, it is reflected at the same angle as it hits. The diagram shows the light from the lamp striking the mirror at a 25° angle. It reflects into the girl's eye at the same angle.

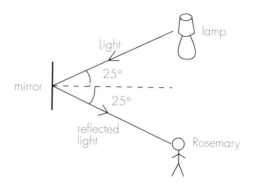

Paul knows about the reflective angles of a mirror. He decides to use this property to measure the height of a building to see how long a ladder he'll need to reach the roof. Paul has a mirror and a meter stick.

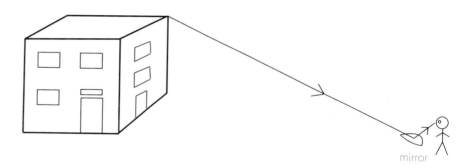

24. Draw the similar triangles in the problem.

25. What measurements does Paul need to make?

26. Assign some possible measures to your drawing, with x being the height of the building.

27. Write the proportion, and solve for x.

CHAPTER SUMMARY

4.1 Triangles *ABC* and *DFE* are similar.
Vertices: $A \leftrightarrow D$, $B \leftrightarrow F$, $C \leftrightarrow E$
Angles: $\angle A \leftrightarrow \angle D$, $\angle B \leftrightarrow F$, $\angle C \leftrightarrow \angle E$
Sides: $\overline{AB} \leftrightarrow \overline{DF}$, $\overline{BC} \leftrightarrow \overline{FE}$, $\overline{AC} \leftrightarrow \overline{DE}$

4.2 A ratio is one number divided by another number that is not zero.
$$\frac{6}{7} \qquad \frac{x}{3} \qquad \frac{(y-5)}{13}$$
A proportion is an equation consisting of two equal ratios.
$$\frac{300}{25} = \frac{15}{x} \qquad \text{(Multiply both sides by } 25x\text{)}$$
$$300x = 375$$
$$x = 1.25$$

4.3 Triangles *RAP*, *YES*, and *TUC* are similar.
$RAP \sim YES$ by SAS
$YES \sim TUC$ by SSS
$RAP \sim TUC$ by AA

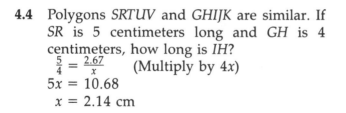
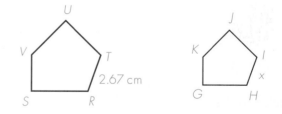

4.4 Polygons *SRTUV* and *GHIJK* are similar. If *SR* is 5 centimeters long and *GH* is 4 centimeters, how long is *IH*?
$$\frac{5}{4} = \frac{2.67}{x} \qquad \text{(Multiply by } 4x\text{)}$$
$$5x = 10.68$$
$$x = 2.14 \text{ cm}$$

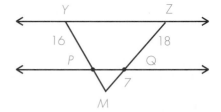

4.5 Lines *YZ* and *PQ* are parallel. What is the length of *PM*?
$$\frac{PM}{YP} = \frac{QM}{ZQ} \qquad \text{(By Theorem 4-5-2)}$$
$$\frac{PM}{16} = \frac{7}{18}$$
$$PM = 6\frac{2}{9}$$

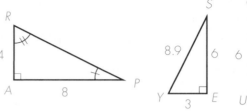

4.6 How wide is (what is the diameter of) the water tower?
$\frac{18}{45} = \frac{20}{50} = \frac{2}{5}$ and $m\angle WTA = m\angle RTE$
$\triangle WAT \sim \triangle RET$ (By SAS)
$$\frac{TR}{WT} = \frac{ER}{WA}$$
$$\frac{20}{50} = \frac{16}{x}$$
$$x = 40 \text{ ft}$$

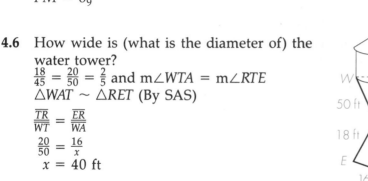

COMPUTER ACTIVITY

Variables can be used when defining LOGO procedures. A variable in LOGO is identified by a colon preceding the variable's name. The variable :SIDE is used in the EQUITRI procedure below to represent the length of a side of the equilateral triangle.

```
TO EQUITRI
RT 30
FD :SIDE
REPEAT 2[RT 120 FD :SIDE]
END
```

To execute the procedure, you must type EQUITRI and then a number to indicate the length of a side. For example, EQUITRI 30 would assign the value 30 to :SIDE and thus tell the turtle to draw an equilateral triangle whose sides are 30 spaces long.

To draw two similar equilateral triangles, simply use the EQUITRI procedure twice.

```
TO SIMTRI1
PU FD 30 PD
EQUITRI 20
HOME
EQUITRI 10
END
```

The output is shown below.

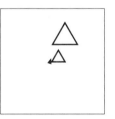

The equilateral triangle on the top has sides 20 spaces long.
The equilateral triangle on the bottom has sides 10 spaces long.

The lengths of the sides of an equilateral triangle are given in Exercises 1–3 below. Choose the length of the sides of an equilateral triangle that is similar to each given triangle. Then modify the SIMTRI1 procedure to tell the turtle to draw the triangles.

1. 15   **2.** 28   **3.** 42

More than one variable can be used in a procedure. The TRI1 procedure will draw one angle and one side of an acute triangle. In the procedure, the variables :ANGLE and :SIDE are used to represent the measure of an angle and the length of a side, respectively. Notice that to find the measure of the angle the turtle will need to turn through, you will need to use the operation symbol for subtraction (−).

```
TO TRI1 :ANGLE :SIDE
RT 180 − :ANGLE
FD :SIDE
END
```

This procedure can now be used in another procedure to draw a complete triangle.

```
TO TRI3
TRI1 30 20
TRI1 65 10
TRI1 85 18
END
```

The output is shown below.

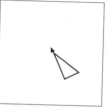

Use the TRI1 procedure in a procedure named SIMTRI2 to tell the turtle to draw the similar triangles given in Exercises 4 and 5 below.

4.

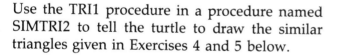

5.

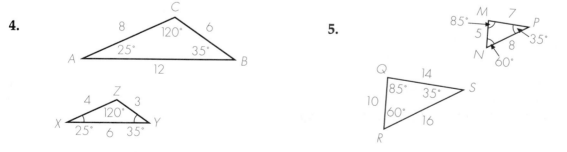

6. In △ABC, m∠A = 50°, m∠B = 60°, m∠C = 70°, AB = 20, BC = 16, and AC = 18. Find the lengths of the sides of a triangle that is similar to △ABC. Use the SIMTRI2 procedure to tell the turtle to draw the triangles.

For each pair of triangles, determine why they are similar and the ratio of the sides. Map the triangles.

1.
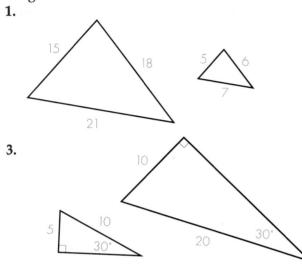

2.
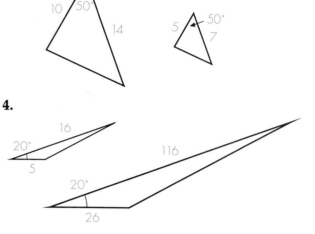

3.

4.

5. In the figure to the right, $l_1 \parallel l_2$ with transversals $\overline{AD}$ and $\overline{BC}$. Are the triangles similar? If so, why?

6. If 7 dictionaries cost $213.50, then what will 12 dictionaries cost at the same price per book?

In $\triangle ABC$, $m\angle C = 90°$, $\overline{CD} \perp \overline{AB}$, and $m\angle A = 25°$.

7. $m\angle B =$

8. $m\angle ADC =$

9. $m\angle ACD =$

10. $m\angle BCD =$

11. How many triangles are in the figure?

12. Are the triangles similar? If so, why?

13. Refer to $\triangle QRP$. How long is ST?

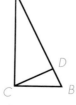

14. In the drawing at the right, the scale is $\frac{1}{4}$ in. = 3 ft. What are the true measurements of the space?

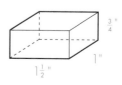

1. A school has a ratio of 1 teacher for every 17 students. There are 650 students in the school with an expected increase of 81 students next year. How many teachers will the school need next year?

2. A school has a ratio of 1 coach to every 14 players. If the school has 3 full-time and 1 half-time coach, then how many players are there?

3. A road map has a scale of 1 inch to 39 miles. Two cities are $5\frac{3}{4}$ inches apart on the map. How many miles are they apart?

4. Solve the proportion for x. Use your calculator and round the answer to two decimal places. $\frac{(3x-4)}{12} = \frac{14}{23}$

Which of the pairs of triangles are similar and why?

5.

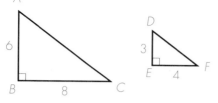

6.

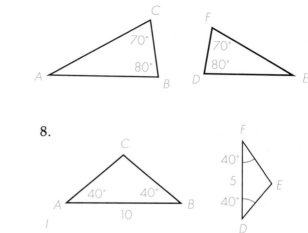

7.

8.

9. What is the length of $\overline{LM}$ in $\triangle GHI$?

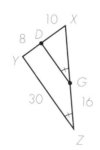

42 cm

Given a right triangle ABC with C the right angle. $\overline{CD} \perp \overline{AB}$.

10. Draw the figure.
11. Map the three similar triangles.
12. Complete: $\overline{AC}/\overline{AD} = \frac{?}{?} = \frac{?}{?}$

Refer to $\triangle XYZ$. $m\angle XGD = m\angle XZY$.
13. $XG =$
14. $DG =$

Example: Evaluate $x^2 + y$ using $x = 16$ and $y = -13$.

$16^2 + (-13)$

$256 - 13$

243

Example: Solve for x.

$21 - 3(2x - 7) = 15$

$21 - 6x + 21 = 15$

$-6x + 42 = 15$

$-6x = -27$

$x = 4.5$

Evaluate each expression using $x = -76$ and $y = 38$.

1. $x + y$

2. $x - y$

3. xy

4. $\frac{x}{y}$

5. $y - x$

6. $|x + y|$

7. $|x - y|$

8. $\frac{y}{x}$

9. $\frac{1}{x} + \frac{1}{y}$

Solve each equation for x.

10. $14x + 65 = 2x - 7$

11. $5(2x - 6) = 3x + 12$

12. $16 - 4(3x - 12) = 112$

13. $2.5x + 3 = 3(1.5x - 5)$

14. $1.6x + 6 = 2(1.8x - 9)$

15. $2.3x + 4 = 3(1.6x - 5)$

A student rode on her bike for 3 hours and traveled 135 miles. After a half-hour rest, she rode another 2 hours and traveled 75 miles more to her campsite.

16. If she left at 8 a.m., when did she arrive at her campsite?

17. What was her average speed during the first three hours?

18. What was her average speed during the last two hours?

19. What was the average speed for the entire trip (including the rest time)?

20. What was the average speed counting only the time she was traveling?

5

CONGRUENT TRIANGLES

He [Lincoln] studied and nearly mastered the six books of Euclid since he was a member of Congress.

Abraham Lincoln

165

SECTION 5.1

Sets and Subsets

After completing this section, you should understand
▲ what a set is.
▲ what Venn diagrams are.
▲ what set operations are.
▲ how to solve problems using sets.

Before the study of similar triangles can be extended to congruent triangles, some terms from the study of set theory must be defined.

WHAT IS A SET?

You can tell by looking at a plate whether it belongs to a certain set of dishes.

NOTEBOOK

DEFINITION 5-1-1   **A SET is a well-defined collection.**

Well-defined means that if you are given an object, you can tell whether or not it is a member, or element, of the set. Members of a set are enclosed in braces: { }. For example, $S = \{1, 2\}$ is read S is the set of numbers 1 and 2.

Example: The set of counting numbers between 1 and 10 = $\{2, 3, 4, 5, 6, 7, 8, 9\}$.

Example: $N = \{1, 3, 5, 7, \ldots\}$ = the set of odd counting numbers. The three dots (. . .) indicate that the terms continue in the given pattern, until the last member is reached. When no last member is given, the set continues forever, or to infinity.

Example: The set of all geometry students in your school is well-defined. Given a name or an ID number, you can determine whether or not a student is taking geometry.

DEFINITION 5-1-2 **A SUBSET is any set contained in the given set.**

A is a subset of *U* is written as $A \subset U$.

$\{1, 2\} \subset \{1, 2, 3, 4, 5\}$

Notice that a set can be a subset of itself. It can also be a subset of a larger set, called a universal set.

DEFINITION 5-1-3 **The UNIVERSE or UNIVERSAL SET is all the possible members in the well-defined set.**

The set of all geometry students in your school could be a universal set. Your geometry class would be a subset of the universe since your class is a part of the set of all geometry students.

CLASS ACTIVITY

Let $U = \{0, 1, 2, 3, 4, 5, 6, 7, 8, 9\}$. Write true or false about each statement.
1. $\{0, 2, 4, 6, 8\} \subset U$
2. $\{10\} \subset U$
3. $\{1, 3, 5, 7, 9\} \subset U$
4. The set of all even numbers $\subset U$
5. Describe *U* in words.
6. Write the one-member subsets for $\{a, b, c\}$.
7. Write the two-member subsets for $\{2, 4, 6, 8\}$.

VENN DIAGRAMS

Venn diagrams are often used to help show how sets relate to each other. A rectangle represents the universal set for a given situation. Sets are then represented by circles within the rectangle.

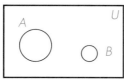
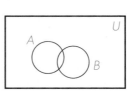
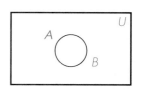

SET OPERATIONS

There are two operations in set theory: union and intersection.

DEFINITION 5-1-4 **The UNION of *n* sets is the set consisting of all the members in the *n* sets.**

The shaded area shows the union of two sets, *A* and *B*.

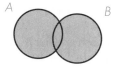

A union *B* is written as $A \cup B$.

DEFINITION 5-1-5 **The INTERSECTION of *n* sets is the set consisting of the members common to the *n* sets.**
The shaded area shows the intersection of two sets, *A* and *B*.

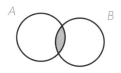

A intersection *B* is written as $A \cap B$.

Example: Let $T = \{3, 6, 9, 12\}$ and $E = \{2, 4, 6, 8, 10, 12\}$. Draw a Venn diagram to show $T \cup E$ and to show $T \cap E$. Write the sets for $T \cup E$ and $T \cap E$.

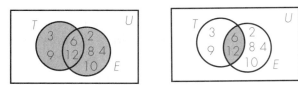

$T \cup E = \{2, 3, 4, 6, 8, 9, 10, 12\}$ $T \cap E = \{6, 12\}$
The elements are in *T* **or** *E*.   The elements are in *T* **and** *E*.

Example: Let $N = \{0, 1, 2\}$ and $M = \{4, 5, 6\}$. Find $N \cap M$.

There are no elements in common to both sets. The intersection is empty. This set is called the empty set.

THEOREM 5-1-1 The empty set is a subset of every set. The symbol is either ∅ or { }.

Example: List all of the subsets of $A = \{1, 2, 3\}$.

$\{1, 2, 3\}$	A set is a subset of itself.
$\{1, 2\}, \{1, 3\}, \{2, 3\}$	Subsets having 2 elements.
$\{1\}, \{2\}, \{3\}$	Subsets having 1 element.
$\{ \}$	Subsets having no elements, or empty set.

CLASS ACTIVITY

1. Draw a Venn diagram illustrating the following:
   U: the set, or number, of students in your school
   F: the set of freshmen   *S:* the set of sophomores
   J: the set of juniors   *R:* the set of seniors
2. Using Exercise 1, if there are 300 in *F*, 275 in *S*, 254 in *J*, and 199 in *R*, then how many are in *U*?

A survey was taken of a group of students.
10 were taking English (*E*).   8 were taking geometry (*G*).
7 were taking history (*H*).   4 were taking *E* and *G*.
3 were taking *G* and *H*.   5 were taking *H* and *E*.
2 were taking all three subjects.

3. Draw a Venn diagram showing the information.
4. How many were taking only geometry? only English? only history?
5. How many were taking history or English? geometry or history? geometry or English?
6. How many students were interviewed?

Use a computer and a geometric drawing tool to complete Exercises 7–10.

7. Draw a line. Can you draw a ray on the line? Is a ray a subset of a line?
8. Draw an isosceles triangle. Can you make the isosceles triangle into a rectangle? Are isosceles triangles a subset of rectangles?
9. Draw the union of two rays with a common endpoint. What figure is formed?
10. Draw two parallel lines. Is there a point common to both lines? What set describes the intersection of these two lines?

HOME ACTIVITY

Write each set using braces, or set notation.

1. The first four counting numbers divisible by 2 and 3
2. The first four counting numbers divisible by 2 or 3
3. The set of vowels of the alphabet
4. The set of integers whose absolute value is less than 3
5. The set of counting numbers greater than 100

Write whether each set is a subset of $A = \{1, 2, 3, 4, 5, \ldots, 12\}$.

6. $B = \{2, 4, 6\}$
7. $C = \{\,\}$
8. $D = \{2, 4, 6, 8, \ldots\}$
9. $E = \{1, 3, 5, \ldots, 11\}$

A school's population was surveyed about extracurricular activities.
110 were out for sports (S).
83 were out for dramatics (D).
36 were out for D and S.
12 were out for all three.

95 were out for band (B).
41 were out for D and B.
56 were out for B and S.
150 were not out for any of the three.

10. Draw a Venn diagram showing the information.
11. How many students were taking only sports? only band? only dramatics?
12. How many were out for S or B? for B or D? for S or D?
13. How many students were in the school?

CRITICAL THINKING

14. Copy and complete the following table.

Set	Number of Elements	Subsets	Number of Subsets
$\{a\}$	1	$\{a\}, \{\,\}$	2
$\{a, b\}$	2	$\{a, b\}, \{a\}, \{b\}, \{\,\}$	4
$\{a, b, c\}$	____		____
$\{a, b, c, d\}$	____		____

15. How many subsets are there for a set of n elements?

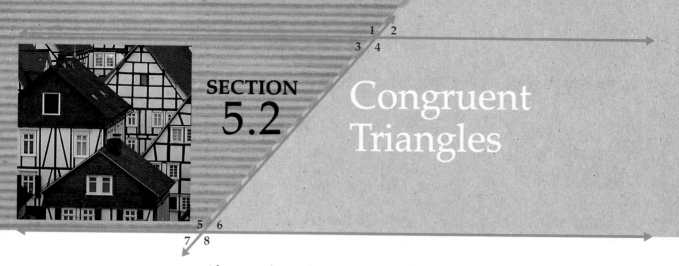

SECTION 5.2

Congruent Triangles

After completing this section, you should understand
▲   the meaning of congruent triangles.
▲   how to prove triangles congruent.

In Chapter 4, you solved problems involving similar triangles. You used a reversible definition for similarity, with conditions involving angle-angle (AA), side-side-side (SSS), and side-angle-side (SAS).

DISCOVERY ACTIVITY

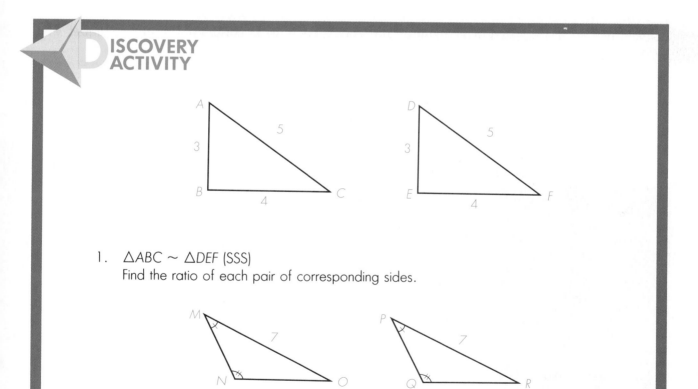

1.   △ABC ~ △DEF (SSS)
     Find the ratio of each pair of corresponding sides.

2.   △MNO ~ △PQR (AA)
     Find MO/PR.

3. What appears to be true about the pairs of triangles?

WHAT ARE CONGRUENT TRIANGLES?

The triangles of the Discovery Activity have not only the same shape but also the same size. In both pairs of triangles, the ratio of the lengths of the corresponding sides is one. The triangles are congruent.

DEFINITION 5-2-1 **If two triangles are similar and the ratio of corresponding sides is one, then the triangles are CONGRUENT.**

All congruent triangles are similar, but not all similar triangles are congruent. A Venn diagram can be used to display this.

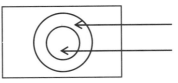

Set of similar triangles
Set of congruent triangles

The symbol for congruency is a combination of the equal and similarity symbols, ≅.

$$\triangle ABC \cong \triangle DEF$$

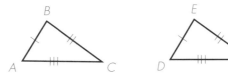

Example: Are the following triangles congruent?

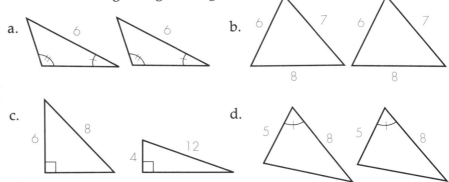

a.

b.

c.

d.

a. Yes; the triangles are similar by AA, and $\frac{6}{6} = 1$.
b. Yes; the triangles are similar by SSS, and $\frac{6}{6} = \frac{7}{7} = \frac{8}{8} = 1$.
c. No; $\frac{6}{4} \neq \frac{8}{12} \neq 1$.
d. Yes; the triangles are similar by SAS, and $\frac{5}{5} = \frac{8}{8} = 1$.

CLASS ACTIVITY

1. Use a computer and a geometric drawing tool to draw $\triangle AED$ and $\triangle BEC$ so that E is the midpoint of $\overline{AB}$, $AD = BC$, $\overline{DA} \perp \overline{AB}$, and $\overline{CB} \perp \overline{AB}$. Does $\triangle AED$ appear to be congruent to $\triangle BEC$? Justify your answer.

2. The gable end of a house is built by making $AC = BC$ and $AD = BD$. Is $\triangle ADC \cong \triangle BDC$? If so, why?

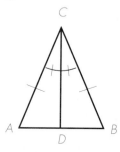

PROVING TRIANGLES CONGRUENT

When a theorem is justified, it is justified for a *general case*. For example, if the theorem is about triangles having a right angle, it is justified for *all* triangles that have a right angle.

Example: Given: $\triangle ABC$ with $AC = BC$,
$\overline{CD}$ bisects $\angle ACB$.
Prove: $\triangle ACD \cong \triangle BCD$

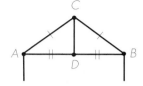

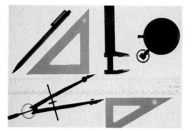

Statements	Reasons
1. $AC = BC$	1. Given
2. $\overline{CD}$ bisects $\angle ACB$.	2. Given
3. $CD = CD$	3. Identity
4. $m\angle ACD = m\angle BCD$	4. Definition of angle bisector
5. $AC/BC = CD/CD$	5. Algebra
6. $\triangle ACD \sim \triangle BCD$	6. SAS
7. $AC/BC = 1$	7. Algebra
8. $\triangle ACD \cong \triangle BCD$	8. Definition of congruent triangles

This is a general case since it applies to all triangles with two equal sides. Therefore, it will be stated as a theorem.

THEOREM 5-2-1 **If a triangle has two equal sides and the included angle is bisected, then the triangle is divided into two congruent triangles.**

CLASS ACTIVITY

1. Use a computer and a geometric drawing tool. Draw △MON so that MN = MO. Draw P on $\overline{ON}$ so that NP = 6 and m∠NMP = m∠OMP. Find NO. Justify your answer.

2. In △FGH, FG = HG and $\overline{GI}$ bisects ∠FGH. If m∠F = 50°, find m∠H.

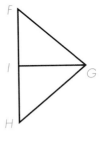

HOME ACTIVITY

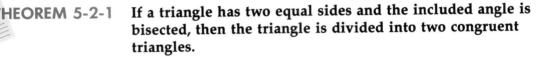

State whether each pair of triangles are congruent.

1.

2.

 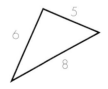

3.

4.

Refer to triangles *ABC* and *DEF*.

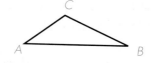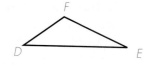

$AB = DE$, $AC = DF$, $m\angle B = m\angle E$

5. Are the triangles similar?

6. Are the triangles congruent?

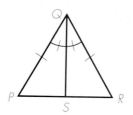

7. In $\triangle PQR$, $PQ = RQ$ and $m\angle PQS = m\angle RQS$. What do you know about $\triangle PQS$ and $\triangle RQS$? Explain your answer.

8. Given: $\triangle ABM$ and $\triangle DCM$ with *M* the midpoint of $\overline{AD}$ and $\overline{BC}$. Explain why $\triangle AMB \cong \triangle DMC$.

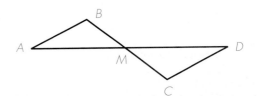

9. Determine the height of the flagpole. $AB = 41$ ft

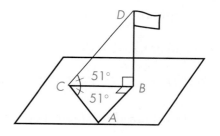

10. In $\triangle ABC$, $AC = BC$ and *D* is the midpoint of $\overline{AB}$. Explain why $\triangle ACD \cong \triangle BCD$.

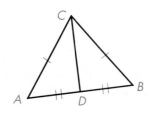

11. Given: $\triangle AEC$ and $\triangle DFB$, with $AE = DF$, $AB = CD$, and $m\angle A = m\angle D$
    Justify: $\triangle AEC \cong \triangle DFB$ (Use a paragraph proof.)

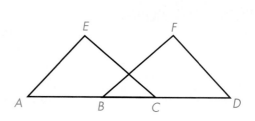

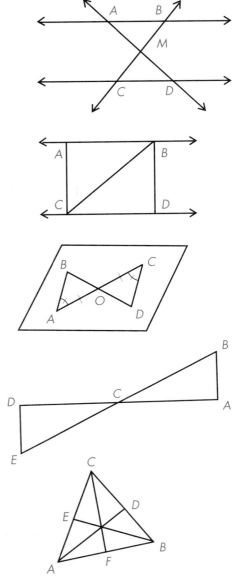

12. In the figure at the right, $\overleftrightarrow{AB} \parallel \overleftrightarrow{CD}$, $\overleftrightarrow{AD}$ and $\overleftrightarrow{BC}$ are transversals, and $CM = BM$. Explain why $\triangle AMB \cong \triangle DMC$.

13. Rectangle $ABDC$ has $\overleftrightarrow{AB} \parallel \overleftrightarrow{CD}$, $\overline{CA} \perp \overline{AB}$, $\overline{DB} \perp \overline{CD}$. Explain why $\triangle ABC \cong \triangle DCB$.

14. In the figure at the right, $\overline{AOC}$ and $\overline{BOD}$ are line segments. Explain why $\triangle AOB \cong \triangle COD$.

15. Given: $AC = DC$, $m\angle A = m\angle D = 90°$
    Justify: $\triangle ACB \cong \triangle DCE$ (Write a paragraph proof.)

$\triangle ABC$ is equilateral. $\overline{AD}$, $\overline{BE}$, and $\overline{CF}$ are angle bisectors.

16. Is $\triangle ABD \cong \triangle ACD$? Explain your answer.

17. Is $\triangle BCE \cong \triangle BAE$? Explain your answer.

18. Is $\triangle CAF \cong \triangle BCF$? Explain your answer.

19. Plot the points for each triangle on the Cartesian plane. Use your ruler and protractor to determine if the triangles appear to be congruent.
    Triangle 1: $(0,0)$, $(0,3)$, $(4,0)$;
    Triangle 2: $(-1,0)$, $(-1,4)$, $(-4,0)$

CRITICAL THINKING

20. A student found an easy way to remember a certain year in history. The year is a four-digit number, such as 1988. The four digits ad up to 17. When the first two digits are turned upside down, the second two digits result. What is the year? (There are two answers.)

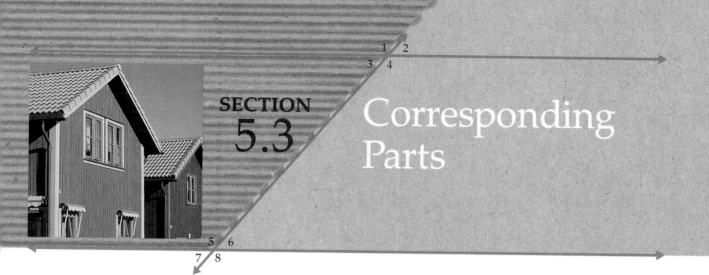

SECTION 5.3

Corresponding Parts

After completing this section, you should understand
▲ corresponding parts of congruent figures.
▲ how to determine lengths of segments and measurements of angles.

If you had two road maps of your state that had the same scale, or ratio, you could say that they were in proportion. If you measured a segment on one map, it would represent an equal length on the other map.

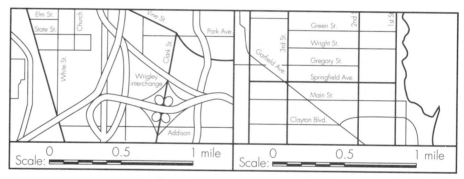

The measured segments on one map would correspond identically to the corresponding segment on the other map. The concept of corresponding, identical parts is widely used in industry. It would be very difficult (and costly!) to replace some items, such as house doors, if they did not come in standard (identical) sizes.

CORRESPONDING PARTS OF CONGRUENT FIGURES

In order for triangles to be congruent, they must be similar and the ratio of the corresponding sides must be one. Recall that triangles are similar if one of three situations is present.

1. Two angles in each triangle are equal (AA).
2. The ratios of all pairs of corresponding sides are equal (SSS).
3. The ratios of two pairs of corresponding sides are equal and the included angles are equal (SAS).

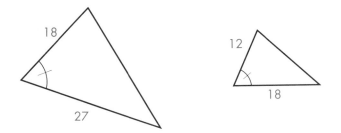

Thus, for congruent triangles, both corresponding sides and corresponding angles are equal. This can be extended to other figures, as well.

DEFINITION 5-3-1 **Corresponding parts of congruent figures are equal. This may be abbreviated as CPCF.**

To this point, you have worked primarily with triangles. However, the definition holds for all geometric figures. Since definitions are reversible, if two figures have all corresponding parts equal, then the figures are congruent.

DISCOVERY ACTIVITY

1. Draw a 2-inch line segment on a sheet of paper, labeling the segment *AB*.

2. From *A* and from *B*, use your protractor to draw ∠*BAC* and ∠*ABC* so that the two angles are equal and the rays intersect at *C*.

3. Measure the lengths of $\overline{AC}$ and $\overline{BC}$.

4. What conclusion do you reach about measures of $\overline{AC}$ and $\overline{BC}$?

SIDES OPPOSITE EQUAL ANGLES

The class should all have reached the same conclusion, that $AC = BC$. The following will justify the conclusion.

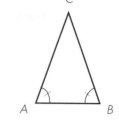

Given: $\triangle ABC$ with m$\angle A$ = m$\angle B$
Justify: $AC = BC$

From the given, we know that m$\angle A$ = m$\angle B$. We can draw the angle bisector from C because there is a number assigned to the measure of $\angle ACB$, and that number can be divided by 2. Let the angle bisector intersect $\overline{AB}$ at D.

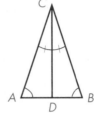

m$\angle ACD$ = m$\angle BCD$. Then $\triangle ACD \sim \triangle BCD$ by AA. $CD = CD$, so $CD/CD = 1$. Therefore, $\triangle ACD \cong \triangle BCD$. Definition 5-3-1 tells us that the other corresponding parts are equal. So $AC = BC$.

THEOREM 5-3-1   **If a triangle has two equal angles, then the sides opposite the equal angles are equal.**

The above theorem applies to isosceles triangles. For an equiangular triangle, Theorem 5-3-1 can be applied twice.

THEOREM 5-3-2   **If a triangle is equiangular, then it is equilateral.**

CLASS ACTIVITY

1. Use a computer and a geometric drawing tool to draw an equiangular triangle. Measure each side. Repeat this activity several times, using different angle measures each time. Do your results show that Theorem 5-3-2 is true?

Are the pairs of triangles congruent?

2.

3.

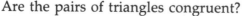

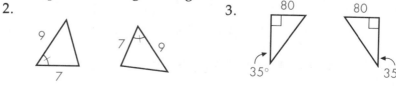

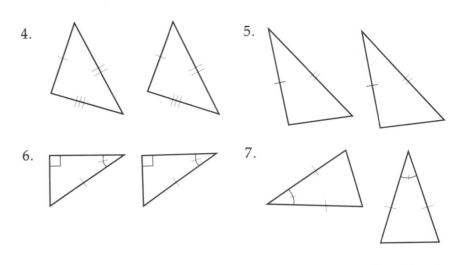

4.

5.

6.

7.

Use a computer and a geometric drawing tool to complete Exercises 8–10.

8. Draw a triangle with two equal angles. Measure each side. Repeat this activity several times, using different angle measures each time. If a triangle has two equal angles, is it isosceles?

9. Draw two triangles that have equal corresponding angles. Measure each side. Repeat this activity several times, using different angle measures each time. If two triangles have equal corresponding angles, are the triangles congruent?

10. Draw an acute triangle and a right triangle. Measure each side and each angle. Repeat this activity several times, using different angle measures and different side lengths each time. Is an acute triangle ever congruent to a right triangle?

HOME ACTIVITY

Refer to △ABC and △DEF.

1. Why are the triangles congruent?

2. Why does $AB = DE$?

3. Does $m\angle A = m\angle D$?

4. Find $m\angle A$.

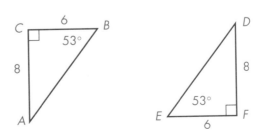

Refer to △MNO and △PQR.

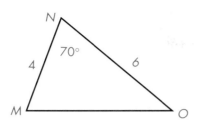

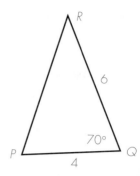

5. Are the two triangles similar? If so, why?

6. Are the two triangles congruent? If so, why?

7. The length of $\overline{MO}$ is equal to the length of which side?

8. Is m∠N = m∠P?

Refer to △STU and △VWX.

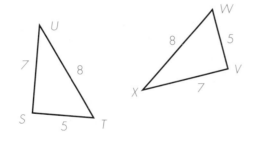

9. Are the triangles congruent? If so, why?

10. Is m∠U = m∠W?

11. What angle is ∠S equal to?

12. Classify the triangles according to the lengths of their sides.

Tell whether each statement is always, sometimes, or never true.

13. If a triangle has two equal angles, then it is isosceles.

14. If two triangles have corresponding angles equal, the triangles are congruent.

15. If two lines intersect, then the vertical angles are supplementary.

16. An acute triangle and a right triangle are congruent.

17. Given: $AB = BC = CD = DA$
Justify: $m\angle A = m\angle C$

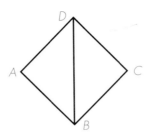

Statements	Reasons
1. $AB = BC = CD = DA$	1.
2. $BD = BD$	2.
3. $\frac{AB}{CB} = \frac{AD}{CD} = \frac{BD}{BD}$	3.
4. $\triangle ABD \sim \triangle CBD$	4.
5. $\frac{AB}{CB} = 1$	5.
6. $\triangle ABD \cong \triangle CBD$	6.
7. $m\angle A = m\angle C$	7.

18. In $\triangle ABC$, $AC = BC$, and D, E, and F are midpoints. Write a paragraph proof to show that $m\angle DFA = m\angle EFB$.

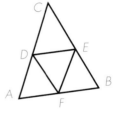

19. Given: Isosceles $\triangle ABC$ with $AC = BC$. $\overline{CD}$ bisects $\angle ACB$. Point X is on $\overline{CD}$.
Justify: $\triangle AXB$ is isosceles.

Statements	Reasons
1. Isosceles $\triangle ABC$ with $AC = BC$	1.
2. $\overline{CD}$ bisects $\angle ACB$.	2.
3. $m\angle ACD = m\angle BCD$	3.
4. $CX = CX$	4.
5. $\frac{AC}{BC} = \frac{CX}{CX}$	5.
6. $\triangle ACX \sim \triangle BCX$	6.
7. $\frac{CX}{CX} = \frac{AC}{BC} = 1$	7.
8. $\triangle ACX \cong \triangle BCX$	8.
9. $AX = BX$	9.
10. $\triangle AXB$ is isosceles.	10.

CRITICAL THINKING

20. Without using a calculator, determine which of these four numbers is the greatest: 3^6, 6^3, 4^5, and 5^4. (Hint: Factor each number.)

The Converse of a Theorem

After completing this section, you should understand
▲ the meaning of converse.
▲ how a statement and its converse are related.

You have used conditional statements many times so far in this course. The following statement is an example.

If <u>two lines are parallel</u>, then the two lines do not intersect.

Recall that the underlined part of the statement is the hypothesis, or given, and that the circled part is the conclusion.

CONVERSES

You form the converse of a conditional statement by switching the hypothesis and the conclusion.

If <u>two lines do not intersect</u>, then they are parallel.

The two statements are not the same. In the first, you are given the fact that two lines are parallel and conclude that the lines do not intersect. In the second, you are given the fact that two lines do not intersect and conclude that the lines are parallel.

DEFINITION 5-4-1 **The CONVERSE of an "If A, then B" statement is "If B, then A."**

Example: Statement: If it rains, then I carry an umbrella.
Converse: If I carry an umbrella, then it rains.

When writing the converse of a statement, you may have to change the wording of the hypothesis and conclusion slightly to make the converse read more clearly.

Example: Statement: If Jim has a dog, then he has a pet.
Converse: If he has a pet, then Jim has a dog.
Reworded converse: If Jim has a pet, then he has a dog.

The reworded converse does not change the meaning; it simply restates the converse in a way that makes it more clear.

CLASS ACTIVITY

Give the converse of each of the following statements.

1. If you are late to class, then you are tardy.

2. If you are a safe driver, then you do not have accidents.

3. If you watch the news on TV, then you are informed.

4. If a ray is an angle bisector, then it divides the angle into two equal angles.

5. If a triangle is isosceles, then it is equilateral.

6. If a triangle has two equal angles, then the sides opposite these angles are equal.

DISCOVERY ACTIVITY

1. Tell whether each statement is true or false.
Statement A: If you own a wristwatch, then you own a watch that you wear on your wrist.
Statement B: If a person skis in the Olympics, then that person is an excellent skier.
Statement C: If Jack owns a piece of jewelry, then he owns a ring.
Statement D: If a teenager owns a bicycle, then the teenager is a girl.
2. Write the converse of each statement above, and tell whether it is true or false.

3. Copy and complete the following tables, using the results from parts 1 and 2. Entries for statement A and its converse have been done for you.

	Statement	Converse
A	true	true
B		

	Statement	Converse
C		
D		

4. Study the tables and write a generalization that seems to be true.

You may have discovered that even when you know that a statement is true or false, you cannot use this to tell whether the converse is true or false.

CLASS ACTIVITY

Consider each of the following statements to be true. Write the converse of each and tell whether the converse is true or false.

1. If John goes swimming, then the temperature is above 75°F.

2. If Joan rides her bicycle to school, then she does not walk.

3. If Maria plays tennis, then it is not snowing.

4. If it is raining, then I carry an umbrella.

Consider each of the following statements to be false. Write the converse of each and tell whether the converse is true or false.

5. If Alex is twelve years old, then he is a teenager.

6. If a person is a girl, then the person owns a horse.

7. If Juan owns an automobile, then he owns a convertible.

8. If a dog is brown, then it is a male.

A SPECIAL CONVERSE

In the last section, it was proved that if a triangle has two equal angles, then the sides opposite these angles are equal (Theorem 5-3-1).

THE CONVERSE OF THEOREM 5-3-1 IS

If two sides of a triangle are equal, then the angles opposite the two sides are equal.

Just because Theorem 5-3-1 was proved to be true, you cannot assume its converse is true. You must prove the converse true separately.

Provide a reason for each statement in the proof.

Given: $\triangle ABC$ with $AC = BC$
Justify: $m\angle A = m\angle B$

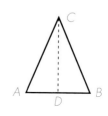

Statements		Reasons
1.	$AC = BC$	1.
2.	Draw the bisector of $\angle C$.	2.
3.	$m\angle ACD = m\angle BCD$	3.
4.	$CD = CD$	4.
5.	$\triangle ACD \sim \triangle BCD$	5.
6.	$\frac{AC}{BC} = 1$	6.
7.	$\triangle ACD \cong \triangle BCD$	7.
8.	$m\angle A = m\angle B$	8.

NOTEBOOK

THEOREM 5-4-1 **If a triangle is isosceles, then the angles opposite the equal sides are equal.**

CLASS ACTIVITY

1. Write the reasons to complete the following proof.

Given: $AB = DB$ and
$AC = DC$
Justify: $m\angle BAC = m\angle BDC$

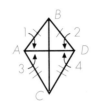

Statements		Reasons
1.	$AB = DB$ and $AC = DC$	1.
2.	$m\angle 1 = m\angle 2$ and $m\angle 3 = m\angle 4$	2.
3.	$m\angle 1 + m\angle 3 = m\angle 2 + m\angle 4$	3.
4.	$m\angle BAC = m\angle BDC$	4.

2. Use a computer and a geometric drawing tool to draw an isosceles triangle. Measure each angle. Repeat this activity several times, using different angle measures and different side lengths each time. Do your results show that Theorem 5-4-1 is true?

HOME ACTIVITY

Write the converse of each of the following statements.

1. If it is a cat, then it is a Siamese.

2. If it is summer, then I'll go swimming.

3. If he is a judge, then he is honest.

4. If you study, then you earn good grades.

5. If the triangle is isosceles, then it has two equal angles.

Consider each of the following statements to be true. Write the converse of each and tell whether the converse is true or false.

6. If I have a quarter, then I have 25 cents.

7. If Barb is not walking, then she is riding her horse.

8. If clothes are washed, then they are clean.

9. If two lines are perpendicular, then they meet at right angles.

10. If two angles are complementary, then the sum of their measures is 90°.

Consider each of the following statements to be false. Write the converse of each and tell whether the converse is true or false.

11. If it is winter in New York, then the date is January 31.

12. If James is not at home, then he is in school.

13. If an animal is a dog, then it has two legs.

14. If a figure has four sides, then it is a triangle.

15. If two angles are congruent, then they are right angles.

16. Write the reasons to complete the proof.
Equilateral triangles are equiangular.
Given: △ABC with AB = BC = CA
Prove: m∠A = m∠B = m∠C

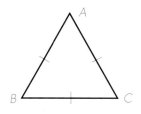

Statements	Reasons
1. AB = CA	1.
2. m∠B = m∠C	2.
3. BC = CA	3.
4. m∠A = m∠B	4.
5. m∠A = m∠B = m∠C	5.

17. Given: △ABC with AC = BC. $\overline{CD}$ bisects ∠ACB.
Prove: $\overline{CD} \perp \overline{AB}$

18. Given: △ABC with AC = BC. $\overline{AD}$ and $\overline{BE}$ are angle bisectors.
Prove: AE = BD

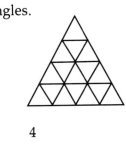

CRITICAL THINKING

19. Observe the patterns of equilateral triangles.

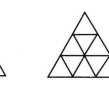

Number of rows	1	2	3	4
Number of small △'s	1	4	9	16

Find a pattern. Predict the number of small triangles in a pattern with 15 rows.

Applications

After completing this section, you should understand
▲ applications of congruent triangles.
▲ applications of the converse.

You saw in the last section that no matter if a statement is true (or false), you must look at the converse itself to decide if it is true or false. The advertising world makes use of this. A true statement is made. Some people may believe that the converse is also true.

APPLICATIONS OF THE CONVERSE

Example: "Smart shoppers shop at store X."

In an if-then form, this statement is
If you are a smart shopper, then you shop at store X.

This can be a true statement. A smart shopper may well go to store X because the store has some good buys. The store is hoping that people will think that the converse is true: if you shop at store X, then you are a smart shopper. They are trying to appeal to a person's pride—we all like to think we are smart shoppers. But going to a particular store doesn't necessarily mean a person knows how to look for the best buy!

REPORT 5-5-1   Consult with several local merchants as to

1. the time (day of week, month, season) when they order their merchandise;
2. how they select the merchandise; and
3. how they develop advertising to attract the market.

Bring to class some examples of local advertising. When possible, write the advertisement in an if-then form. Write the converse of the statement, and check it for accuracy.

CLASS ACTIVITY

In a magazine directed at high school athletes, an ad shows an NBA star wearing a certain brand of basketball shoes.

1. Write your interpretation of the ad in if-then form.

2. Write the converse of the statement in Exercise 1.

3. What do you think the ad wants you to conclude?

Assume that each statement is true. Write the converse. Is the converse true?

4. If two triangles are congruent, then they are similar.

5. If two sides of a triangle are equal in length, then the angles opposite the equal sides are equal in measure.

6. If the sum of the lengths of two sides of a triangle is 13, then the length of the third side is less than 13.

APPLICATION OF CONGRUENT FIGURES

Mass production plays an important role in our society. The blueprints for houses involve drawings that are similar figures of the houses. The construction of the houses can involve thousands of congruent pieces.

Triangles are often used in construction, as triangular braces are very strong.

Example: The Williams family wants to build a fence around their backyard. They drew a diagram of two sections of fence.

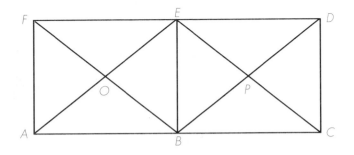

The following information is given in the diagram.

$\overline{AF} \parallel \overline{BE} \parallel \overline{CD}$
$\overline{FD} \parallel \overline{AC}$
$\overline{AE} \parallel \overline{BD}$
$\overline{FB} \parallel \overline{EC}$
$\overline{FA} \perp \overline{FD}, \overline{FA} \perp \overline{AC}$
$\overline{EB} \perp \overline{FD}, \overline{EB} \perp \overline{AC}$
$\overline{DC} \perp \overline{FD}, \overline{DC} \perp \overline{AC}$
$FA = EB = DC = 5$ ft
$AB = BC = FE = ED = 6$ ft

1. Find all of the triangles that are congruent to $\triangle EFA$.
    $\overline{FA} \perp \overline{FD}$, so $\angle EFA$ is a right angle. $\overline{EB} \perp \overline{FD}$, so $\angle FEB$ is a right angle. Then m$\angle EFA$ = m$\angle FEB$. $FE = FE$. $FA = 5$ ft and $EB = 5$ ft, so $FA = EB$. Then $\triangle FEB \cong \triangle EFA$. By similar reasoning, $\triangle BAF$, $\triangle ABE$, $\triangle DEB$, $\triangle CBE$, $\triangle EDC$, and $\triangle BCD$ are congruent to $\triangle EFA$.

2. Find all of the triangles that are congruent to $\triangle FEO$.
    $\overline{FD} \parallel \overline{AC}$, so $\angle EFB \cong \angle ABF$ since they are alternate interior angles. Also, $\angle FEA \cong \angle BAE$ for the same reason. $FE = AB$. Therefore, $\triangle FEO \cong \triangle BAO$. $\overline{AE} \parallel \overline{BD}$. $\angle EAC \cong \angle DBC$ since they are corresponding angles (transversal $\overline{AC}$). $\overline{FB} \parallel \overline{EC}$. $\angle FBA \cong \angle ECA$ since they are corresponding angles (transversal $\overline{AC}$). $AB = BC$. Therefore, $\triangle BAO \cong \triangle BCP$. By use of alternate interior angles, $\triangle EDP \cong \triangle CBP$. Then $\triangle FEO \cong \triangle BAO \cong \triangle EDP \cong \triangle CBP$.

Example: The figure shows the pattern for the Jack in the Pulpit quilt. Each shape is created by cutting and sewing in a piece of fabric in a particular color. If you wanted a quilt large enough to cover a double bed, you would need to create 20 squares like this and sew them together.

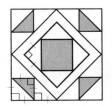

1. Are the triangular pieces congruent?

Yes, they each contain a right angle enclosed by two equal legs. Therefore, they are similar by SAS, and the ratio of two corresponding sides is one.

2. How many unshaded congruent triangles would you need to cut for the entire quilt?

There are 16 unshaded congruent triangles. You would need $16 \cdot 20 = 320$ triangles of this shape and color for your quilt.

HOME ACTIVITY

1. Write your interpretation of the ad in if-then form.

2. Write the converse of the statement in Exercise 1.

3. Do you think the statement in Exercise 1 is true?

4. Do you think the converse in Exercise 2 is true?

Step out in style.

5. Write your interpretation of this ad in if-then form.

6. Write the converse of the statement in Exercise 5.

7. Do you think the statement in Exercise 5 is true?

8. Do you think the converse in Exercise 6 is true?

The following ad was placed by a summer camp.

> We promote
> —Positive self-image
> —Personal confidence
> —Self-reliance
> —Academic achievement

9. Write your interpretation of the above ad in if-then form.

10. Write the converse of the statement you wrote for Exercise 9.

11. Do you think the statement in Exercise 9 is true?

12. Do you think the converse in Exercise 10 is true?

13. In figure $ABCD$, $AB = BC = CD = DA$. $m\angle D = m\angle C = 90°$. M is the midpoint of $\overline{DC}$. Explain why $AM = BM$, and why $m\angle 1 = m\angle 2$.

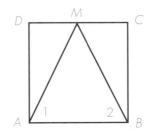

14. In this figure, $JL = LM$, $KL = LN$, and $NM = 54$ feet. What length is JK?

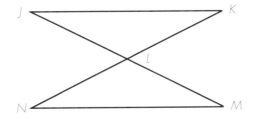

15. In this diagram of a garage roof, $AM = BM$. If $\angle PMA$ is a right angle, then how do you know that $AP = BP$?

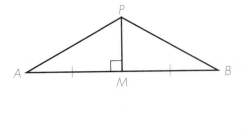

16. In this diagram of a gable roof, $AC = BC$, M is the midpoint of $\overline{AB}$. How do you know that $\overline{CM} \perp \overline{AB}$?

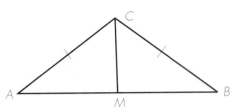

17. A designer created the following design for an ad. $AC = AE$, $AB = AD$, and $m\angle CAB = m\angle EAD$. Show that $CB = ED$.

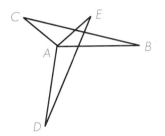

18. In this drawing of a park, $\overline{DE} \parallel \overline{AB}$. $AB = 40$ m, $AC = 25$ m, $BC = 35$ m, and $AD = 10$ m. What is the length of $\overline{CE}$?

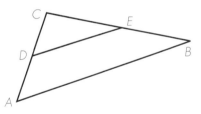

CRITICAL THINKING

19. Find a pattern in the table. Use the pattern to predict the number of diagonals in a figure having 24 sides.

Number of Sides	Number of Diagonals
3	0
4	2
5	5
6	9
7	14
8	20

5.1 Write all the subsets of $A = \{1, 4, 9\}$. $\{1\}, \{4\}, \{9\}, \{1, 4\}, \{1, 9\}, \{4, 9\}, \{1, 4, 9\}, \{\ \}$

$B = \{1, 4, 7, 10, 13\}$ $C = \{2, 4, 6, 8, 10\}$

Draw Venn diagrams showing $B \cup C$ and $B \cap C$.

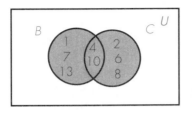 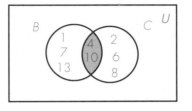

5.2 SQ bisects $\angle PQR$. Is $\triangle PQS$ congruent to $\triangle RQS$?

$PQ = QR$, $QS = QS$, and $m\angle PQS = m\angle RQS$.

$\triangle PQS \sim \triangle RQS$ (by SAS) and $\frac{SQ}{SQ} = \frac{PQ}{RQ} = 1$.

Therefore, $\triangle PQS \cong \triangle RQS$.

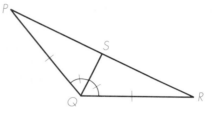

5.3 Find x in $\triangle DRY$.

$\triangle DRY \sim \triangle TON$ (by AA) and $\frac{DR}{TO} = 1$

Therefore $\triangle DRY \cong \triangle TON$, and $\frac{DY}{TN} = 1$.

$\frac{x}{10} = 1$

$x = 10$

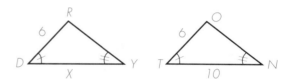

5.4 This is a true statement: If a triangle is equilateral, then it is isosceles. Is the converse true?

Converse: If a triangle is isosceles, then it is equilateral; false.

This is a true statement: If a triangle has two equal angles, then the sides opposite these angles are equal. Is the converse true?

Converse: If a triangle has two equal sides, then the angles opposite these sides are equal; true.

The truth of a statement does not determine the truth of its converse.

You have learned that all right triangles contain one right angle. So we automatically know that two right triangles have one pair of angles with equal measures. Thus, to prove that two right triangles are similar, we must only prove either that another pair of angles have the same measure (AA) or that the pairs of legs of the right triangles have the same lengths (SS).

The LOGO procedure in the example below can be used to show that two right triangles are similar if one pair of the acute angles have equal measures.

Example: Write a LOGO procedure to draw two right triangles, each with one acute angle whose measure is 25. Do the two right triangles appear to be similar? Justify your answer.

First write a procedure to draw one right triangle with an acute angle with a measure of 25.

```
TO RTTRI1
BK 80
RT 90
FD 40
LT 155
FD 70
END
```

Then use the RTTRI1 procedure in a procedure to draw both of the right triangles.

```
TO RTTRI2
RTTRI1
PU HOME LT 90 FD 100 RT 90 PD
RTTRI1
END
```

The output of RTTRI2 is shown below.

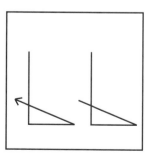

The two right triangles do appear to be similar.

The triangles can be proven to be congruent by AA and showing that the ratio of the corresponding sides is $\frac{40}{40}$ or one.

1. Write a LOGO procedure named RTTRI3 to draw two right triangles, each with one leg with a length of 25 spaces and one leg with a length of 35 spaces. Do the two right triangles appear to be congruent? Justify your answer.

Modify the RTTRI1 procedure to tell the turtle to draw each right triangle. Then modify the RTTRI2 procedure to tell the turtle to draw a right triangle that is congruent to the given triangle.

2.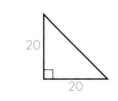

3.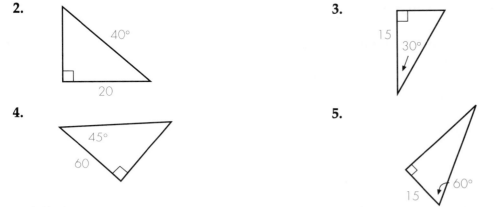

4.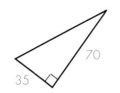

5.

Modify the RTTRI3 procedure to tell the turtle to draw each right triangle and a right triangle that is congruent to it.

6.

7.

8.

9.

Write a LOGO procedure to tell the turtle to draw the figure.

10.

11.

1. Explain the significance of the congruence symbol.

Classify each statement as true or false.

2. All similar triangles are congruent.

3. All congruent triangles are similar.

4. Some similar triangles are congruent.

5. If the ratio of the corresponding sides for similar triangles is equal to one, then the triangles are congruent.

Refer to figures a–h.

6. Which triangles are similar?

7. Which triangles are congruent?

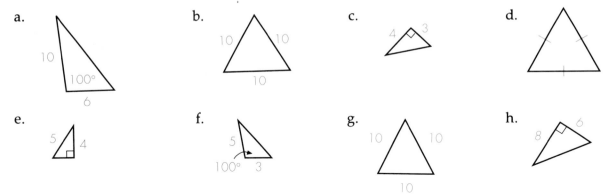

a.   b.   c.   d.

e.   f.   g.   h.

This information was collected for the after-school geometry class picnic.

    10 voted for pizza (P).   17 voted for hot dogs (H).
    6 voted for P and H.   2 voted for neither P or H.

8. Draw the Venn diagram illustrating this information.

9. How many students are in this geometry class?

Refer to $\triangle ABC$.

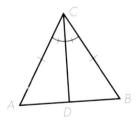

10. Is $\triangle ACD \sim \triangle BCD$? If so, why?

11. Is $\triangle ACD \cong \triangle BCD$? If so, why?

12. Give the converse of the statement: If you want to be a better thinker, then study geometry.

Let $A = \{1, 4, 9, 16, 25\}$.

1. Is $\{36\} \subset A$?

2. List all the three-element subsets of A.

Refer to figure $ABCD$. $AB = CD = AD = BC$, $AE = CF$. $\overline{AC}$ is a straight line segment.

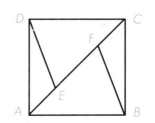

3. Is $m\angle DAC = m\angle DCA$? Explain.

4. Justify that $\triangle AED \cong \triangle CFB$.

5. Is $DE = BF$? Explain.

6. Show that $m\angle DEF = m\angle BFE$.

A car manufacturer places an ornament on the hood of the new model. The hood ornament is a triangle, $\triangle$. The ad copy is

$$\triangle \quad \text{A SYMBOL OF QUALITY}$$

7. Write in if-then form what you think the advertisement wants you to conclude.

8. Write the converse of your statement in Exercise 7.

Use the survey information for Exercises 9 and 10.

| 12 people like grapes (G). | 15 people like nectarines (N). |
| 5 people like both G and N. | 3 people like neither G or N. |

9. Draw a Venn diagram showing this information.

10. How many people were surveyed?

Example: First-degree equation

$$3x - 4 = 5(2x - 7)$$
$$3x - 4 = 10x - 35$$
$$3x + 31 = 10x$$
$$31 = 7x$$
$$x = 4\tfrac{3}{7}$$

Example: Second-degree equation

$$2x^2 - 3x - 35 = 0$$

Use the quadratic formula, $x = \dfrac{-b \pm \sqrt{b^2 - 4ac}}{2a}$

$$x = \dfrac{-(-3) \pm \sqrt{(-3)^2 - 4(2)(-35)}}{2(2)}$$

$$x = \dfrac{3 \pm \sqrt{9 + 280}}{4}$$

$$x = \dfrac{3 \pm \sqrt{289}}{4}$$

$$x = \dfrac{3 \pm 17}{4}$$

$$x = 5 \text{ or } x = -\tfrac{7}{2}$$

Solve each equation for x.

1. $3x - 45 = 90$

2. $x^2 = 36$

3. $5x^2 = 125$

4. $17x - 5(3x - 50) = 68$

5. $x^2 + x - 6 = 0$

6. $5x - 29 = 41$

7. $2(x - 3) = 3(2x + 6)$

8. $2x^2 - 2x - 12 = 0$

9. $x^2 + 15x + 26 = 0$

10. $4x + 6 = 7x - 9$

11. $3x^2 - 5x - 2 = 0$

12. $3(2x - 7) = 4(3x + 9)$

13. $5(4x + 6) = 3(5x - 4)$

14. $x^2 - 7x + 6 = 0$

15. $x^2 - 5x - 9 = 0$

16. $3(7x - 9) = 4(2x + 5)$

17. $3x + 17 = 5x - 9$

18. $3x^2 + 6x + 4 = 0$

19. $2x^2 - 7x + 6 = 0$

20. $7(2x - 4) = 8(3x + 5)$

21. A counselor asked a new student her age. The student replied that two times her age plus the square root of 36 resulted in a perfect square number. The counselor knew that the student was between 10 and 18, and that today was her birthday. How old is the new student?

Tell what kind of figure each of the following is. Use symbols to name it.

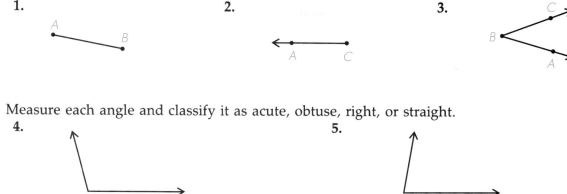

1.   **2.**   **3.**

Measure each angle and classify it as acute, obtuse, right, or straight.

4. 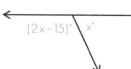   **5.**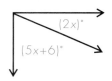

6. What is the measure of each angle formed by the bisector of a 90° angle?

Each of the following pairs of angles is either complementary or supplementary. Find the measure of each angle.

7.   **8.**

$(2x-15)° \quad x°$ $(2x)°$

$(5x+6)°$

Refer to the number line at the right.

9. What are the coordinates of points A, B, C, and D?

10. What are the lengths of $\overline{AC}$ and $\overline{BD}$?

Suppose C is between B and G. Point L is the midpoint of $\overline{BC}$. $BG = 10$ and $BL = 1\frac{1}{2}$.

11. Draw a diagram to show how B, L, C, and G are situated.

12. What is the measure of $\overline{LC}$? of $\overline{CG}$?

13. Write LOGO commands for drawing a 110° angle and its bisector.

Use the diagram. Identify the following.

14. Corresponding angles

15. Alternate interior angles

16. Vertical angles

17. Alternate exterior angles

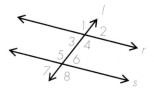

In the diagram, $\overleftrightarrow{ML}$ and $\overleftrightarrow{DH}$ are in plane P and do not intersect. Also, $m\angle RTM = 90°$. Identify the following.

18. Intersecting lines
19. Parallel lines
20. A transversal
21. Skew lines
22. Coplanar lines
23. Three collinear points
24. Three noncollinear points

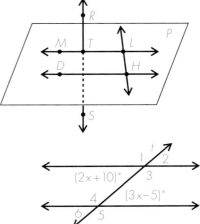

In the diagram, $r \parallel s$ and t is a transversal.

Give the degree measures of each angle.

25. $\angle 1$ 26. $\angle 2$ 27. $\angle 3$
28. $\angle 4$ 29. $\angle 5$ 30. $\angle 6$

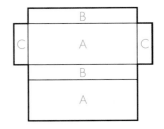

Measure the segment as indicated. State the error in each measurement.

31. to the nearest $\frac{1}{2}$ inch
32. to the nearest $\frac{1}{16}$ inch
33. Draw the closed three-dimensional figure this layout would form.
34. "If a person studies the manual carefully, that person will pass the driver's written test. Kim studied the manual carefully." If both statements are true, what can you conclude about how Kim will do on the test?

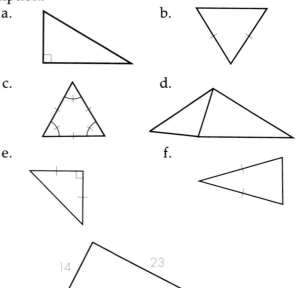

Write the letter of each triangle that fits the description.

35. Acute triangle

36. Scalene triangle

37. Equilateral triangle

38. Obtuse triangle

39. Right triangle

40. Isosceles triangle

41. Equiangular triangle

42. The length of the unknown side in the triangle is between what two numbers?

43. What is the measure of ∠C?

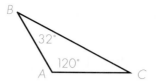

Consider the figure and the information given.
Find the measure of each angle.

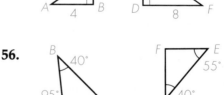

44. ∠1   **45.** ∠2
46. ∠3   **47.** ∠4
48. ∠5   **49.** ∠6

50. Plot the following points on the *x-y* plane: (6,1), (0,4), and (6,4). Identify the type of triangle that is formed when the points are connected by line segments.

51. Two cities are $4\frac{1}{4}$ in. apart on a road map. If the map uses a scale of 1 in. = 30 mi, what is the actual distance between the cities?

52. A factory wants to have a ratio of 2 managers for every 15 trainees. If 90 people are in the training program, how many managers will it need?

Identify the pairs of similar triangles. Tell how you know they are similar.

53.

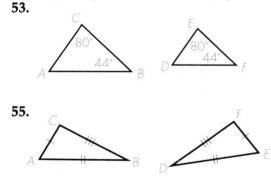

54.

55.

56.

57. What is the length of $\overline{RS}$ in △ABC?

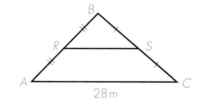

Refer to the diagram. $m\angle SQR = m\angle SVT$. Find the length of each segment.
58. SV =
59. TV =

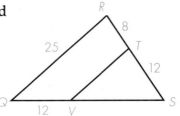

Classify each statement as true or false.
60. No similar triangles are congruent.

61. If two triangles are congruent, then the ratio of two corresponding sides is equal to 1.

Refer to figures a – f.
62. Which triangles are similar?   **63.** Which triangles are congruent?

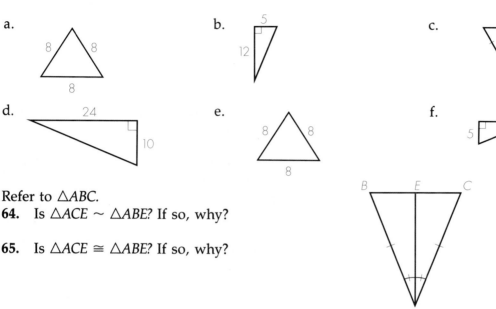

a.

b.

c.

d.

e.

f.

Refer to △ABC.
64. Is △ACE ∼ △ABE? If so, why?

65. Is △ACE ≅ △ABE? If so, why?

The members of a geometry class were asked what after-school clubs they belong to.
12 are in the computer club   8 are in the photography club
5 are members of both clubs   6 are in neither club

66. Draw a Venn diagram illustrating this information.

67. How many students are in this geometry class?

68. In the figure at the right, $m\angle 1 = m\angle 2$,
    $PA = PB$, and $PC = PD$.
    Justify: $\overline{PB}$ bisects $\angle APD$

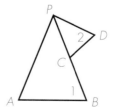

Give the converse of each of the following statements.
69. If you read the newspaper, then you are informed.

70. If you got into the theatre, then you had a ticket.
71. If a person has visited London, then the person has been to England.

6

CONSTRUCTIONS

Those who hear, forget.
Those who see, remember.
Those who do, understand.

Anonymous

SECTION
6.1

Bisection of Line Segments and Angles

After completing this section, you should understand
▲ how to copy a line segment using the tools of geometry.
▲ how to copy an angle using the tools of geometry.
▲ how to bisect a line segment.
▲ how to bisect an angle.

In Chapter 1, you learned that the tools for geometry include a compass and a straightedge (or ruler). In this chapter, you will learn how to use these tools and how to solve some problems using these tools.

COPYING A LINE SEGMENT

Given a segment, $\overline{AB}$, copy the segment onto line l.

$\overline{}$ $\longleftrightarrow l$
A   B

Place a point on l.

$\longleftrightarrow l$
C

Place the point of your compass on point A. Open the setting on the compass until the pencil point is on point B. Move the point of the compass over to the point on line l. Make an arc. You have just copied a line segment.

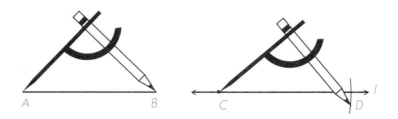

COPYING AN ANGLE

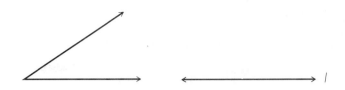

Copy the angle. Place the point of the compass on the vertex of the angle. Mark an arc on each ray of the angle.

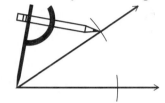

Mark a point on line l. Using the same setting on the compass, mark one arc on line l, and another above it.

Place the point of your compass where an arc intersects one ray of the angle. Open the setting until the point of the pencil touches the intersection of the arc and the other ray.

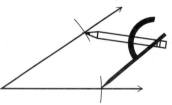

Keeping this setting, place the point of the compass at the point on line l where the arc intersects l. Make a small arc.

Complete the angle.

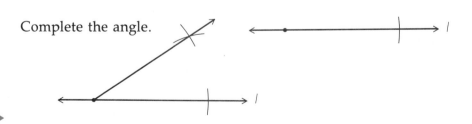

BISECTING A LINE SEGMENT

With an unruled straightedge, you can't measure a segment in order to find the midpoint. You must bisect the segment in order to find the midpoint.

Bisect $\overline{AB}$.

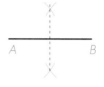

To bisect a segment, open your compass to a setting that is just a little more than half the length of $\overline{AB}$. Mark two arcs from A and two from B, as shown below.

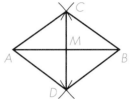

Connect the points where the arcs intersect. This segment crosses $\overline{AB}$ at its midpoint, thus bisecting it.

Show that M is really the midpoint of $\overline{AB}$.

$\overline{AC}$, $\overline{BC}$, $\overline{AD}$, and $\overline{BD}$ were constructed so that they were all the same length. Then $\triangle ACD \cong \triangle BCD$, by SSS and the ratio of corresponding sides is 1. This means $\angle ACM \cong \angle BCM$, and since $AC = BC$ and $CM = CM$, $\triangle ACM \cong \triangle BCM$. Then $AM = BM$ by CPCF. By the definition of midpoint, the construction is justified.

CLASS ACTIVITY

1. Use a computer and a geometric drawing tool to draw any line segment. Use the principles from the construction for bisecting a line segment to bisect the segment.

BISECTING AN ANGLE

Place the point of your compass on the vertex of the angle. Mark an arc on both rays of the angle.

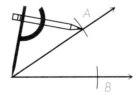

Using the same setting, place the point on A, and mark an arc in the center. Do the same from B. Label the vertex C and the intersecting arcs D. $\overrightarrow{CD}$ is the angle bisector.

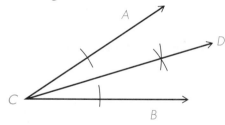

REPORT 6-1-1   The compass we use today can hold a setting until it is changed. Euclid used a collapsible compass. He lost the setting every time he picked the compass up to move it! Research how Euclid could bisect

1.   a line segment.   2.   an angle.

Suggested reference:
Cajori, F. *A History of Elementary Mathematics.* New York: Macmillan Publishing Co., 1905.

CLASS ACTIVITY

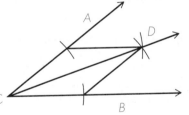

1.   Show that $\overline{CD}$ is really the bisector of $\angle ACB$.   (Hint: Show that $\triangle CAD \sim \triangle CBD$.)

2.   Use a computer and a geometric drawing tool to draw any angle. Use the principles from the construction for bisecting an angle to bisect the angle.

Copy each line or angle. Then use a compass and straightedge to bisect each.

3.   4.

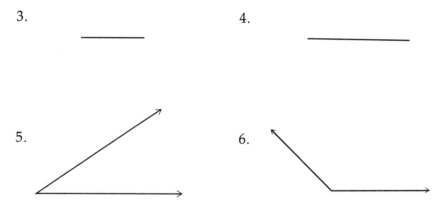

5.   6.

HOME ACTIVITY

1. Draw a $6\frac{1}{2}$-inch line segment. Bisect it.

2. Draw a 5-inch line segment. By construction, divide the segment into four equal segments.

3. On l_1, look at line segments $\overline{AB}$ and $\overline{CD}$. Construct $\overline{MN}$ on a line so that $MN = AB + CD$. (HINT: Use your compass as a walking tool)

4. Use your compass to show that $AB + AC > BC$ in $\triangle ABC$.

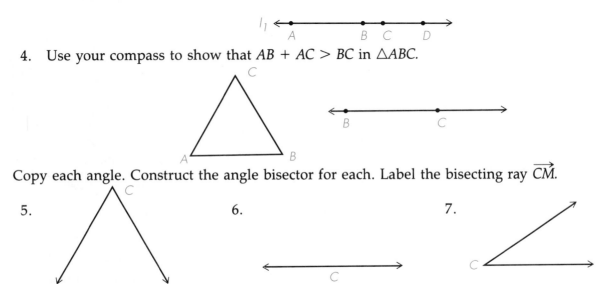

Copy each angle. Construct the angle bisector for each. Label the bisecting ray $\overrightarrow{CM}$.

5.   6.   7.

8. Draw a 60° angle. Use your compass and straightedge to divide the angle into four equal angles.

9. Copy the three segments. Copy $\overline{AC}$ onto line w, then use the other two segments to construct a triangle that has sides with lengths of the given segments.

         A —————— B

         C —————— B

         A ———————— C   ←————————————→ w

10. Construct a triangle so that $AB = 4$ cm, $AC = 6$ cm, and $m\angle A = 50°$. (Use your protractor to draw $\angle A$.)

CRITICAL THINKING

11. How many points are needed to divide a line segment into 100 equal segments? Try a simpler case first, and look for a pattern.

SECTION 6.2

Perpendiculars

After completing this section, you should understand
▲ how to construct a perpendicular to a line from a point on the line.
▲ how to construct a perpendicular to a line from a point not on the line.

Recall that perpendicular lines form a 90° angle, or a right angle. The symbol for perpendicularity is ⊥.

PERPENDICULARS TO A LINE FROM A POINT ON THE LINE

Given line $\overleftrightarrow{AB}$ and point C on $\overleftrightarrow{AB}$, construct a line through C that is perpendicular to $\overleftrightarrow{AB}$.

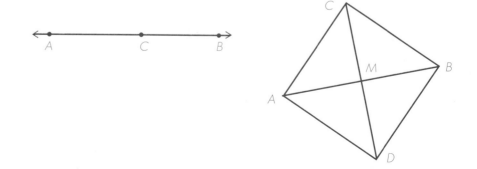

Recall that in the last section, we showed that $\triangle ACM \cong \triangle BCM$, where $AC = BC = AD = BD$. Then by CPCF, $\angle CMA \cong \angle CMB$. Since $m\angle AMB = 180°$, both $\angle CMA$ and $\angle CMB$ are right angles. Then $\overline{CM} \perp \overline{AB}$. Thus, the bisector constructed was perpendicular to the segment $\overline{AB}$.

In given line $\overleftrightarrow{AB}$, C is not the midpoint of $\overline{AB}$. If C were a midpoint, you could construct a perpendicular at C as you did in the last section.

DISCOVERY ACTIVITY

1. Copy line $\overleftrightarrow{AB}$ on page 211 onto a sheet of paper.

2. Open the setting on your compass so that it is less than AC or BC. Without changing the setting, place the point of the compass on C and make a mark to the left and to the right of C. Label the points of intersection D and E.

3. What do you know about CD and CE?

4. What name can be given to point C?

You may have found that by using the same compass setting, you can make C the midpoint of a line segment. Now you can construct a perpendicular at C.

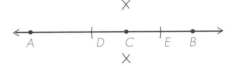

Open the compass setting so that it is slightly more than CD or CE. From D, make a mark above the line and below the line. Do the same from E.

Connect the intersections of the arcs to form the perpendicular to $\overleftrightarrow{AB}$ through C.

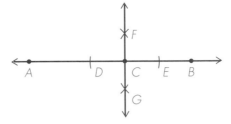

To justify that $\overleftrightarrow{FG} \perp \overleftrightarrow{AB}$, you can show that $\angle BCF$ is a right angle.

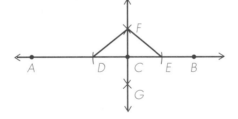

Connect F to D and F to E. You are able to do this because two points determine a line. $EF = DF$ since these were marked off without changing the setting on the compass. This reasoning is called "by construction."

Steps	Reasons
1. $CF = CF$	1. Identity
2. $CD = CE$	2. By construction
3. $DF = EF$	3. By construction
4. $\triangle DEF$ is isosceles.	4. Definition of isosceles triangle
5. $m\angle CDF = m\angle CEF$	5. Angles opposite the equal sides of an isosceles triangle are equal.
6. $CF/CF = CD/CE = 1$	6. Algebra
7. $\triangle DFC \sim \triangle EFC$	7. SAS
8. $\triangle DFC \cong \triangle EFC$	8. Definition of congruent triangles
9. $m\angle DCF = m\angle ECF$	9. CPCF
10. $m\angle DCE = 180°$	10. Definition of straight angle
11. $m\angle ECF + m\angle DCF = 180°$	11. Supplementary angles
12. $2(m\angle ECF) = 180°$	12. Substitution
13. $m\angle ECF = 90°$	13. Algebra
14. $m\angle BCF = 90°$	14. Same angle
15. $\angle BCF$ is a right angle.	15. Definition of right angle

CLASS ACTIVITY

1.   Use a computer and a geometric drawing tool to draw any line with point C between points A and B. Use the principles from the construction of a perpendicular to a line from a point on the line to draw a perpendicular to the line through point C.

2. Draw a 3-inch line segment, $\overline{AB}$, on your paper. Mark point C 1 inch from B. Construct the perpendicular to $\overleftrightarrow{AB}$ at C.

3. Draw a 12-centimeter line segment, $\overline{AB}$, on your paper. Mark a point C 5 centimeters from A. Construct the perpendicular to $\overline{AB}$ at C.

PERPENDICULARS TO A LINE FROM A POINT NOT ON THE LINE

The next exercise shows how to construct a perpendicular to a line from a point not on the line.

DISCOVERY ACTIVITY

1. Copy line $\overleftrightarrow{AB}$ onto a sheet of paper.

2. Place the point of your compass on point C and open the setting so that you can make two arcs on $\overleftrightarrow{AB}$. Label the intersections of the arcs and $\overleftrightarrow{AB}$ D and E.

3. What do you know about CD and CE?

4. Keeping the same setting, place the point of your compass on D and make an arc below $\overleftrightarrow{AB}$. Do the same from E. Label the intersection of the arcs F.

5. Connect C and F. What appears to be true?

The justification for $\overleftrightarrow{CF} \perp \overleftrightarrow{AB}$ in the Discovery Activity is similar to the proof for a perpendicular from a point on the line. Try the justification on your own.

CLASS ACTIVITY

1. Use a computer and a geometric drawing tool to draw any line. Draw points A and B on the line and point C above the line and between points A and B. Use the principles from the construction of a perpendicular to a line from a point not on the line to draw a perpendicular to the line through point C.
2. Draw a 5-inch segment, $\overline{AB}$, and a point C 2 inches from A and 2 inches above $\overline{AB}$. Construct the perpendicular from C to the line segment.

3. Draw a 10-centimeter segment and mark a point below the segment. Construct a perpendicular to the segment from the point.

HOME ACTIVITY

Draw the perpendicular at or from C to AB.

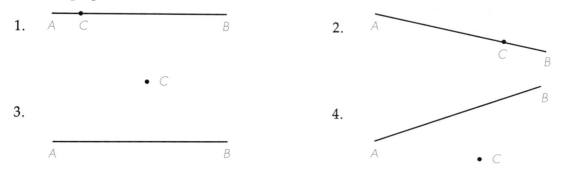

1. A   C   B

2. A   C   B

3. A   B

4. A   • C   B

5. Draw $\overline{AB}$, a 6-inch line segment, and mark C 1.5 inches from A. Construct the perpendicular to $\overline{AB}$ from C.
6. Draw $\overline{AB}$, a 9-centimeter line segment, and mark C 3.5 centimeters from A. Construct the perpendicular to $\overline{AB}$ from C.
7. Draw $\overline{AB}$, a 60-millimeter line segment, and mark C 20 millimeters from A and 15 millimeters below $\overline{AB}$. Construct the perpendicular to $\overline{AB}$ from C.
8. Draw a 5-inch line segment $\overline{AB}$, and mark a point D 1 inch above the segment and 3 inches from A. Construct the perpendicular to $\overline{AB}$ from D.
9. Draw $\overline{MN}$, a 7-inch line segment, and mark P $3\frac{1}{4}$ inches from N. Construct the perpendicular to $\overline{MN}$ from P.
10. Draw a 5-centimeter line segment $\overline{XY}$, and mark a point Z 2.4 centimeters from Y and 1.8 centimeters above $\overline{XY}$. Construct the perpendicular to $\overline{XY}$ from Z.
11. Draw a $4\frac{1}{2}$-inch line segment $\overline{PQ}$, and mark a point R $1\frac{1}{4}$ inches from Q. Construct the perpendicular to the segment from R.
12. Draw a 3.2-centimeter line segment $\overline{HI}$, and mark a point J 1.6 centimeters from H and 2.1 centimeters above $\overline{HI}$. Construct the perpendicular to $\overline{HI}$ from J.
13. Draw a 43-millimeter line segment $\overline{DE}$, and mark a point F 26 millimeters from E. Construct the perpendicular to $\overline{DE}$ from F.
14. Draw a $3\frac{1}{4}$-inch line segment $\overline{GK}$, and mark a point T $1\frac{1}{2}$ inches from K and $\frac{3}{4}$ inch below $\overline{GK}$. Construct the perpendicular to $\overline{GK}$ from T.
15. Draw an angle and pick a point between the two rays. From that point construct perpendiculars to the two rays.

16. A pipeline is to be constructed from the oil well to the river. By construction, indicate the path that is perpendicular to the river.

• oil well

17. A surveyor knows line $\overleftrightarrow{AB}$ on a map is a north-south line. Show, by construction, how to determine an east-west line through city *P*.

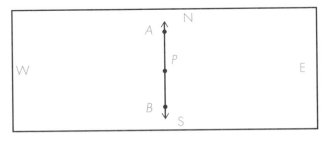

18. Perform the following directions:
    a. Draw line segment $\overline{AB}$ 5 inches in length.
    b. Mark point *C*, 2 inches from *A*, and *D*, 2 inches from *B*.
    c. From *C* and *D*, construct perpendiculars.
    d. From *C*, mark a point *E*, 1 inch above *C* on the perpendicular.
    e. From *E*, construct a perpendicular to $\overline{CE}$. Call it $\overline{GH}$.
    f. Extend $\overline{GH}$ to intersect the perpendicular at *D*. Label the point *M*.
    g. Darken 1.5 inches from *C* through *E*, and do the same from *D* through *M*. Now darken $\overline{EM}$.
    h. What does the darkened figure remind you of?
19. How would you construct an angle of 45°? All you are given to start with is a line on your paper.

CRITICAL THINKING

20. Which printed capital letters in the alphabet consist of perpendicular line segments? Form one of these letters by construction.

SECTION
6.3

Constructing a Line Parallel to Another Line

After completing this section, you should understand
▲ how to construct a line parallel to another line.

In blueprints or floor plans you can find many examples of parallel lines. How are they constructed?

NUMBER OF PARALLEL LINES THROUGH A POINT

Recall Theorem 3-1-1: if the figure is a triangle, then the sum of the angles is 180°. The proof of this theorem was based on the assumption (Postulate 3-1-1) that, through a point not on a line, there is only one line parallel to a given line.

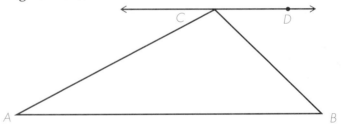

Mathematicians in the past have investigated other possibilities that could be true if Postulate 3-1-1 did not hold. The other possibilities are

1. there are many lines through a point parallel to the given line; and

2. there are no lines through a point parallel to the given line.

These possibilities may be true in non-Euclidean geometry, which plays an important role in navigating a ship or plane. For the work in this book, Postulate 3-1-1 will hold; we assume there is exactly one line parallel to a given line through a point not on the line.

217

REPORT 6-3-1 Mathematicians developed non-Euclidean geometry to deal with the possibilities of either no parallel lines or more than one parallel line through a point. Use the references below to write a brief report on non-Euclidean geometry.

Suggested sources:
Bergamini, David. *Mathematics.* New York: Time-Life Books, 1963.

CONSTRUCTING A LINE PARALLEL TO A GIVEN LINE

To construct a parallel line, you first need to recall how to copy an angle.

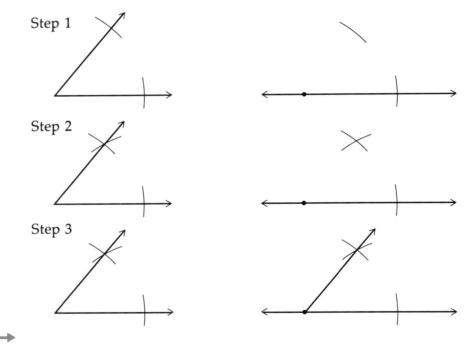

Step 1

Step 2

Step 3

JUSTIFYING THE METHOD FOR COPYING AN ANGLE

Here is the justification of the method of copying an angle.

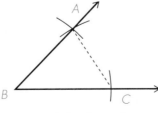

original angle

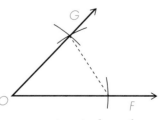

constructed angle

Steps	Reasons
1. Draw $\overline{AC}$, $\overline{GF}$.	1. Two points determine a line.
2. $BC = OF$, $BA = OG$, $AC = GF$	2. By construction
3. $BC/OF = BA/OG = AC/GF$	3. Algebra
4. $\triangle ABC \sim \triangle GOF$	4. SSS
5. $BC/OF = BA/OG = AC/GF = 1$	5. Algebra
6. $\triangle ABC \cong \triangle GOF$	6. Definition of congruent triangles
7. $\angle ABC \cong \angle GOF$	7. CPCF

DISCOVERY ACTIVITY

1. On a sheet of paper, draw line $\overleftrightarrow{AB}$. Choose a point above the line, and label the point P.

2. Choose a point on $\overleftrightarrow{AB}$. Label the point F. Draw $\overleftrightarrow{FP}$.

3. Copy $\angle PFB$ to form $\angle FPN$. $\angle FPN$ is to be the alternate interior angle for $\angle PFB$.

4. What appears to be true about $\overleftrightarrow{NP}$ and $\overleftrightarrow{AB}$?

5. Make a second copy of line $\overleftrightarrow{AB}$ and point P. Choose a different point F on $\overleftrightarrow{AB}$.

6. Copy $\angle PFB$ as before. Does your answer to question 4 change?

If two lines are intersected by a transversal so that the alternate interior angles are equal, then the two lines appear to be parallel. This is indeed the case.

JUSTIFYING THE CONSTRUCTION

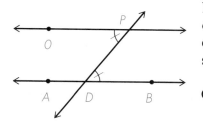

Recall Theorem 2-3-2: if two parallel lines are intersected by a transversal, then the alternate interior angles are equal. In the Discovery Activity, you constructed alternate interior angles to be equal. Your conclusion is that the lines are parallel. This is the converse of Theorem 2-3-2, so we must justify this conclusion separately.

Given: Lines $\overleftrightarrow{AB}$ and $\overleftrightarrow{OP}$ with transversal $\overleftrightarrow{PD}$ so that $\angle OPD \cong \angle PDB$.

Justify: Lines $\overleftrightarrow{AB}$ and $\overleftrightarrow{OP}$ are parallel.

Indirect reasoning will be used for this justification.

Only two possibilities exist: either $\overleftrightarrow{OP}$ is parallel to $\overleftrightarrow{AB}$ or $\overleftrightarrow{OP}$ is not parallel to $\overleftrightarrow{AB}$. Assume that $\overleftrightarrow{OP}$ is not parallel to $\overleftrightarrow{AB}$. If the lines are not parallel, then $\overleftrightarrow{AB}$ and $\overleftrightarrow{OP}$ intersect. This means a triangle is formed, as indicated in the figure below.

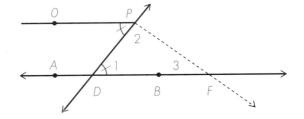

It was given that $\angle OPD \cong \angle PDB$. Because $\overleftrightarrow{OP}$ is a line, $m\angle 1 + m\angle 2 = 180°$. In $\triangle DPF$, $m\angle 1 + m\angle 2 + m\angle 3 = 180°$. But this means that $m\angle 3 = 0°$. This is impossible. Since the assumption led to a contradiction, the assumption must be false, or invalid. This means that $\overleftrightarrow{OP} \parallel \overleftrightarrow{AB}$.

THEOREM 6-3-1   **If two lines are intersected by a transversal so that the alternate interior angles are equal, then the lines are parallel.**

Thus, the method presented in the Discovery Activity has been justified as a way of constructing parallel lines.

CLASS ACTIVITY

1. Use a computer and a geometric drawing tool to draw any two lines intersected by a transversal so that the alternate interior angles are equal. What appears to be true about the lines? Repeat this activity several times. Do your findings support Theorem 6-3-1?

Construct a line parallel to the given line through the given point.

2.   3.

4. Draw a line segment 15 centimeters long. Label it $\overline{KL}$. Mark a point Z, 6 centimeters from K and 3 centimeters above $\overline{KL}$. Construct a line parallel to $\overline{KL}$ through Z.

5. Draw a 4-inch line segment. Choose two points, A and B, so that A is 1 inch above the line segment and B is 2 inches below the line segment. Construct a line through A and one through B that are parallel to the segment.

HOME ACTIVITY

Construct a line parallel to the given line through the given point(s).

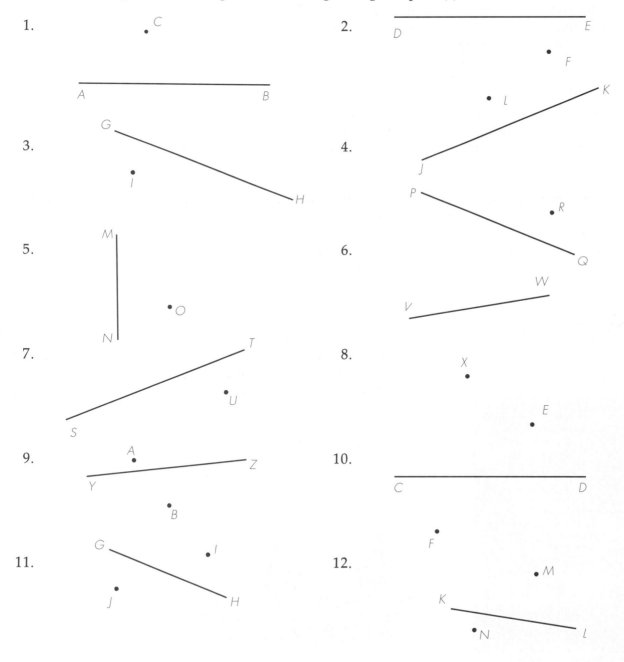

13. Draw a line segment $\overline{DE}$ 4.5 inches long. Choose a point H so that H is 2.25 inches from D and 1.25 inches above $\overline{DE}$. Construct the line through H that is parallel to $\overline{DE}$.

14. Draw a line segment $\overline{PQ}$, 130 millimeters long. Choose a point T that is 64 millimeters from Q and 57 millimeters below $\overline{PQ}$. Construct the line through T that is parallel to $\overline{PQ}$.

15. Draw a line segment $\overline{AB}$, 6 inches long. Choose a point C 2 inches from A and $1\frac{1}{2}$ inches above $\overline{AB}$, and a point D $1\frac{1}{2}$ inches from B and 1 inch below $\overline{AB}$. Draw a line through each of these points that is parallel to $\overline{AB}$.

16. Draw a line segment $\overline{MN}$, 180 millimeters long. Choose a point P 53 millimeters from M and 26 millimeters above $\overline{MN}$, and a point Q, 74 millimeters from N and 36 millimeters below $\overline{MN}$. Draw a line segment through each of these points that is parallel to $\overline{MN}$.

17. Draw an acute triangle and label it $\triangle SWC$. Construct the line through C parallel to $\overline{SW}$.

18. Construct a right angle, then draw a right triangle ACB, with C the vertex of the right angle. Construct a line through A parallel to $\overline{BC}$, and a line through B parallel to AC, and a line through C parallel to $\overline{AB}$. Extend the constructed lines until they intersect. How many triangles are formed?

Draw an acute triangle. Bisect each angle and extend the bisectors until each intersects the opposite side.

19. What do you observe about the three angle bisectors?

20. How many triangles are there in the final figure?

CRITICAL THINKING

21. Use indirect reasoning to solve the following: Sara, Bob, and Cheryl are three bowlers with different averages. If only one of the following statements is true, then who has the highest average?

Bob has the highest average.

Sara does not have the highest average.

Cheryl does not have the lowest average.

SECTION 6.4

Constructions Involving Triangles

After completing this section, you should understand
▲ the medians of a triangle.
▲ the altitudes of a triangle.
▲ the angle bisectors of a triangle.
▲ the perpendicular bisectors of the sides of a triangle.

So far in this chapter, you have learned how to bisect line segments, how to bisect angles, how to construct perpendiculars, and how to construct a line parallel to a given line.

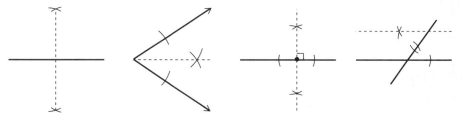

This section incorporates these constructions into applications with triangles. Some new terminology will be introduced.

MEDIANS OF A TRIANGLE

You may already be familiar with medians because of the median of a highway. The median divides a highway into two equal portions.

The median of a triangle divides a side of the triangle into two equal portions.

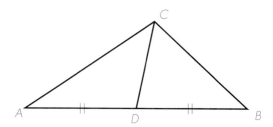

$\overline{CD}$ is one of the medians of $\triangle ABC$. $\overline{CD}$ divides $\overline{AB}$ into two equal portions or lengths.

DEFINITION 6-4-1 **A MEDIAN of a triangle is a line segment joining the vertex of an angle and the midpoint of the opposite side.**

To draw a median, you must locate the midpoint of a side of the triangle. Connect this point to the vertex opposite it.

DISCOVERY ACTIVITY

1. Use a ruler and a protractor to draw $\triangle ABC$ on a sheet of paper. In $\triangle ABC$, let $AB = 3$ inches, $m\angle A = 40°$, and $m\angle B = 60°$.

2. By construction, locate the midpoints of $\overline{AB}, \overline{AC}$, and $\overline{BC}$. Label the midpoints $D, E,$ and F, respectively.

3. Connect points A and F, B and E, and C and D. What name is given to $\overline{AF}, \overline{BE},$ and $\overline{CD}$?

4. How many medians does a triangle have?

5. Write a statement that appears to be true about the medians that you constructed.

You may have found that the medians all intersect at the same point. Lines that do this are called concurrent lines.

DEFINITION 6-4-2 **CONCURRENT lines are lines that intersect at the same point.**

The point at which the lines intersect is the point of concurrency. The point of concurrency for the medians of a triangle is the centroid.

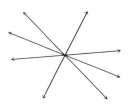

DEFINITION 6-4-3 **The point where the medians of a triangle intersect is the CENTROID, or the center of gravity, of the triangle.**

The definition mentions an important fact about the centroid—it is the center of gravity of a triangle. This idea is explored in the following project.

PROJECT 6-4-1 Cut a large triangle from a piece of cardboard. Draw the medians, and label the point where the medians intersect M. Using the point of a nail or a pencil, try to support the triangle by placing the point on M. Does the triangle balance?

Move the nail or point away from M and try to balance the triangle. Does the triangle balance?

Try this experiment with triangles of different shapes, and write a conclusion. Present your experiment and findings to the class.

You may have found that you could balance the triangle at its centroid but not anywhere else. The triangle balances at its center of gravity.

PERPENDICULAR BISECTORS OF THE SIDES OF A TRIANGLE

The three perpendicular bisectors of the sides of a triangle also have a special property, as you will discover.

DISCOVERY ACTIVITY

1. Use a ruler and a protractor to draw $\triangle ABC$ on a sheet of paper. Let $AB = 10$ cm, $m\angle A = 40°$, and $AC = 12$ cm.

2. Construct the perpendicular bisectors of each side of the triangle. Label the midpoints of the sides D, E, and F. Extend the perpendicular bisectors. What do you observe?

You may have discovered that the perpendicular bisectors of the sides of a triangle are concurrent.

ANGLE BISECTORS OF A TRIANGLE

Next, you will investigate the angle bisectors of a triangle.

DISCOVERY ACTIVITY

1. On a sheet of paper, draw a large acute triangle.
2. Construct the angle bisector for each angle.
3. Extend each angle bisector until it intersects the opposite side.
4. Write a conclusion that appears to be true about the bisectors.

You may have observed that the three angle bisectors are concurrent.

DEFINITION 6-4-4 **An ANGLE BISECTOR of a triangle is a line segment from a vertex of a triangle to the point where the angle bisector of that angle intersects the opposite side.**

ALTITUDES OF A TRIANGLE

The altitude of a triangle also involves a perpendicular segment.

DISCOVERY ACTIVITY

1. On a sheet of paper, draw a large acute triangle.

2. From each vertex, construct a line segment that is perpendicular to the opposite side.

3. Write a conclusion that appears to be true about the altitudes.

You may have discovered that the three perpendiculars from the vertices to the opposite sides are concurrent. These line segments are called altitudes.

DEFINITION 6-4-5 An **ALTITUDE** of a triangle is a perpendicular line segment from a vertex of the triangle to the opposite side or to the line determined by the opposite side.

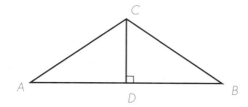

$\overline{CD}$ is the altitude from C.

You used an acute triangle in the Discovery Activity. Observe what happens when the triangle is obtuse.

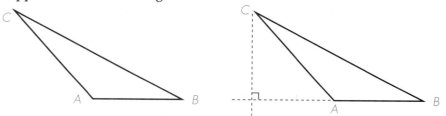

In order to draw the altitude from C, you must extend $\overline{AB}$.

You may have noticed that in each construction, the line segments intersect at a common point. Only one, the centroid, or center of gravity, was discussed. The usefulness of the other points of concurrency will be discussed in Section 6.6.

CLASS ACTIVITY

Copy each triangle four times. For each triangle, construct the medians, the perpendicular bisectors of the sides, the angle bisectors, and the altitudes. Label each example.

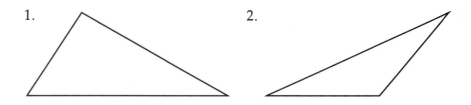

1.

2.

3. Draw $\triangle ABC$ with $AB = 4$ inches, $BC = 3.5$ inches, and $m\angle B = 50°$. Construct the medians and label the centroid M.

4. Construct $\triangle DEF$ with $DE = 3$ inches, $DF = 2.5$ inches, and $EF = 4.5$ inches. Construct the perpendicular bisectors of the sides. Label the point of concurrency P. Classify the triangle according

to its sides and its angles.

5. Construct △ABC, given that AB = 12.5 cm, BC = 10 cm, and AC = 15 cm. Construct the three angle bisectors and label the point of concurrency F. Classify the triangle according to its sides and its angles.

6. Construct △ABC with AB = 4 inches, BC = 4 inches, and CB = 4 inches. Construct the three altitudes. Label the point of concurrency P. Classify the triangle according to its sides and its angles.

Use a computer and a geometric drawing tool to complete Exercises 7 and 8.

7. Draw any triangle and the three medians of the triangle. Repeat this activity several times. Is the centroid ever outside the triangle? Explain your answer.

8. Draw any triangle and the three altitudes of the triangle. Repeat this activity several times. Do the altitudes ever intersect outside the triangle? Explain your answer.

HOME ACTIVITY

Copy each triangle four times. For each triangle, construct the medians, the perpendicular bisectors of the sides, the angle bisectors, and the altitudes. Label each example.

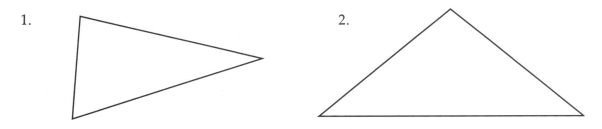

1.

2.

3. Construct △PQR with PQ = 12 cm, QR = 9 cm, and PR = 15 cm. Construct the medians and label the centroid C.

4. Construct △ABC with AB = 3 inches, BC = 5 inches, and AC = 4 inches. Construct the perpendicular bisectors of the sides. Label the point of concurrency O. Classify △ABC according to its sides and angles.

5. Draw △PQR with PQ = 14 cm, m∠P = 80°, and m∠Q = 40°. Construct the perpendicular bisectors of the sides. Label the point of concurrency B.

6. Construct an equilateral triangle with sides measuring 2 inches. Construct the three medians, then construct the three perpendicular bisectors of the sides. What do you observe?

7. Construct △ABC, with AB = 1 inch, and AC = BC = 2 inches. Construct the median from C, and construct the perpendicular bisector of $\overline{AB}$. What do you observe?

8. Draw △DHK with DK = 8 cm, m∠D = 100°, and m∠K = 50°. Construct the angle bisectors. Label the point of concurrency P.

9. Draw △THM with TM = 10 cm, m∠T = 110°, and m∠M = 35°. Construct the three altitudes. Label the point of concurrency K.

10. Construct △ABC, with AB = 3 inches, CA = 4 inches, and BC = 5 inches. Construct the three altitudes, and label the point of concurrency P. What do you observe about the point P?

11. Construct an isosceles right triangle. Construct the three perpendicular bisectors of the sides. Where is the point of concurrency?

12. Draw a large equilateral triangle. Construct each of the following: the three medians, the three angle bisectors, the three altitudes, and the three perpendicular bisectors of the sides. What do you observe?

CRITICAL THINKING

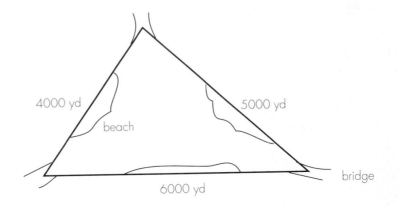

You live on a triangular island that has a beach on each side and a bridge at each vertex. The measurements of the island are 4000 yards by 6000 yards by 5000 yards.

13. You plan to build a cabin equidistant from the three beaches. Would you use the medians, the altitudes, the angle bisectors, or the perpendicular bisectors of the sides to help you decide where to build?

14. If you were to build your cabin equidistant from each bridge, would you use the medians, the altitudes, the angle bisectors, or the perpendicular bisectors of the sides to help you find the spot to build?

SECTION 6.5

Dividing a Line Segment into *n* Equal Parts

After completing this section, you should understand
▲ how to divide a line segment into n equal parts.

Section 6.1 not only introduced you to the tools and basic constructions of geometry, but also suggested a challenging problem. In that section, you learned to bisect a line segment.

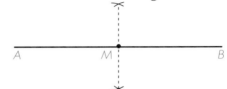

Applying the method of bisecting a line segment to $\overline{AM}$ and $\overline{MB}$ would result in four equal line segments. The process could continue, resulting in a sequence of equal line segments.

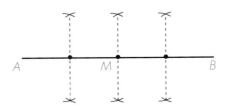

This method will work for dividing a line segment into 2^x congruent parts. However, a new method of dividing the segment is needed for a number of equal parts that is not a power of two.

DIVIDING A LINE SEGMENT INTO THREE EQUAL PARTS

The object is to find two points on a line segment that will divide it into three equal parts.

$$AC = CD = DB$$

Divide $\overline{AB}$, 10 centimeters long, into three equal parts.

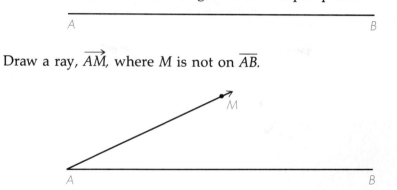

Draw a ray, $\overrightarrow{AM}$, where M is not on $\overline{AB}$.

Open a compass setting to about 1.5 centimeters. With the point of the compass on A, mark an arc that intersects $\overrightarrow{AM}$. Label the intersection of the arc and $\overrightarrow{AM}$ point R. Now place the point of the compass on R, and make an arc on $\overrightarrow{AM}$, labeling the intersection S. Repeat the process, with the third point labeled T. Since the setting on the compass was not changed, $AR = RS = ST$.

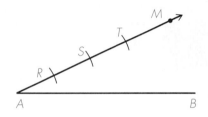

Connect points T and B with a line segment.

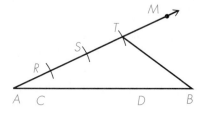

Construct a line through point S parallel to $\overline{TB}$. (See Section 6.3 if necessary.) Label the intersection with $\overline{AB}$ as D.

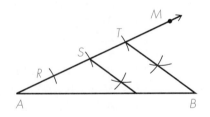

Construct a line through point R parallel to $\overline{TB}$. Label the intersection with $\overline{AB}$ as C.

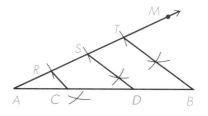

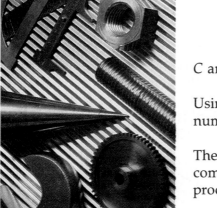

C and D divide $\overline{AB}$ into three equal parts, or line segments.

Using the same method, you can divide a line segment into any number of equal parts, or n equal parts.

The construction method used should be justified. Refer to the completed construction, and supply the reasons for each step in the proof.

	Steps	Reasons
1.	$\angle A \cong \angle A$	1.
2.	$\angle ARC \cong \angle ASD \cong \angle ATB$	2.
3.	$\triangle ARC \sim \triangle ASD \sim \triangle ATB$	3.
4.	$AR/AS = AC/AD$, and $AR/AT = AC/AB$	4.
5.	$AR = RS = ST$	5.
6.	$AS = 2(AR)$, and $AT = 3(AR)$	6.
7.	$AR/AS = AC/AD = \frac{1}{2}$, and $AR/AT = AC/AB = \frac{1}{3}$	7.
8.	$AD = 2(AC)$, and $AB = 3(AC)$	8.
9.	$AD = AC + CD$	9.
10.	$2(AC) = AC + CD$	10.
11.	$AC = CD$	11.
12.	$AB = AD + DB$	12.
13.	$3(AC) = 2(AC) + DB$	13.
14.	$AC = DB$	14.
15.	$AC = CD = DB$	15.

PHYSICAL MODEL OF DIVIDING A SEGMENT INTO EQUAL PARTS

Construction workers or carpenters may need to divide a piece of wood into congruent segments. These people probably wouldn't carry a compass, but they can use a method that is similar to the one in this section for dividing the piece of wood.

DISCOVERY ACTIVITY

1. On a sheet of paper, draw a line segment, $\overline{AB}$, 4 inches long. This will be the carpenter's piece of wood, a two by four, 8 feet long.

2. Place your ruler so that it is on A. Select a point M that is not on $\overline{AB}$, and draw $\overrightarrow{AM}$.

3. Use the width of your ruler, rather than your compass, to mark points R, S, and T on $\overrightarrow{AM}$ so that $AR = RS = ST$.

4. Draw $\overline{TB}$.

5. Now, use your ruler to draw parallels to $\overline{TB}$ through points S and R. Remember, the carpenter is using another board, rather than a ruler. The carpenter would mark on the second board two points, X and Y, when $\overline{TB}$ was drawn. Then the carpenter would slide the board to S so that X and Y that lie on $\overrightarrow{AM}$ are still on $\overrightarrow{AM}$, and draw the next line through S. Do the same with your ruler. Label the points of intersection with $\overline{AB}$ C and D.

This method is probably faster than using a compass to construct the parallel line segments but may not be quite as accurate.

CLASS ACTIVITY

1. Use a computer and a geometric drawing tool to draw any line segment. Use the principles from the construction for dividing a line segment into n equal parts to divide the line segment into three equal parts.
2. Draw a line segment 3 inches long. Bisect the segment using the method of this section.
3. Draw a line segment 10 centimeters long. Divide the line segment into three equal parts.
4. Draw a line segment 8 centimeters long. Divide the line segment into five equal parts.

HOME ACTIVITY

1. Draw a line segment 7 centimeters long. Bisect the segment using the method of this section.

2. Draw a line segment 5 inches long. Bisect the segment using the method of this section.

3. Draw a line segment 8 centimeters long. Divide the segment into three equal parts.

4. Draw a line segment $3\frac{1}{2}$ inches long. Divide the segment into three equal parts.

5. Draw a line segment 9 centimeters long. Divide the segment into four equal parts using the method of this section.

6. Draw a line segment $4\frac{3}{4}$ inches long. Divide the segment into four equal parts using the method of this section.

7. Draw a line segment 12 centimeters long. Divide the segment into five equal parts.

8. Draw a line segment $4\frac{1}{2}$ inches long. Divide the segment into five equal parts.

9. Draw a line segment 15 centimeters long. Divide the segment into six equal parts.

10. Draw a line segment 7 inches long. Divide the segment into six equal parts.

11. Draw a line segment 4 inches long. Bisect the segment using the method in the Discovery Activity.

12. Draw a line segment 8 centimeters long. Divide the segment into three equal parts using the method of this section.

13. Draw a line segment 5 inches long. Divide the segment into four equal parts using the method in the Discovery Activity.

14. Draw a line segment 12 centimeters long. Divide the segment into five equal parts using the method in the Discovery Activity.

CRITICAL THINKING

Copy the angle to the right. Construct a line segment connecting the rays of the angle at points equidistant from R. Trisect (divide into three equal parts) the line segment you constructed. Draw line segments to R from the points that trisect the line.

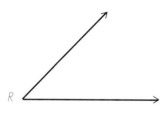

15. Do these connecting segments trisect the angle?

16. Would your answer to Exercise 15 be the same for any angle?

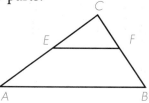

SECTION 6.6

Theorems from Constructions

After completing this section, you should understand
▲ how constructions lead to conclusions.

In Section 6.4, you learned that the medians of a triangle are concurrent at the centroid, the center of gravity of the triangle. The other constructions that you did lead to some important theorems about triangles.

THE TRIANGLE MIDPOINT THEOREM

For △ABC, construct the bisectors of $\overline{AC}$ and $\overline{BC}$. Label the midpoints E and F. Draw $\overline{EF}$.

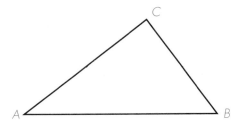

$\overline{EF}$ appears to be parallel to $\overline{AB}$. Measuring their lengths, EF appears to be half of AB.

THEOREM 6-6-1 **If the midpoints of two sides of a triangle are joined, then the line segment determined or formed is parallel to the third side and is equal to $\frac{1}{2}$ its length.**

The justification of this theorem is in two parts.

Given: △ABC with E and F midpoints by construction
Justify: a. $EF = \frac{1}{2}(AB)$
b. $\overline{EF} \parallel \overline{AB}$

235

Steps	Reasons
Part a	
1. $EC = \frac{1}{2}(AC)$	1. Midpoint divides a segment into two equal parts.
2. $FC = \frac{1}{2}(BC)$	2. Same as reason 1
3. $\angle C \cong \angle C$	3. Identity
4. $\triangle ABC \sim \triangle EFC$	4. SAS
5. $EF = \frac{1}{2}(AB)$	5. Sides of similar triangles are in equal ratio.
Part b	
6. $\angle A \cong \angle CEF$	6. Corresponding angles of similar triangles are equal.
7. $\overline{EF} \parallel \overline{AB}$	7. If two lines are intersected by a transversal so that the corresponding angles are equal, then the lines are parallel.

CLASS ACTIVITY

1. Use a computer and a geometric drawing tool to draw any $\triangle ABC$. Find the midpoint of $\overline{AC}$ and label it E. Find the midpoint of $\overline{BC}$ and label it F. Draw line segment $\overline{EF}$. Find the measures of $\overline{AB}$, $\overline{EF}$, $\angle CEF$, and $\angle CAB$. Repeat this activity several times. How does the measure of $\overline{EF}$ relate to the measure of $\overline{AB}$? What do the measures of $\angle CEF$ and $\angle CAB$ tell you about $\overline{EF}$ and $\overline{AB}$? Do your findings support Theorem 6-6-1?

Find the length of the line segment.

2. $\overline{PQ}$ 3. $\overline{XY}$

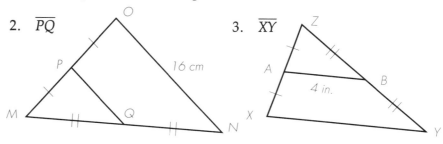

PERPENDICULAR BISECTORS OF TRIANGLE SIDES

Recall that the perpendicular bisectors of the sides of a triangle are concurrent.

DISCOVERY ACTIVITY

1. On a sheet of paper, draw a large acute triangle, *ABC*.

2. Construct the perpendicular bisectors of all three sides, and label the midpoints *D*, *E*, and *F*. Label the point of concurrency *O*.

3. Place the point of your compass on *O*. Open the setting on your compass until the point of the pencil is on *A*. Rotate the compass until the pencil point is back on *A*. What do you observe?

You may have found that the arc passes through all three vertices.

DEFINITION 6-6-1   **A CIRCLE is the set of all points on a plane equidistant from a point called the center.**

In the Discovery Activity, *O* was the center of the circle. This is justified below.

Given: $\overline{MN}$, with constructed perpendicular bisector, *K* the midpoint of $\overline{MN}$, and *P* any point besides *K* on the perpendicular bisector.

Justify: $PM = PN$

Steps	Reasons
1. Draw $\overline{PM}$, $\overline{PN}$.	1. Two points determine a line.
2. $MK = NK$	2. Midpoint divides a segment into two equal parts.
3. $PK = PK$	3. Identity
4. $m\angle MKP = m\angle NKP$	4. By construction (perpendicular bisector)
5. $\triangle MKP \sim \triangle NKP$	5. SAS
6. $MK/NK = 1$	6. Algebra
7. $\triangle MKP \cong \triangle NKP$	7. Similar triangles and ratio of sides = 1
8. $MP = NP$	8. CPCF

THEOREM 6-6-2   **If a point is on the perpendicular bisector, then it is equidistant from the endpoints of the segment.**

Thus, in the Discovery Activity, $OA = OB = OC$, and O is the center of the circle. The circle goes around the outside of the triangle, and is called the circumcircle. Point O is the circumcenter of the triangle.

CLASS ACTIVITY

1. Copy $\triangle DEF$. Construct the circumcenter.

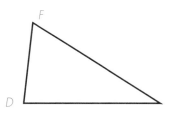

THE ANGLE BISECTORS OF A TRIANGLE

Recall that the angle bisectors of a triangle are also concurrent.

DISCOVERY ACTIVITY

1. On a sheet of paper, draw a large triangle. Bisect the angles. Label the point of concurrency of the angle bisectors T.

2. From T, construct the perpendiculars to the sides of the triangle. Label the points of intersection of the perpendiculars and the sides X, Y, and Z. Measure TX, TY, and TZ. What do you observe?

3. Place the point of your compass on T and the pencil point on X. Construct a circle. What do you observe?

You may have found that the circle you constructed passes through X, Y, and Z. The circle is inside the triangle and so is called the in-circle.

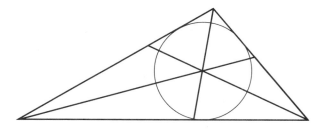

NOTEBOOK

THEOREM 6-6-3　**If a point is on the angle bisector, then the point is equidistant from the sides of the angle.**

Given: $\angle ABC$ with $\overrightarrow{BT}$ the angle bisector, $\overline{TD} \perp \overrightarrow{BC}$, and $\overline{TE}$ $\perp \overrightarrow{BA}$

Justify: $TD = TE$

Supply the missing reasons.

Steps	Reasons
1. $\angle TBD \cong \angle TBE$	1.
2. $BT = BT$	2.
3. $\overline{TD} \perp \overline{BC}, \overline{TE} \perp \overline{BA}$	3.
4. $\angle TEB$ and $\angle TDB$ are right angles.	4.
5. $\angle TEB \cong \angle TDB$	5.
6. $\triangle BTD \sim \triangle BTE$	6.
7. $BT/BT = 1$	7.
8. $\triangle BTD \cong \triangle BTE$	8.
9. $TD = TE$	9.

CLASS ACTIVITY

1. Construct the in-circle of a triangle whose sides measure 3 centimeters, 5 centimeters, and 6 centimeters.

2. Construct the in-circle of a right triangle whose sides measure 3 inches, 4 inches, and 5 inches.

HOME ACTIVITY

1. Draw an obtuse triangle with two sides measuring 8 cm and 6 cm, and the included angle equal to 110°. Construct the perpendicular bisectors of the sides. Label the circumcenter O. Construct the circumcircle.

2. Draw an acute triangle ABC, with m$\angle A = 60°$, $AB = 3$ inches, and m$\angle B = 40°$. Construct the angle bisectors. Label the incenter V. Construct the in-circle.

3. D and E are midpoints of two sides of $\triangle MNO$. If $MN = 32$ mm, what is DE?

In △*ABC*, *D*, *E*, and *F* are midpoints.

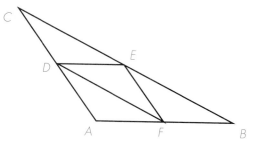

4. If *AC* = 8, *AB* = 10, and *BC* = 15, what does *DF* + *DE* + *EF* equal?

5. If *DE* = 6, *DF* = 8, and *EF* = 4, what does *AB* + *AC* + *BC* equal?

6. Construct a triangle with sides of length 3 inches, 3 inches, and 2 inches. Construct the circumcenter of the triangle.

7. Draw a triangle with sides of length 3 inches and 4 inches, and the included angle equal to 115°. Construct the circumcircle of the triangle.

8. Construct a triangle with sides of length 8 cm, 10 cm, and 12 cm. Construct the in-circle of the triangle.

9. Draw a triangle with sides of length 4 inches and $5\frac{1}{2}$ inches, and included angle equal to 75°. Construct the incenter of the triangle.

10. Construct an equilateral triangle with sides of length 4 inches. Construct the circumcenter and the incenter. What do you observe?

11. Draw an isosceles triangle with two of the angles measuring 70° and the base equal to 6 centimeters. Construct the three medians, labeling the centroid *M*. Construct the three altitudes, labeling the point of concurrency *A*. Construct the three angle bisectors, labeling the incenter *B*. What do you observe about *M*, *A*, and *B*?

12. Construct an isosceles right triangle, *ABC*, with ∠*C* the right angle. Construct the median from ∠*C* to $\overline{AB}$. Label the point *O* where the median intersects $\overline{AB}$. Measure *OA*, *OB*, and *OC*. What do you observe?

CRITICAL THINKING

13. A surveyor was asked to determine the distance from *A* to *B*. The direct line is blocked by the building, as indicated in the figure. Using a theorem from this section, devise a method to help the surveyor.

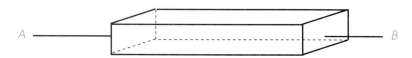

6.1 Bisect the line segment.   Bisect the angle.

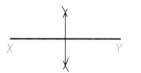

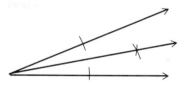

6.2 Construct a perpendicular to $\overline{AB}$ from C.

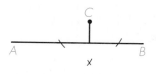

6.3 Construct a line parallel to $\overline{EF}$ through D.

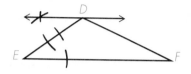

6.4 In $\triangle GHI$, find the incenter R and the circumcenter S of GHI.

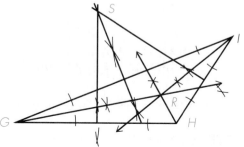

6.5 Divide $\overline{JK}$ into three equal parts.

6.6 $\overrightarrow{NP}$ is the bisector of $\angle MNO$. Find x. Because $\overrightarrow{NP}$ is an angle bisector, $MP = OP$. So $x = 20$ cm.

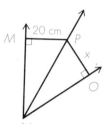

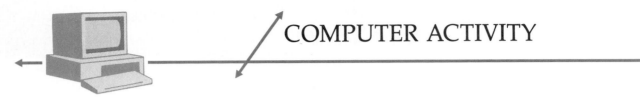
The three vertices of a right triangle have been graphed on the *x*–*y* plane at the right. In order to draw the medians of this triangle, you must find the coordinates of the midpoint of each side.

Consider the side whose endpoints have coordinates of (1, 2) and (5, 2). We can count 4 units between these points, so it follows that the midpoint would divide the segment into two segments that are each 2 units long. So the midpoint must have coordinates (3, 2). Notice that these coordinates can be computed by adding the *x*-coordinates (1 + 5 = 6) and then dividing the sum by 2 (6 ÷ 2 = 3), and by adding the *y*-coordinates (2 + 2 = 4) and then dividing the sum by 2 (4 ÷ 2 = 2). The general rule for finding the coordinates of the midpoint of a line segment is given below.

The coordinates of the midpoint of the line segment whose endpoints have coordinates (x_1, y_1) and (x_2, y_2) are

$$\left(\frac{x_1 + x_2}{2}, \frac{y_1 + y_2}{2}\right).$$

You can use LOGO and this rule to tell the turtle to draw triangles and their medians on the *x*–*y* plane.

Example: The coordinates of the three vertices of a triangle are (10, 5), (10, 15), and (30, 5). Write a procedure named MEDIANS that tells the turtle to draw the triangle and its three medians.

First, compute the coordinates of the midpoint of each side.

The side whose endpoints have coordinates of (10, 5) and (10, 15) has a midpoint whose coordinates are $\left(\frac{10 + 10}{2}, \frac{5 + 15}{2}\right)$, or (10, 10). Thus, the first median will start at (10, 10) and go to (30, 5). The side whose endpoints have coordinates of (10, 15) and (30, 5) has a midpoint whose coordinates are $\left(\frac{10 + 30}{2}, \frac{15 + 5}{2}\right)$, or (20, 10). Thus, the second median will start at (20, 10) and go to (10, 5). The side whose endpoints have coordinates (10, 5) and (30, 5) has a midpoint with coordinates $\left(\frac{10 + 30}{2}, \frac{5 + 5}{2}\right)$ or (20, 5). Thus, the third median will start at (20, 5) and go to (10, 15).

Now, write the procedure.

```
TO MEDIANS
PU SETXY 10 5 PD
SETXY 10 15                    First, draw the triangle.
SETXY 30 5
SETXY 10 5
PU SETXY 10 10 PD              Draw the first median.
SETXY 30 5
PU SETXY 20 10 PD             Draw the second median.
SETXY 10 5
PU SETXY 20 5 PD              Draw the third median.
SETXY 10 15
END
```

The output is shown below.

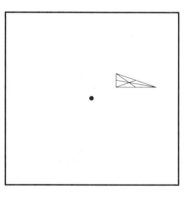

The coordinates of the three vertices of a triangle are given below. Modify the MEDIANS procedure to tell the turtle to draw each triangle and its three medians.

1. (0, 0), (10, 0), (0, 10)

2. (5, 5), (15, 5), (5, 20)

3. (0, −6), (8, −6), (0, 2)

4. (−6, −4), (−6, 6), (0, −4)

5. (−25, −25), (−25, −10), (−5, −25)

6. (−19, 18), (−19, 12), (−13, 18)

7. Modify the MEDIANS procedure to tell the turtle to draw a triangle of your choice and its three medians.

CHAPTER REVIEW

1. Copy $\overline{AB}$. Bisect $\overline{AB}$, and label the midpoint M.

        A   B

2. Draw an acute angle, $\angle ABC$, and copy the angle onto $\overline{EF}$ with the vertex at E.
3. Copy the angle and bisect it.

4. Copy the figure, and construct a line perpendicular to $\overleftrightarrow{AB}$ through C.

5. Copy the figure from Exercise 4. Construct a line parallel to $\overleftrightarrow{AB}$ through C.
6. Draw a line segment 4 inches long. By construction, divide the segment into three equal parts.

7. Draw a large obtuse triangle ABC, with $\angle C$ the obtuse angle. Construct the angle bisector of $\angle C$, the median from B, and the altitude from A.
8. What is the measure of $\overline{KL}$ in $\triangle ABC$? K and L are midpoints.

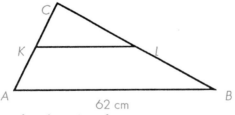

62 cm

9. Construct the circumcenter for the triangle.

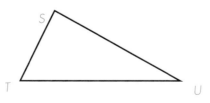

10. Construct the incenter for the triangle.

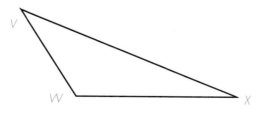

244

1. Draw a line segment $2\frac{3}{4}$ inches long. Bisect the segment.
2. Copy the figure. Construct the line perpendicular to $\overline{MN}$ through W.
3. Copy the figure from Exercise 2. Construct the line parallel to $\overline{MN}$ through W.
4. Draw a line segment 9 centimeters long. By construction, divide the segment into three equal parts.
5. Copy and then bisect the angle.

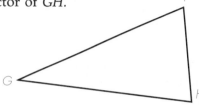

6. Draw an obtuse angle measuring 105°. Copy the angle by construction onto $\overline{VW}$ with the vertex at V.
7. Copy $\triangle HGI$. Construct the angle bisector for $\angle H$. From G, construct the median. From I, construct the altitude. Construct the perpendicular bisector of $\overline{GH}$.

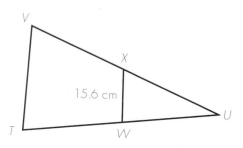

8. X and W are midpoints. Find the length of $\overline{VT}$.

15.6 cm

9. Construct the circumcenter for the triangle.

10. Construct the incenter for the triangle.

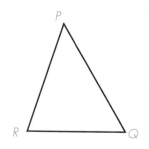

ALGEBRA SKILLS

The square roots of 9 are 3 and -3. Because 9 is a perfect square, $\sqrt{9}$ is a rational number. The square roots of numbers that are not perfect squares are irrational numbers: the roots have a decimal that is nonterminating and nonrepeating. Irrational roots may be expressed as decimal approximations or in square-root form.

Example: $\sqrt{5} = ?$ Using a calculator, enter 5, and press the square-root key $\boxed{\sqrt{}}$.

$5\boxed{\sqrt{}} \approx 2.236068$

Example: $\sqrt{500} = ?$ Factor the number, looking for perfect squares.
$\sqrt{500} = \sqrt{5 \cdot 100} = 10\sqrt{5}$
You can leave the answer in square-root form, or use the table of square roots on page 562. Locate 5 in the leftmost column, headed n. The column headed $\sqrt{n}$ gives 2.236 for the approximate value of $\sqrt{5}$.
$10\sqrt{5} \approx 10(2.236) = 22.36$

Example: $x^2 = 16$
$\pm\sqrt{x^2} = \pm\sqrt{16}$ Take the square root of both sides of the equation.
$x = 4 \text{ or } x = -4$

Example: $x^2 + 8x + 16 = 169$ Reformulate the left side as a
$(x + 4)^2 = 169$ square. Take the square root
$x + 4 = \pm13$ of both sides.
$x = 9 \text{ or } x = -17$

Find the square roots to two decimal places, using the table on page 562.

1. $\sqrt{7}$
2. $\pm\sqrt{36}$
3. $\pm\sqrt{90}$
4. $\pm\sqrt{300}$

5. $\sqrt{30}$
6. $\sqrt{(9)^2}$
7. $\sqrt{(-5)^2}$
8. $\sqrt{-36}$

Solve for x. If x is not a perfect square, leave the answer in square-root form.

9. $x^2 = 64$
10. $2x^2 = 8$
11. $(x - 4)^2 = 125$

12. $x^2 = (3)^2 + (4)^2$
13. $(x + 3)^2 = 56$

14. $x^2 - 4x + 4 = 100$
15. $3x^2 + 6x + 3 = 48$

Solve for x, to the nearest two decimal places. Use the table of square roots or your calculator.

16. $x^2 = 3$
17. $5x^2 = 180$
18. $(x + 2)^2 = 33$

19. $(2x)^2 = (20)^2 + (15)^2$
20. $x^2 = 2(17)^2$
21. $x^2 - 6x + 9 = 32$

7

FUNDAMENTAL THEOREM OF PYTHAGORAS

Pythagoras then imported proof into mathematics. This is his greatest achievement . . . [his] second outstanding mathematical contribution . . . [was irrational numbers].

E.T. Bell

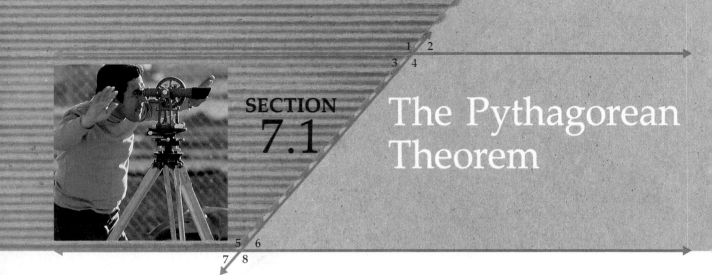

The Pythagorean Theorem

After completing this section, you should understand

▲ the Pythagorean Theorem.

▲ squares and square roots of numbers.

You are familiar with the three types of triangles shown below.

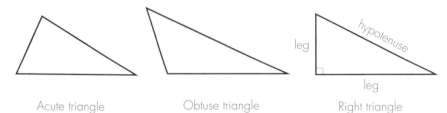

Acute triangle   Obtuse triangle   Right triangle

RIGHT TRIANGLES

Notice that there are special names for the three sides of the right triangle. The side opposite the right angle is the *hypotenuse*. The other two sides are the *legs* of the right triangle.

DEFINITION 7-1-1: **The HYPOTENUSE of a right triangle is the side opposite the right angle.**

DISCOVERY ACTIVITY

1. Find the length of the hypotenuse of each right triangle.

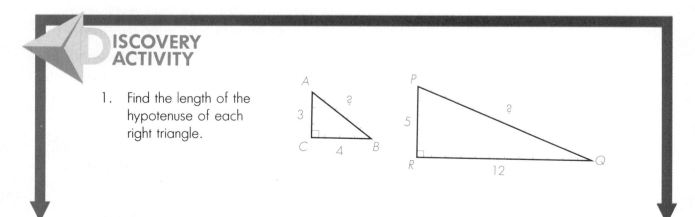

2. Measure the sides of each right triangle to the nearest tenth of a centimeter.

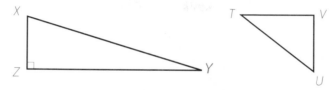

3. Copy and complete the following table for the triangles in Exercises 1 and 2. The entries for △ABC have been done for you.

	Length of first leg	Length of second leg	Length of hypotenuse	Square of first leg	Square of second leg	Square of hypotenuse
△ABC	3	4	5	9	16	25
△PQR						
△XYZ						
△TUV						

4. Study the table in Exercise 3. What do you notice about how the squares of the three sides of each right triangle are related? Write a generalization that seems to be true.

5. Construct a triangle with sides of lengths 100 mm, 96 mm, and 28 mm. Use the methods of Section 6.4.

6. Square the length of each side of the triangle you constructed. How are the squares related?

7. Measure the largest angle of the triangle you constructed. What kind of angle is it?

8. Write a generalization based on your observations concerning the triangle you constructed.

Based upon your observations in Exercises 1–4 of the Discovery Activity, you might conclude that the following statement seems to be true: If a triangle is a right triangle, then the sum of the squares of its two legs is equal to the square of the hypotenuse. Based upon your observations in Exercises 5–8 of the Discovery Activity, you might conclude that the *converse* of the above statement also seems to be true: If the sum of the squares of the lengths of two sides of a triangle is equal to the square of the length of the third side, then the triangle is a right triangle.

CLASS ACTIVITY

Each of the following is the set of lengths of three sides of a triangle. Is the triangle a right triangle?

1. 6, 8, 10   2. 7, 10, 12   3. 26, 10, 24

Each of the following triangles is a right triangle. Find the missing length. Choose from the following: 12, 12.5, 14, 15, 15.5, 16.

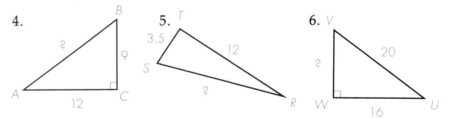

4.

5.

6.

THE PYTHAGOREAN THEOREM

In the figure at the right, a and b are the legs of the right triangle ABC. The hypotenuse is c. The right angle is $\angle C$. Earlier in this section you learned how the three sides of the triangle are related. You saw that the following statement was true: $a^2 + b^2 = c^2$. You can prove that the statement is true.

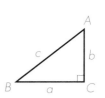

Given: Right triangle ABC with
 $\qquad$ m$\angle C = 90°$ and $\overline{CD} \perp \overline{AB}$
Justify: $a^2 + b^2 = c^2$
First, show that $\triangle ABC \sim \triangle ACD$
and $\triangle ABC \sim \triangle CBD$.

Step		Reason	
1.	In triangles ABC and ACD, m$\angle A$ = m$\angle A$.	1.	Identical angles
2.	In triangles ABC and ACD, m$\angle C$ = m$\angle ADC$.	2.	All right angles are equal.
3.	$\triangle ABC \sim \triangle ACD$	3.	AA
4.	In triangles ABC and CBD, m$\angle B$ = m$\angle B$.	4.	Identical angles
5.	In triangles ABC and CBD, m$\angle C$ = m$\angle CDB$.	5.	
6.	$\triangle ABC \sim \triangle CBD$	6.	
7.	$\frac{b}{x} = \frac{c}{b}; \frac{a}{y} = \frac{c}{a}$	7.	If two figures are similar, then the corresponding sides are in equal ratio.
8.	$b^2 = cx; a^2 = cy$	8.	Property of algebra
9.	$b^2 + a^2 = cx + cy$	9.	Property of algebra
10.	$b^2 + a^2 = c(x + y)$	10.	
11.	$x + y = c$	11.	
12.	$b^2 + a^2 = c \cdot c, = c^2$	12.	Substitution; step 10

THEOREM 7-1-1 **Pythagorean Theorem: If a right triangle has sides of lengths a, b, and c, where c is the hypotenuse, then $a^2 + b^2 = c^2$.**

The converse of the Pythagorean Theorem is also true. Its proof is not given.

THEOREM 7-1-2 **Converse of the Pythagorean Theorem: If a triangle has three sides of lengths a, b, and c, such that $a^2 + b^2 = c^2$, then the triangle is a right triangle.**

When you know the lengths of two sides of a right triangle, you can use the Pythagorean Theorem to find the third side. When you need to find the square root of a number to find the missing length, use a calculator or the table of squares and square roots on page 562.

Example: $c^2 = a^2 + b^2$
$= 3^2 + 5^2$
$= 9 + 25$, or 34
$c = \sqrt{34}$, or 5.831

CLASS ACTIVITY

Find the missing length. Use a calculator or the table of squares and square roots on page 562. Round lengths to the nearest tenth.

1.

2.

3.

Copy and complete the following table. The lengths of the legs of right triangle ABC are a and b. The length of the hypotenuse is c. Use a calculator or the table of squares and square roots on page 562. Round lengths to the nearest tenth.

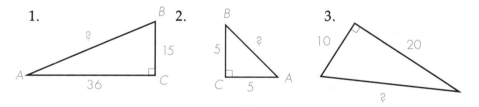

	a	b	c	$a^2 + b^2 = c^2$
4.	8	6	_____	$8^2 + 6^2 = c^2$
5.	3.6	1.5	_____	$3.6^2 + 1.5^2 = c^2$
6.	10	12	_____	_____ + _____ = c^2
7.	8	_____	12	$8^2 + b^2 = 144$

HOME ACTIVITY

Each of the following is the set of lengths of three sides of a triangle. Is the triangle a right triangle?
1. 20, 48, 52
2. 0.6, 0.8, 1
3. 4, 5, 6
4. 27, 364, 365

Find the missing length. Use a calculator for squares and square roots. Round answers to the nearest tenth.

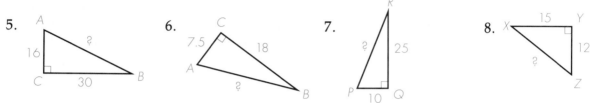

5.
6.
7.
8.

Copy and complete the following table. The lengths of the legs of right triangle *ABC* are *a* and *b*. The length of the hypotenuse is *c*. Use a calculator or the table of squares and square roots on page 562. Round lengths to the nearest tenth.

	a	b	c	$a^2 + b^2 = c^2$
9.	8	15	_____	$8^2 + 15^2 = c^2$
10.	24	7	_____	$24^2 + 7^2 = c^2$
11.	6	7	_____	_____ + _____ $= c^2$
12.	15	_____	20	$15^2 + b^2 = 20^2$

13. A classroom measures 25 ft by 30 ft. To the nearest tenth of a foot how far apart are diagonally opposite corners of the floor?

14. A regulation baseball diamond has the shape of a square. The distance from home plate to first base is 90 ft. Find the distance between first base and third base to the nearest tenth of a foot.

15. A brace is needed to reinforce the gate shown below. To the nearest hundredth of a foot, what length of lumber must be bought?

16. In the figure, find the length of $\overline{AB}$. (Hint: Find the length of another segment first.)
$AB =$

CRITICAL THINKING

The piece of luggage shown is to carry an umbrella. Will an umbrella 30 in. long fit inside?

SECTION
7.2

The Isosceles Right Triangle

After completing this section, you should understand
▲ the properties of isosceles right triangles.
▲ how to solve problems using isosceles right triangles.

The three triangles below are all isosceles. However, just one is both isosceles and right.

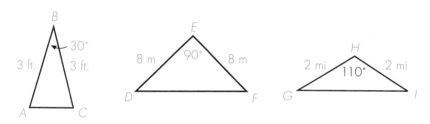

The triangle in the middle, △*DEF,* is an *isosceles right* triangle.

Examples: △*RST* is a *right* triangle but is not isosceles.

△*UVW* is an *isosceles* triangle but is not right.

△*XYZ* is a triangle that is both *right* and *isosceles.*

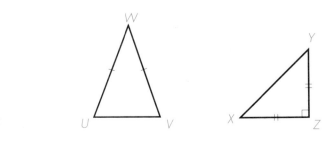

THE 45°-45° RIGHT TRIANGLE

You can prove that each of the acute angles of an isosceles right triangle has an angle measure of 45°. Provide a reason for each step in the proof.

Given: Triangle ABC
with $AC = BC$
Justify: $m\angle A = 45°$
and $m\angle B = 45°$

Step	Reason
1. $m\angle A + m\angle B + 90° = 180°$	1. If a figure is a triangle, then the sum of the angles is 180°.
2. $m\angle A = m\angle B$	2.
3. $m\angle A + m\angle A + 90° = 180°$	3. Substitution
4. $2m\angle A + 90° = 180°$	4. Property of algebra
5. $2m\angle A = 180° - 90°$, or 90°	5. Property of algebra
6. $m\angle A = 90° \div 2$, or 45°	6.
7. $m\angle B = 45°$	7.

THEOREM 7-2-1 **If a triangle is an isosceles right triangle, then the acute angles are each 45°.**

DISCOVERY ACTIVITY

1. Use a centimeter ruler to measure the hypotenuse of each isosceles right triangle below. Measure to the nearest tenth of a centimeter.

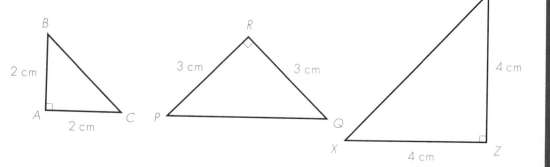

2. Copy and complete the following table. Use the results of Exercise 1.

	Length of leg	Length of hypotenuse	Hypotenuse ÷ leg
△ABC	2 cm	_____ cm	_____
△PQR	_____ cm	_____ cm	_____
△XYZ	4 cm	_____ cm	_____

3. Study the table and write a generalization that seems to be true.

From the table, you might conclude that the ratio $\frac{\text{hypotenuse}}{\text{leg}}$ has a value of approximately 1.4. If this is so, then the relationship can also be written as *hypotenuse* ≈ 1.4 × *leg*. (The symbol ≈ means *is approximately equal to*.) You can use algebra to show that this relationship *is* true for all isosceles right triangles.

Here is the proof. The three sides of right triangle *ABC* are s, s, and c, where c is the length of the hypotenuse. By the Pythagorean Theorem, $s^2 + s^2 = c^2$. Solve for c in terms of s.

Use a calculator or the table of squares and square roots on page 562 to find the value of $\sqrt{2}$: $\sqrt{2} \approx$ 1.414, or about 1.4. Thus, in right triangle *ABC*, $c \approx$ 1.414s.

THEOREM 7-2-2 **The length of the hypotenuse of an isosceles right triangle is $\sqrt{2}$ (about 1.414) times the length of either leg. In symbols this is written as $c = s\sqrt{2}$, or $c \approx 1.414s$.**

The formula of Theorem 7-2-2 is written as $c = s\sqrt{2}$ rather than as $c = \sqrt{2}s$ to stress that only the 2 (not 2s) is under the radical symbol.

Example: 1. To find the length of the hypotenuse of isosceles right triangle *DAN*, *multiply* the length of a leg by $\sqrt{2}$, or 1.414.

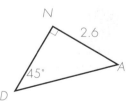

$n = 2.6\sqrt{2}$
$\approx 2.6(1.414)$
≈ 3.6764, or about 3.7

Example: 2. To find the length of one leg of isosceles right triangle *KIM, divide* the length of the hypotenuse by $\sqrt{2}$, or 1.414.

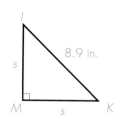

$$s = 8.9 \div \sqrt{2}$$
$$\approx 8.9 \div 1.414$$
$$\approx 6.29, \text{ or about } 6.3 \text{ in.}$$

CLASS ACTIVITY

In this activity, use a calculator or the table of squares and square roots on page 562. Answer to the nearest tenth.

Find the length of the hypotenuse of each isosceles right triangle.

1.

2.

Find the length of one leg of each isosceles right triangle.

3.

4.

You can use the Pythagorean Theorem to find the missing length of a side of an isosceles right triangle. Study the examples shown.

Example:
$$a^2 + b^2 = c$$
$$11^2 + 11^2 = c^2$$
$$121 + 121 = c^2$$
$$242 = c^2$$
$$\sqrt{242} = c$$
$$c = \sqrt{242}$$
$$c \approx 15.6 \text{ ft}$$

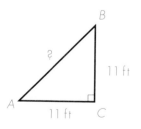

Example: $a^2 + b^2 = c^2$
$s^2 + s^2 = (5\sqrt{2})^2$
$2s^2 = 5^2 \cdot 2$
$2s^2 = 50$
$s^2 = 25$
$s = 5$

Use the Pythagorean Theorem to find the missing length of a side.

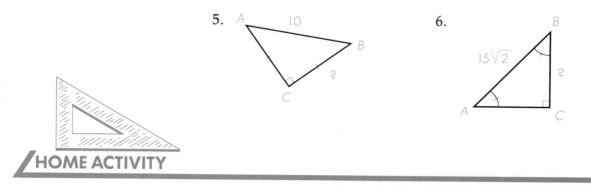

5.

6.

HOME ACTIVITY

In these exercises, use a calculator for squares and square roots or use the table on page 562.
Find the length of the hypotenuse of each isosceles right triangle.

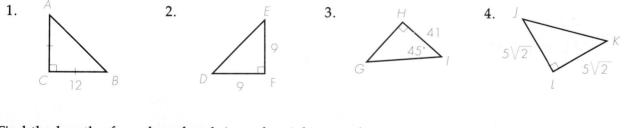

1.

2.

3.

4.

Find the length of one leg of each isosceles right triangle.

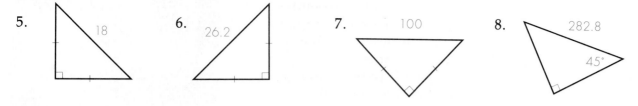

5.

6.

7.

8.

Use the Pythagorean Theorem to find the missing length of a side. Answer to the nearest tenth.

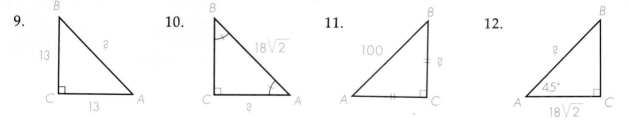

9.

10.

11.

12.

Copy and complete the table below. Answer to the nearest tenth.

	Length of one leg of an isosceles right triangle	Length of hypotenuse
13.	____	$1\sqrt{2}$, or 1.4
14.	2	$2\sqrt{2}$, or ____
15.	3	$3\sqrt{2}$, or ____
16.	____	$5\sqrt{2}$, or 7.071
17.	____	$10\sqrt{2}$, or ____

18. Write a Logo procedure named ISOSRTTRI to draw an isosceles right triangle. Use variables in the procedure so that the lengths of the sides can be different each time the procedure is called.

19. The roof of a shed forms a right angle at its peak. About how many feet long is the beam that runs from point A to point B?

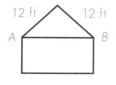

20. Alice jogged from A to C and then on to B. Returning she jogged from B to A. About how much shorter was the return trip?

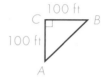

CRITICAL THINKING

In the basement of a building, two water pipes are parallel and 25 ft apart. If a furnace were not in the way, Ms. Karpov could run a connecting pipe from point A on water pipe 1 to the point on water pipe 2 that is closest to A. Instead, she must use a 45° elbow pipe at A to make the connection elsewhere on pipe 2. To the nearest tenth of a foot how long will the connection pipe be?

SECTION 7.3

The 30°-60°-90° Triangle

After completing this section, you should understand

▲ theorems about 30°-60° right triangles.

▲ how to solve problems using 30°-60° right triangles.

There are two special triangles that you will often encounter in geometry and later in trigonometry. You already know about the first such triangle, the isosceles right triangle. The second such special triangle is the *30°-60° right triangle.* A few 30°-60° right triangles are shown below.

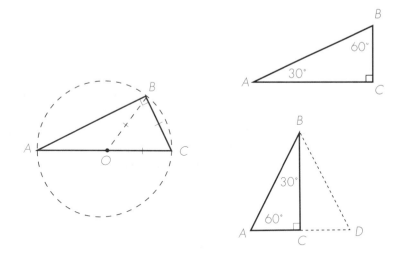

LEG OPPOSITE THE 30° ANGLE

For most right triangles, it is necessary to use trigonometry to find how the sides of a triangle are related to each other and to the angles. For 30°-60° right triangles, you do not need trigonometry to find these relationships.

259

DISCOVERY ACTIVITY

1. On a separate piece of paper, draw line segment $\overline{AC}$ 4 cm long.
2. Use a protractor and a ruler to draw $\angle ACX$ equal to 90°.
3. Use your protractor and a ruler to draw $\angle CAY$ equal to 30°. Use B to label the intersection of $\overrightarrow{CX}$ and $\overrightarrow{AY}$.
4. Measure $\overline{AB}$ to the nearest tenth of a centimeter.
5. Measure $\overline{CB}$ to the nearest tenth of centimeter.
6. Find the ratio $CB : AB$.
7. Repeat steps 1–6 using a 6-cm line segment instead of a 4-cm segment.
8. Based on your experiments, state a generalization that appears to be true.

It appears as though the shorter leg of a 30°-60° right triangle has half the length of the hypotenuse. This is a conclusion based upon *inductive reasoning,* that is, reasoning using the results of a number of experiments or observations.

It is possible to show that the conclusion is correct using *deductive reasoning,* that is, reasoning based upon undefined terms, defined terms, postulates, and previously justified theorems.

The following proof is based upon the fact that an equilateral triangle can be divided into two 30°-60° right triangles.

Equilateral triangle *ABC* has three equal sides, by definition of *equilateral.* It is also *equiangular,* so that the measure of each angle is 60°. (The proof that an equilateral triangle is also equiangular is asked for in the HOME ACTIVITY.)

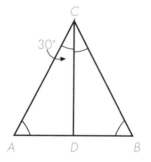

In the figure, $\overline{CD}$ bisects angle C. Therefore, m∠ACD = 30° (half of 60°). You can show that △ACD ≅ △BCD by SAS. Thus m∠ADC and m∠BDC are each 90° by CPCF. So, △ACD is a 30°-60° right triangle.

Also, AD = BD by CPCF. From this you can see that $AD = \frac{1}{2}AB$

Since AB = AC, you can use substitution to conclude that $AD = \frac{1}{2}AC$

THEOREM 7-3-1 **If a triangle is a 30°-60° right triangle, then the length of the leg opposite the 30° angle is half the length of the hypotenuse.**

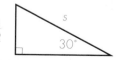

Examples: 1. In △XYZ, $\overline{YZ}$ is opposite the 30° angle. YZ = $x = \frac{1}{2}(20)$, or 10

2. In △NOP, $\overline{OP}$ is opposite the 30° angle. OP = $24 = \frac{1}{2}p$, so $p = 2 \cdot 24 = 48$

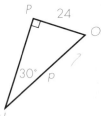

CLASS ACTIVITY

Find the missing lengths.

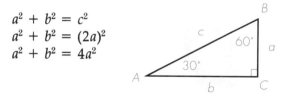

1. 62 in. 30° x

2. 120 cm 30° x

3. 30° x x 20.5 yds

4. 25 mm x 30°

LEG OPPOSITE THE 60° ANGLE

You know that the length of the leg opposite the 30° angle of a 30°-60° right triangle is half the length of the hypotenuse. You can now find out how the leg opposite the 60° angle is related to the hypotenuse. Use the Pythagorean Theorem and algebra. In right triangle ABC, m∠A = 30° and m∠B = 60°. So, $a = \frac{1}{2}c$, or c = 2a. Substitute 2a for c in the Pythagorean Theorem.

$a^2 + b^2 = c^2$
$a^2 + b^2 = (2a)^2$
$a^2 + b^2 = 4a^2$

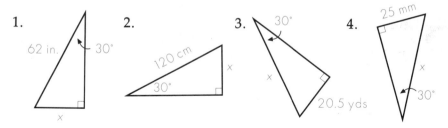

Next, solve for b.

$$a^2 + b^2 - a^2 = 4a^2 - a^2$$
$$b^2 = 3a^2$$

Take the square root of each side of this equation.

$$b = a\sqrt{3}$$

Now substitute $\frac{1}{2}c$ for a.

$$b = \left(\tfrac{1}{2}c\right)\sqrt{3}, \text{ or } c\,\frac{\sqrt{3}}{2}.$$

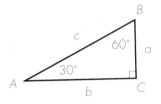

THEOREM 7-3-2 **If a triangle is a 30°-60° right triangle, then the length of the leg opposite the 60° angle is $\frac{\sqrt{3}}{2}$ times the length of the hypotenuse.**

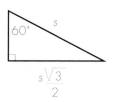

Examples: Find the missing length.

1. $b = c\,\frac{\sqrt{3}}{2}$

   $= (20)\frac{\sqrt{3}}{2}$

   $= 10\sqrt{3}$

   $\approx 10 \cdot 1.732$

   $= 17.32$ in.

2. $8 = z\,\frac{\sqrt{3}}{2}$

   $8 \cdot 2 = z\,\frac{\sqrt{3}}{2} \cdot 2$

   $16 = z\sqrt{3}$

   $\approx z \cdot 1.732$

   $16 \div 1.732 \approx z$

   $9.238 \approx z$, or

   $z \approx 9.238$

CLASS ACTIVITY

Find the value of x. Answer to the nearest tenth. Use 1.732 for $\sqrt{3}$.

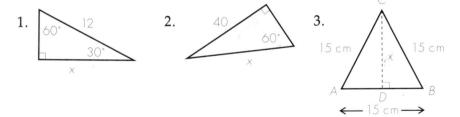

1.

2.

3.

INDIRECT MEASUREMENT

The relationships among the three sides of a 30°-60° right triangle depend on the shape of the triangle but not on its size. They are illustrated in the triangles below.

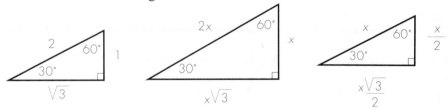

These relationships are the same for all such triangles, large and small. For this reason, you can use 30°-60° right triangles to measure distances indirectly.

Example: Alicia and Susan were surveying a property and needed to find the distance between an oak tree and a large rock on opposite sides of a pond. They paced off several feet along their side of the pond, forming a right angle at C. Next they found a point B such that m$\angle ABC$ = 60°. The distance BC was found to measure 116 ft. They found AB and AC as follows.

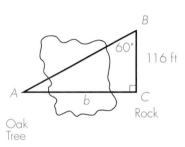

$\frac{1}{2}AB = 116$, so $AB = 232$

$b = 232 \cdot \frac{\sqrt{3}}{2}$ or $116\sqrt{3}$

$b \approx 116\,(1.732)$, or about 201 ft

CLASS ACTIVITY

Find the distance across the pond. Answer to the nearest whole unit. Use 1.732 for $\sqrt{3}$.

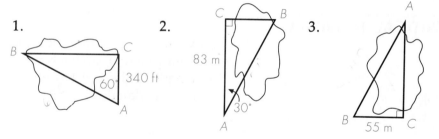

1.

2.

3.

HOME ACTIVITY

Find the missing lengths. Answer to the nearest tenth. Use 1.732 for $\sqrt{3}$.

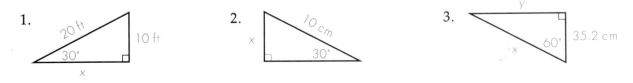

1.

2.

3.

Find the missing measure. Answer to the nearest whole unit. Use 1.732 for $\sqrt{3}$.

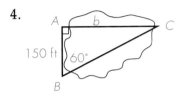

4.

5.

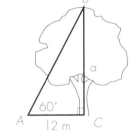

6.

7. Write a Logo procedure named RTTRI3060 to draw a 30°-60° right triangle. Use variables in the procedure so that the lengths of the sides can be different each time the procedure is called.

8. In the figure, $\overline{XY} \parallel \overline{UV}$. Copy the figure. Then bisect angles ABY and BAV. Call the intersection point of the bisectors C. First, find x and y. Then find $m\angle ABC$. Explain how you arrived at your answers.

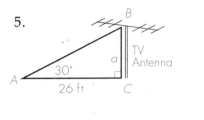

9. Prove that if a triangle is equilateral then it is also equiangular.

CRITICAL THINKING

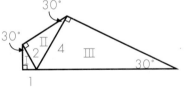

10. Copy the figure at the left and continue the pattern until you have drawn a total of five connected 30°-60° right triangles. Find the length of the last hypotenuse to be drawn when the total number of triangles is:

   1. 1   2. 2   3. 3   4. 4   5. 5   6. 10   7. n

SECTION 7.4

Angle-Side Relationship

After completing this section, you should understand

▲ that the side opposite the greatest angle of a triangle is the longest side.

▲ that the angle opposite the longest side of a triangle is the greatest angle.

▲ applications of triangle inequalities.

Surveyors use triangles and indirect measurement to help them establish the boundaries of large pieces of land. Very often the sides of a triangle are not equal length nor the angles of equal measure. In such cases, how are the sides related to the angles? In this section, you will learn one way in which they are related.

UNEQUAL ANGLES IN A TRIANGLE

In Section 3.2, you learned that the sum of two sides of a triangle is greater than the third side. In this section you will learn about another inequality involving triangles.

DISCOVERY ACTIVITY

1. On a separate piece of paper, draw a triangle such as the one at the right in which m∠C > m∠B.

2. Cut out the triangle and fold it along $\overline{PM}$ so that vertices B and C coincide as shown in the second diagram. Notice that M is the midpoint of $\overline{BC}$.

3. How is AC related to the sum of AP and PC?

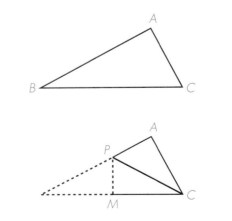

(Hint: Review Theorem 3-2-1.)

4. How is the sum $AP + PC$ related to AB?

5. What can you conclude about the relationship between AB and AC?

From the Discovery Activity, it appears that you can make the following conjecture: If in a triangle, one angle is greater than the other, then the side opposite the greater angle is larger than the side opposite the other angle.

The Discovery Activity also suggests a method of proving the conjecture.

Given: △ABC with
  $m\angle C > m\angle B$

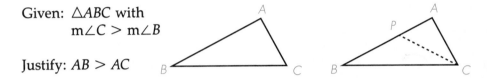

Justify: $AB > AC$

In △ABC, copy $\angle B$ with C as the vertex and $\overline{BC}$ as one of its two sides. The other side of the new angle intersects $\overline{AB}$ in point P. Fill in the reasons in the proof.

Steps	Reasons
1. $PC = PB$	1.
2. $AP + PC > AC$	2.
3. $AP + PB > AC,$ or $AB > AC$	3.

In the above proof, it was assumed that point P lies in the interior of $\angle BCA$.

THEOREM 7-4-1 **If a triangle has one angle greater than another angle, then the side opposite the greater angle is longer than the side opposite the other angle.**

Example: In isosceles triangle *ABC*, m∠*ABC* = 30°. How are the lengths of the three sides related?

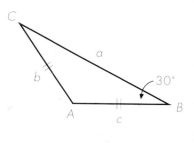

Since △*ABC* is isosceles, *b* = *c*. Also, m∠*B* = m∠*C*, since *in an isosceles triangle the angles opposite the equal sides are equal* (Theorem 5-4-1). So, m∠*C* = 30°. Since the sum of the interior angles of the triangle is 180°, m∠*A* = 120°.

Thus, m∠*A* > m∠*B*. From Theorem 7-4-1 it follows that *a* > *b*. Finally, since *b* = *c*, *a* > *c*.

CLASS ACTIVITY

For each triangle *ABC*, write *a* < *b*, *a* = *b*, or *a* > *b* if sufficient information is provided to justify one of these conclusions. Otherwise write "insufficient information." (The triangles are *not* necessarily drawn to show their proper shapes.)

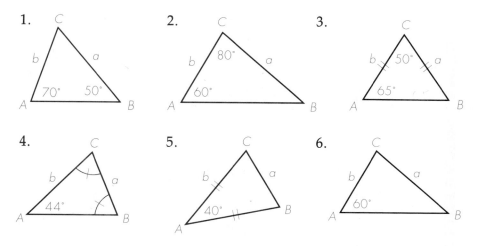

7. Use Theorem 7-4-1 to show that in a right triangle, the hypotenuse must always be the longest side. (Hint: Can either of the non-right angles be equal to 90°? greater than 90°? Why?)

UNEQUAL SIDES IN A TRIANGLE

In Chapter 5 you learned that the converse of a true theorem is not necessarily true. Here is Theorem 7-4-1 and its converse.

Theorem: **If a triangle has one angle greater than another angle, then the side opposite the greater angle is longer than the side opposite the other angle.**

Converse: If a triangle has one side longer than another side, then the angle opposite the longer side is greater than the angle opposite the other side.

Is the converse of the theorem true? The next Discovery Activity will help you answer this question.

DISCOVERY ACTIVITY

1. On a separate piece of paper, draw a scalene triangle such as the one at the right in which $AC > AB$.

2. Cut out the triangle and fold the left part of the triangle in such a way that side AB coincides with a portion of side AC.

3. Unfold the triangle and inspect the fold along line segment AP. What does this line segment bisect?

4. What two triangles are congruent? Why?

5. How is $\angle 1$ related to $\angle 1'$? How do you know?

6. How is $\angle 1'$ related to $\angle 2$? How do you know? (Hint: Review Theorem 3-5-1)

7. What can you conclude about the relationship between $\angle 1$ and $\angle 2$?

From the above Discovery Activity, it appears that the converse of Theorem 7-4-1 *is* true.

THEOREM 7-4-2 **If a triangle has one side longer than another side, then the angle opposite the longer side is greater than the angle opposite the other side.**

The proof of Theorem 7-4-2 is not given.

CLASS ACTIVITY

For each triangle *ABC*, write m∠*A* < m∠*B*, m∠*A* = m∠*B*, or m∠*A* > m∠*B* if sufficient information is provided to justify one of these conclusions. Otherwise write "insufficient information." (The triangles are *not* necessarily drawn to show their proper shapes.)

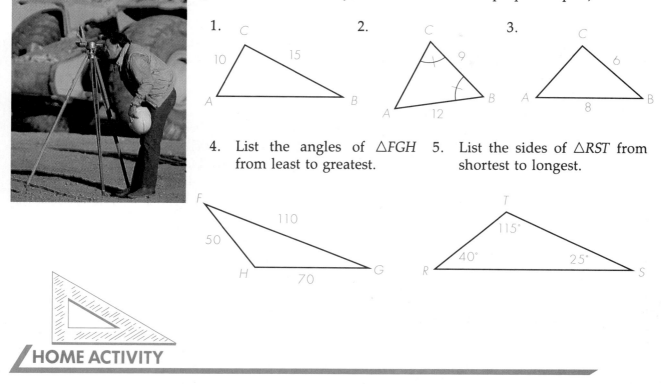

1.

2.

3.

4. List the angles of △*FGH* from least to greatest.

5. List the sides of △*RST* from shortest to longest.

HOME ACTIVITY

For each triangle, write *a* < *b*, *a* = *b*, or *a* > *b* if sufficient information is provided to justify one of these conclusions. Otherwise write "insufficient information." (The triangles are *not* necessarily drawn to show their proper shapes.)

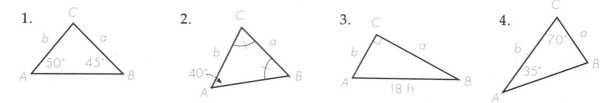

1.

2.

3.

4.

For each triangle, write m∠*A* < m∠*B*, m∠*A* = m∠*B*, or m∠*A* > m∠*B* if sufficient information is provided to justify one of these conclusions. Otherwise write "insufficient information." (The triangles are *not* necessarily drawn to show their proper shapes.)

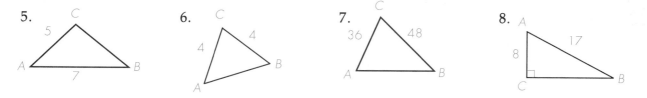

5.

6.

7.

8.

9. List the sides of △PQR from shortest to longest.

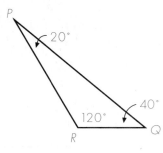

10. List the angles of △WXY from least to greatest.

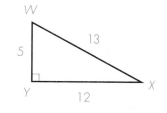

In Exercises 11–14, be prepared to give reasons for your answers.

11. Given: △MPO with
    m∠2 < m∠1 and
    PN < PM
    What can you conclude about m∠2 and m∠3?

12. Given: △WXY with
    m∠1 > m∠2
    What can you conclude about m∠1 and m∠3? about WY and WX? (Hint: Review Theorem 3-5-1.)

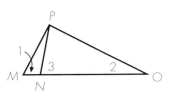

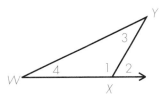

13. Given: △ABC with
    AC > BC
    The bisectors of ∠A and ∠B meet at D. What can you conclude about BD and AD?

14. Given: △DEF is a scalene triangle.
    What can you conclude about the three angles?

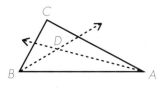

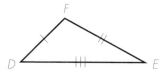

CRITICAL THINKING

15. In △ABC, the bisector of ∠C intersects $\overline{AB}$ in point D. Show that BC > BD.

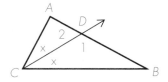

SECTION 7.5

Pythagorean Theorem and Descartes

After completing this section, you should understand
▲ how the Pythagorean Theorem is applied to the coordinate plane.
▲ the distance formula.

You have seen how surveyors can use a 30°-60° right triangle to locate landmarks on property. There are other ways in which landmarks can be located. For example you can give the location of each point in relation to a pre-selected reference point.

The point of reference is called the **origin.** The origin is the intersection of two perpendicular lines of reference, called the **x-axis** and the **y-axis.**

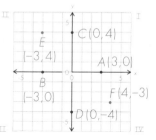

The x- and y-axes divide the plane into four quadrants (I, II, III and IV). A point is represented by a pair of numbers. The first number indicates how far the point is to the right or left of the origin, the second number tells how far it is above or below the origin. For example, point A is (3, 0) and is 3 units to the right of the origin. Point D is (0, −4) and is 4 units below the origin. The pairs for points E and F use the same numbers (−3 and 4) but in a different order. For this reason, a pair such as (4, −3) is called an **ordered pair.** The first number is the x-coordinate and the second number is the y-coordinate. The x- and y-axes are called the **coordinate axes.** The plane that they define is called the **coordinate plane.**

The man who is credited with introducing coordinates into geometry is René Descartes.

REPORT 7-5-1   Prepare a report on René Descartes that includes the following information.

1. The time in history when Descartes lived

2. The important mathematical works that Descartes wrote

3. The names of other important mathematicians of Descartes's era

4. Some major world events during Descartes's lifetime

Sources:
Bell, Eric Temple, *Men of Mathematics.* New York: Simon & Schuster, 1937.
Boyer, Carl B., *A History of Mathematics.* Princeton, NJ: Princeton University Press, 1985.

The study of geometry by means of coordinates is called **coordinate geometry.** The Discovery Activity shows how coordinate geometry can be used to examine distance.

DISCOVERY ACTIVITY

In the figure, $\overline{AB}$ and $\overline{CD}$ are horizontal, $\overline{EF}$ and $\overline{GH}$ are vertical.

1. Count squares to find AB and CD.
2. Count squares to find EF and GH.
3. How could you use the coordinates to find the lengths of the segments?

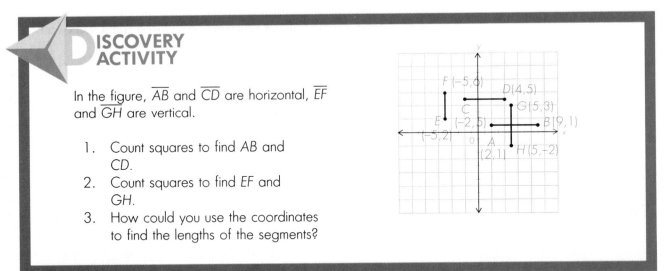

The Discovery Activity suggests the following definition.

NOTEBOOK

DEFINITION 7-5-1   **The LENGTH of the horizontal line segment with endpoints (x_1, y) and (x_2, y) is $|x_2 - x_1|$. For the vertical line segment with endpoints (x, y_1) and (x, y_2), the length is $|y_2 - y_1|$.**

The absolute value symbol (| |) means that you use the positive values of the differences in the definition.

NONHORIZONTAL AND NONVERTICAL LINE SEGMENTS

You can use the Pythagorean Theorem to find a formula for the length of a line segment that is neither horizontal nor vertical.

In $\triangle ABD$,

$$d^2 = a^2 + b^2$$
$$d^2 = (|x_2 - x_1|)^2 + (|y_2 - y_1|)^2$$
$$d^2 = (x_2 - x_1)^2 + (y_2 - y_1)^2$$

So $d = \sqrt{(x_2 - x_1)^2 + (y_2 - y_1)^2}$

THEOREM 7-5-1 **The distance between points $A(x_1, y_1)$ and $B(x_2, y_2)$ is**

$$d = \sqrt{(x_2 - x_1)^2 + (y_2 - y_1)^2} \qquad \textbf{(The Distance Formula)}$$

CLASS ACTIVITY

Mark the points on a coordinate plane. Then find the length of the segment connecting the two points.

1. A (3, 6) 2. C (1, 2) 3. E (−5, −1) 4. M (−4, 5)
    B (6, 2)   D (1, −1)   F (−1, 4)   N (−2, 5)

HOME ACTIVITY

Mark the points on a coordinate plane. Then find the length of the segment connecting the two points.

1. A (−3, 3)   2. C (4, 4)   3. E (−2, −7)   4. G (0, −4)
    B (5, −3)   D (4, 1)   F (6, 8)   H (8, −5)
5. $\triangle ABC$ has vertices $A(0, 1)$, $B(3, 7)$, and $C(6, 1)$. Show that the triangle is isosceles.
6. $\triangle PQR$ has vertices $P(1, 1)$, $Q(4, 4)$, and $R(8, 0)$. Show that the triangle is a right triangle. (Hint: Review Theorem 7-1-2.)

CRITICAL THINKING

7. Which is larger, $\sqrt{10} + \sqrt{17}$ or $\sqrt{53}$? Find the answer without using tables or a calculator. (Hint: From algebra, you know that $(a + b)^2 = a^2 + 2ab + b^2$.)

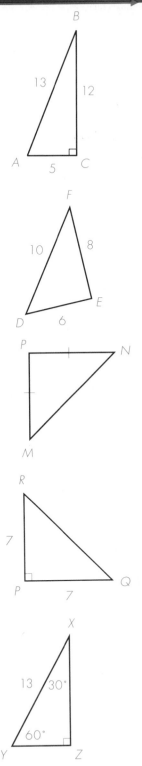

7.1 The hypotenuse of a right triangle is the side opposite the right angle. The other two sides are the legs of the right triangle. In $\triangle ABC$, the right angle is $\angle C$, so the hypotenuse is $\overline{AB}$. The legs are $\overline{AC}$ and $\overline{BC}$.

If a right triangle has sides of lengths a, b, and c, where c is the hypotenuse, then $a^2 + b^2 = c^2$. In right triangle ABC, $5^2 + 12^2 = 13^2$.

If a triangle has three sides of lengths a, b, and c, such that $a^2 + b^2 = c^2$, then the triangle is a right triangle. For $\triangle DEF$, since $8^2 + 6^2 = 10^2$, you know that $\angle DEF$ is a right angle.

7.2 In an isosceles right triangle, the acute angles are each $45°$. In the isosceles right triangle MNP, the acute angles are $\angle M$ and $\angle N$. $m\angle M = 45°$ and $m\angle N = 45°$.

The length of the hypotenuse of an isosceles right triangle is $\sqrt{2}$ (about 1.414) times the length of either leg. $\triangle PQR$ has legs of length 7, so the length of the hypotenuse is: $QR = 7\sqrt{2}$, or about 9.899

7.3 In a $30° - 60°$ right triangle, the length of the leg opposite the $30°$ angle is half the length of the hypotenuse. The length of the leg opposite the $60°$ angle is $\frac{\sqrt{3}}{2}$ times the length of the hypotenuse. In $\triangle XYZ$

$$YZ = \tfrac{1}{2} \cdot 13 = 6\tfrac{1}{2}$$

$$XZ = \tfrac{\sqrt{3}}{2} \cdot 13, \text{ or about } 11.258$$

7.4 If two angles of a triangle are unequal, then the side opposite the greater angle is longer than the side opposite the other angle. In $\triangle ABC$, $m\angle C > m\angle A$. Therefore $AB > BC$.

If two sides of a triangle are unequal, then the angle opposite the longer side is greater than the angle opposite the other side. In $\triangle KLM$, $KL > LM$. Therefore $m\angle M > m\angle K$.

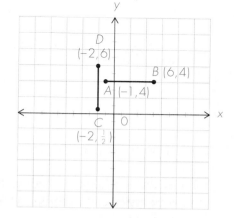

7.5 In a coordinate plane, the length of a horizontal segment equals the absolute value of the difference of the x-coordinates of the endpoints. If the segment is vertical, its length equals the absolute value of the difference of the y-coordinates of the endpoints.

$\overline{AB}$ is horizontal, so

$$AB = |6 - (-1)| = |7| = 7$$

$\overline{CD}$ is vertical, so

$$CD = \left|\tfrac{1}{2} - 6\right| = \left|-5\tfrac{1}{2}\right| = 5\tfrac{1}{2}$$

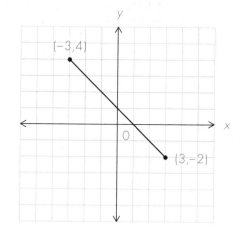

For a segment that is neither horizontal nor vertical and whose endpoints are (x_1, y_1) and (x_2, y_2), the length is the distance d between the endpoints:

$$d = \sqrt{(x_2 - x_1)^2 + (y_2 - y_1)^2}$$

If the endpoints are $(-3, 4)$ and $(3, -2)$, the distance is:

$$d = \sqrt{(3-(-3))^2 + (-2-4)^2}$$

$$d = \sqrt{36 + 36}$$

$$d = \sqrt{72}, \text{ or about } 8.49$$

COMPUTER ACTIVITY

Three positive whole numbers a, b, and c that satisfy the equation $a^2 + b^2 = c^2$ are called a Pythagorean Triple. The ordered triple (a, b, c) is used to represent a Pythagorean Triple, where c is the largest number. The table below shows several Pythagorean Triples, where a is odd and n tells what line in the list the triple is on.

n	a	b	c
1	3	4	5
2	5	12	13
3	7	24	25
4	9	40	41
.	.	.	.
.	.	.	.
.	.	.	.
n	$2n + 1$	$2n^2 + 2n$	$2n^2 + 2n + 1$

The nth Pythagorean Triple in the list is $(2n + 1, 2n^2 + 2n, 2n^2 + 2n + 1)$.

You can write a Logo procedure to find these Pythagorean Triples. In order to do this, you will need to use the operation symbol + for addition, * for multiplication, and ↑ for exponentiation.

For example, $2n^2 + 2n$ is written as $2 * :N \uparrow 2 + 2 * :N$. You will also need to use the OUTPUT command to print the triples. For example, OUTPUT $2 * :N + 1$ will print the result of $2 * :N + 1$ after a value is given for :N.

Example: Write a procedure named PYTHTRIPLE to find the nth Pythagorean Triple in the table above.

```
TO PYTHTRIPLE :N
    OUTPUT 2 * :N + 1
    OUTPUT 2 * :N ↑ 2 + 2 * :N
    OUTPUT 2 * :N ↑ 2 + 2 * :N + 1
END
```

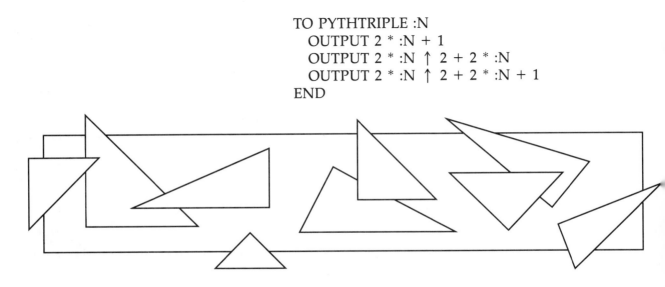

If :N is 5, the output will be:

11

60

61

Use the output from the PYTHTRIPLE procedure to help you fill in the chart below. Use the values in the *n* column for :N in the procedure.

	n	a	b	c
1.	6	___	___	___
2.	7	___	___	___
3.	8	___	___	___
4.	9	___	___	___
5.	10	___	___	___
6.	11	___	___	___
7.	12	___	___	___
8.	13	___	___	___
9.	14	___	___	___
10.	15	___	___	___

11. Write a Logo procedure to draw any right triangle whose sides have lengths $2n + 1$, $2n^2 + 2n$, and $2n^2 + 2n + 1$, where *n* is a positive whole number. Then execute the procedure to draw right triangles whose sides have the lengths given by the Pythagorean Triples from Exercises 1 and 2.

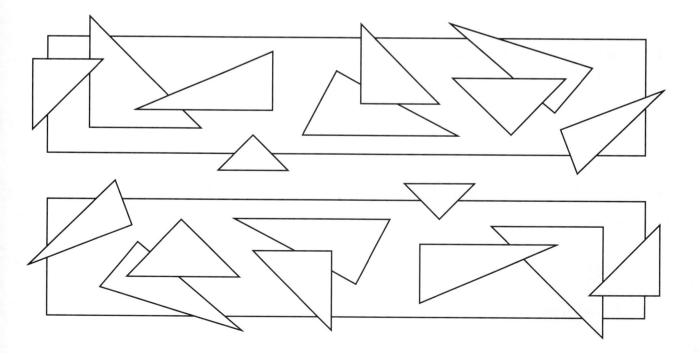

Each of the following is the set of lengths of three sides of a triangle. Is the triangle a right triangle?

1. 7, 8, 9
2. 16, 30, 34
3. 11, 60, 61
4. 9, 10, 12

Find the missing lengths to the nearest tenth. Use a calculator or the table of squares and square roots on page 562.

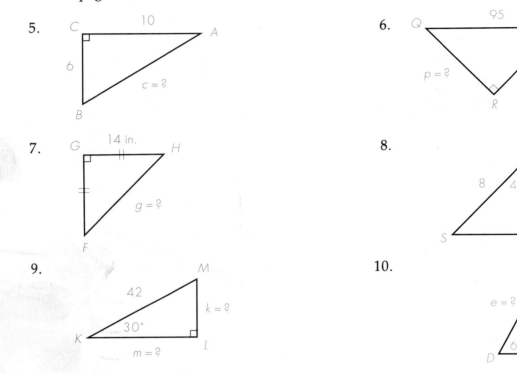

5.

6.

7.

8.

9.

10.

11. List the lengths of the sides of $\triangle XYZ$ from shortest to longest.

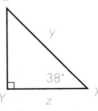

12. Write $m\angle A < m\angle B$, $m\angle A = m\angle B$, $m\angle A > m\angle B$, or "insufficient information."

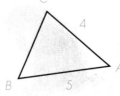

Find the length of the line segment that connects the two points.

13.

14.

15. $(-1, -3)$ and $(-1, 4)$

16. $(-5, 2)$ and $(3, -6)$

You may use a calculator or the table of square and square roots on page 562. Round decimal answers to the nearest tenth.

1. The numbers 13, 84, and 85 are the lengths of the three sides of a triangle. Is the triangle a right triangle?

2. Find the missing length.

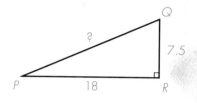

3. A window measures 4 ft by 6 ft. How far is it from one corner to the diagonally opposite corner?

4. When an 18-in. candle just fits in the bottom of a rectangular box 10 in. wide. How long is the box?

Find the missing lengths.

5.

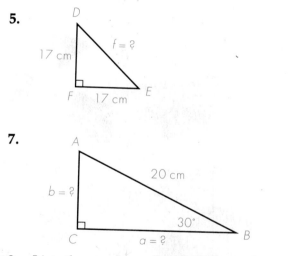

6.

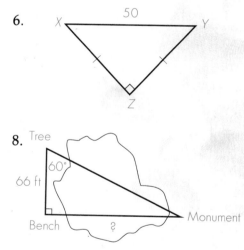

7.

8.

9. List the angles of $\triangle ABC$ from least to greatest.

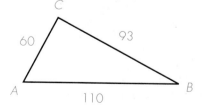

10. Write $x < y$, $x = y$, $x > y$, or "insufficient information."

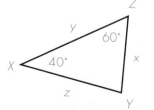

Find the length of the line segment that connects the two points.

11. (2, 4) and (2, −2) 12. (−3, 4) and (3, 2) 13. (2, −6) and (−8, −6) 14. (−1, −1) and (3, 7)

ALGEBRA SKILLS

Example: Solve for x.

$$6(8x - 3) = 5x + 3(5 - x) - 10$$
$$48x - 18 = 5x + 15 - 3x - 10$$
$$48x - 18 = 2x + 5$$
$$48x - 18 - 2x = 2x + 5 - 2x$$
$$46x - 18 = 5$$
$$46x - 18 + 18 = 5 + 18$$
$$46x = 23$$
$$\frac{46x}{46} = \frac{23}{46}$$
$$x = \frac{1}{2}$$

Example: Solve for x.

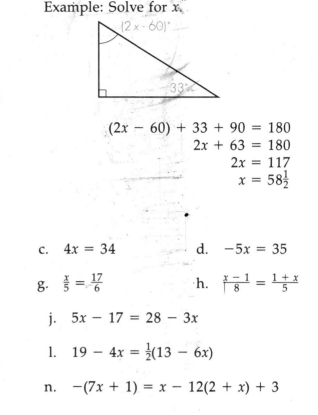

$$(2x - 60) + 33 + 90 = 180$$
$$2x + 63 = 180$$
$$2x = 117$$
$$x = 58\tfrac{1}{2}$$

1. Solve for x.

a. $x - 14 = 27$ b. $x + 33 = 81$ c. $4x = 34$ d. $-5x = 35$

e. $3x - 1 = 50$ f. $53 = 14x + 7$ g. $\frac{x}{5} = \frac{17}{6}$ h. $\frac{x-1}{8} = \frac{1+x}{5}$

i. $0.6x + 1.2 = 4.8$ j. $5x - 17 = 28 - 3x$

k. $5(x + 2) = 7 - x$ l. $19 - 4x = \frac{1}{2}(13 - 6x)$

m. $-6(x - 1) - 8 = 15x + 1$ n. $-(7x + 1) = x - 12(2 + x) + 3$

2. Find the value of the unknown angles.

a.

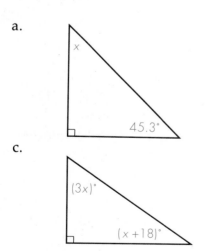

b.

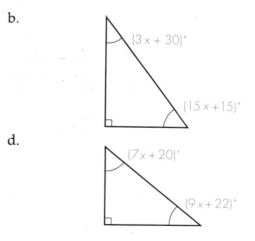

c.

d.

3. On the way to school a student spent half of her money for a snack. At noon she spent half the amount of money she had left. After school she then spent $\frac{1}{3}$ of the money she had left. She finished the day with just $1. How much money did the student begin the day with?

8

QUADRILATERALS

Method consists entirely in properly
ordering and arranging the things to
which we should pay attention.

René Descartes

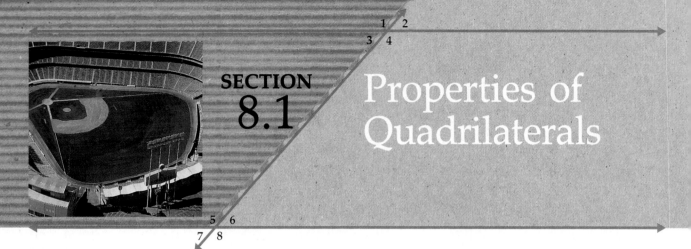

Properties of Quadrilaterals

After completing this section, you should understand
▲ the properties that determine a quadrilateral.
▲ the difference between convex and concave quadrilaterals.

In Chapter 3 you found that a triangle is a figure formed by the line segments that connect three noncollinear, coplanar points. If you connect four coplanar points in order, you will form another kind of figure.

Examples:

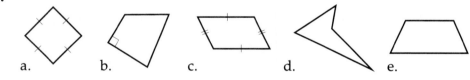

a.   b.   c.   d.   e.

DISCOVERY ACTIVITY

1. Copy and complete the table for figures a–e shown above.

Figure	a	b	c	d	e
Number of sides					
Number of vertices					
At least one pair of opposite sides parallel					
opposite sides congruent					
Opposite sides parallel and congruent					
No parallel sides					
Number of collinear points					

2. Use the information in the table to write a description of the figures that will be true for all of them.

You may have discovered that the figures all have four sides and four vertices.

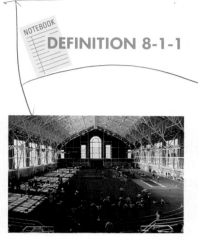

NOTEBOOK

DEFINITION 8-1-1 A QUADRILATERAL is a closed, plane, four-sided figure.

A quadrilateral consists of four line segments that intersect only at their endpoints. It is a *closed* figure because the line segments are connected consecutively. It is a *plane* figure because all four points are on the same plane.

$\overline{JK}$ and $\overline{KL}$ are adjacent sides.
$\overline{KL}$ and $\overline{JM}$ are opposite sides.
K and L are adjacent vertices.
J and L are opposite vertices.
$\angle L$ and $\angle M$ are adjacent angles.
$\angle J$ and $\angle L$ are opposite angles.

Two additional line segments, the diagonals, are also associated with a quadrilateral.

NOTEBOOK

DEFINITION 8-1-2 A DIAGONAL is a line segment determined by two non-adjacent vertices.

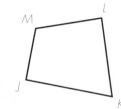

Example: $\overline{PR}$ and $\overline{QS}$ are diagonals of *PQRS*.

CLASS ACTIVITY

Refer to figure *GHJK*.
1. Name the pairs of opposite sides and angles.
2. List the pairs of adjacent sides and angles.
3. Name the diagonals.

Is the figure a quadrilateral? If not, explain why.

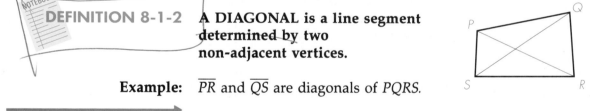

4.   5.   6.   7.

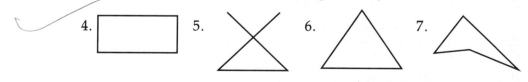

TYPES OF QUADRILATERALS

Compare quadrilateral *ABCD* with quadrilateral *EFGH*. Notice that in *ABCD*, the measure of each angle is less than 180°. In *EFGH*, one angle measures more than 180°. The diagonals $\overline{AC}$ and $\overline{BD}$ in *ABCD* differ from diagonal $\overline{EG}$ in *EFGH* by being inside the figure; $\overline{EG}$ is outside.

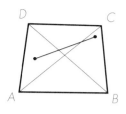

DEFINITION 8-1-3   **A CONVEX quadrilateral is a quadrilateral with the measure of each interior angle less than 180°.**

ABCD is convex because each of its angles is less than 180°, and its diagonals lie inside the figure. In fact, any line segment connecting any two points inside a convex quadrilateral will lie inside the figure.

DEFINITION 8-1-4   **A CONCAVE quadrilateral is a quadrilateral with one interior angle whose measure is greater than 180°.**

EFGH is concave because the measure of interior ∠*H* is greater than 180°, and one of its diagonals, $\overline{EG}$, lies outside the figure. In fact if any line segment connecting any two points inside a quadrilateral intersects the sides of the figure, then the figure is concave.

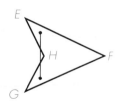

DISCOVERY ACTIVITY

Recall that if a figure is a triangle, then the sum of the measures of its angles is 180° (Theorem 3-1-1).

1. Draw a quadrilateral *ABCD*.
2. Draw the diagonal $\overline{AC}$.
3. How many triangles have you formed?
4. What is the sum of the angles in △*ABC*?
5. Write a conclusion about the sum of the measures of the interior angles of quadrilateral ABCD. Give a reason for your conclusion.
6. Do you think your conclusion will be true for any quadrilateral? Why?

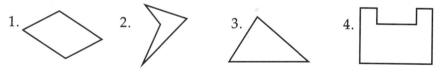

THEOREM 8-1-1 **The sum of the measures of the interior angles of a quadrilateral is 360°.**

CLASS ACTIVITY

Tell whether the figure is convex or concave. Find the sum of the measures of the interior angles.

1. 2. 3. 4.

A SPECIAL PROPERTY OF TRIANGLES

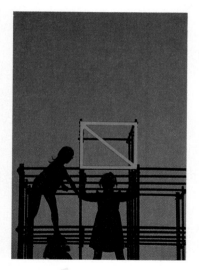

The physical use of diagonals is an important factor in architecture. In the construction of buildings, bridges, and other structures, the diagonal provides support for four-sided figures. Try this experiment.

1. Use four cardboard strips and paper fasteners to construct a convex quadrilateral fastened at the vertices.

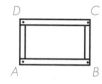

2. Hold the quadrilateral by the sides $\overline{AD}$ and $\overline{BC}$ and move your hands up and down. What happens to the quadrilateral?

3. Cut another cardboard strip to fit from A to C. Attach it at A and C. Repeat step 2.

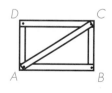

4. What do you observe?

THEOREM 8-1-2 **A triangle is a rigid or non-flexible figure.**

Once the diagonal is attached, the quadrilateral becomes inflexible or rigid. Notice that the diagonal converts the quadrilateral into two triangles. This is why the triangle is called a rigid figure.

HOME ACTIVITY

Refer to figure *LMNO*.
1. Name the diagonals.
2. Name the pairs of opposite sides.
3. Name the pairs of opposite angles.

$180(n-2)$

Refer to figure *QRST*.

4. Name the diagonals.

5. Name the adjacent sides.

6. Name the adjacent angles.

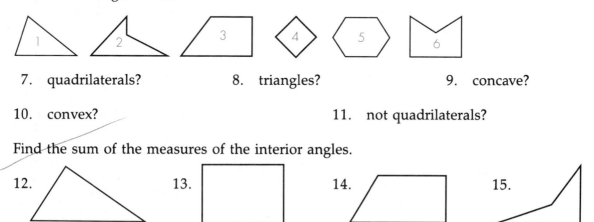

Which of the figures below are

7. quadrilaterals?   8. triangles?   9. concave?

10. convex?   11. not quadrilaterals?

Find the sum of the measures of the interior angles.

12.   13.   14.   15.

 Write Logo commands that tell the turtle to draw each figure described below.

16. Convex quadrilateral   17. Concave quadrilateral

CRITICAL THINKING

18. Copy and complete the table.

Number of Sides	Sum of Interior Angles	Name of Polygon	Number of Sides	Sum of Interior Angles	Name of Polygon
3	180°	triangle	8		octagon
4	360° (2 × 180°)	quadrilateral	10		decagon
5	540° (3 × 180°)	pentagon	12		dodecagon
6	720° (4 × 180°)	hexagon	n		n-gon

SECTION 8.2

Trapezoids

After completing this section, you should understand

▲ the properties of trapezoids and isosceles trapezoids.

▲ how to apply the theorems about isosceles trapezoids and medians of trapezoids.

A trapezoid is a member of the set of quadrilaterals. This means that any property of a quadrilateral also applies to a trapezoid: It is a four-sided closed figure, and the sum of its interior angles is 360°. One more condition is necessary for a quadrilateral to be a trapezoid—one pair of sides must be parallel.

DEFINITION 8-2-1 | **A TRAPEZOID is a quadrilateral with one and only one pair of parallel sides.**

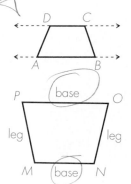

$\overline{AB} \parallel \overline{DC}$

The parallel sides are called **BASES.** A special class of trapezoids has non-parallel sides that are equal in length. These sides are called **LEGS.**

DEFINITION 8-2-2 | **An ISOSCELES TRAPEZOID is a trapezoid with two non-parallel equal sides.**

DISCOVERY ACTIVITY

1. On graph paper, draw several isosceles trapezoids.

2. Use a protractor to measure the interior angles of each trapezoid.

287

3. What do you observe about the measures of the angles in each figure? Compare your observations with your classmates'.

ISOSCELES TRAPEZOIDS

You may have discovered that an isosceles trapezoid has two pairs of angles with equal measures. You can use what you know about similar and isosceles triangles to prove that the measures of the base angles of an isosceles trapezoid are equal.

Given: Trapezoid $ABCD$
with $AD = BC$
Justify: $m\angle DAB = m\angle CBA$ and
$m\angle ADC = m\angle BCD$

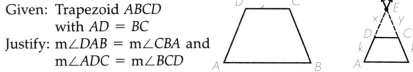

Steps		Reasons	
1.	$AD = BC$	1.	Given
2.	Extend $\overline{AD}$ and $\overline{BC}$ until they intersect at E.	2.	Two non-parallel lines intersect at one point.
3.	$\overline{DC} \parallel \overline{AB}$, so $m\angle DAB = m\angle EDC$ and $m\angle CBA = m\angle ECD$.	3.	Corresponding angles have equal measures.
4.	$\triangle ABE \sim \triangle DCE$	4.	AA
5.	$\frac{x}{(x + k)} = \frac{y}{(y + k)}$	5.	Corresponding sides are in equal ratio.
6.	$x(y + k) = y(x + k)$ $xy + xk = yx + yk$ $xk = yk$ $x = y$	6.	Algebra
7.	$\triangle EDC$ is isosceles.	7.	
8.	$m\angle EDC = m\angle ECD$	8.	
9.	$m\angle DAB = m\angle CBA$	9.	
10.	$m\angle ADC + m\angle EDC = 180°$ and $m\angle BCD + m\angle ECD = 180°$	10.	
11.	$m\angle ADC = m\angle BCD$	11.	

THEOREM 8-2-1 **In an isosceles trapezoid, the measures of each pair of base angles are equal.**

CLASS ACTIVITY

Find the missing angle measures for each isosceles trapezoid.

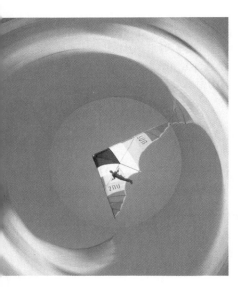

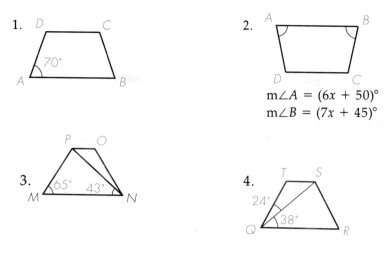

1.

2.

$m\angle A = (6x + 50)°$
$m\angle B = (7x + 45)°$

3.

4.

5. Is a trapezoid concave or convex? Why?

NOTEBOOK

DEFINITION 8-2-3 The **MEDIAN** of a trapezoid is the line segment joining the midpoints of the two non-parallel sides.

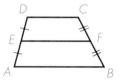

ISCOVERY ACTIVITY

1. On graph paper, draw several *ABCD* trapezoids with *AB* ∥ *DC*.

2. Use a compass to locate the midpoints of the two non-parallel sides for each figure. Label the midpoints *E* and *F*.

3. Draw the medians.

4. Measure to find the length of the median and two bases for each figure. *EF* = ?
   AB = ?   *DC* = ?   *AB* + *DC* = ?

5. How does the length of a median compare to the sum of the lengths of the two bases?

6. Write a conclusion stating what you have discovered.

MEDIANS OF TRAPEZOIDS

You may have found that the length of the median of a trapezoid is one half the sum of the lengths of the bases. You may also have noticed that the median of a trapezoid appears to be parallel to its bases.

Given: Trapezoid $ABCD$ with $\overline{AB} \parallel \overline{DC}$
   E is the midpoint of $\overline{AD}$.
   F is the midpoint of $\overline{BC}$.
Justify: $EF = \frac{1}{2}(AB + DC)$;
   $\overline{EF} \parallel \overline{AB}$ and $\overline{EF} \parallel \overline{DC}$

Steps	Reasons
1. $\overline{AB} \parallel \overline{DC}$; E and F are midpoints of $\overline{AD}$ and $\overline{BC}$.	1. Given
2. Draw the median $\overline{EF}$.	2. Two points determine a line.
3. Extend $\overline{AB}$ and draw $\overleftrightarrow{DF}$ to intersect $\overline{AB}$ at G.	3. Two non-parallel lines intersect at one point.
4. $m\angle BFG = m\angle CFD$	4. Vertical angles are equal.
5. $m\angle FBG = m\angle FCD$	5.
6. $CF = FB$	6. Definition of midpoint
7. $\triangle BFG \cong \triangle CFD$	7.
8. $\overline{DC} \cong \overline{BG}$ and $\overline{DF} \cong \overline{FG}$	8.
9. For $\triangle ADG$, $\overline{EF} \parallel \overline{AG}$, and $EF = \frac{1}{2}AG$.	9. Line joining midpoints of a triangle is parallel to the third side and equal to $\frac{1}{2}$ its length.
10. $AG = AB + BG$	10. Definition of line segment
11. $AG = AB + DC$	11.
12. $EF = \frac{1}{2}(AB + DC)$	12.

THEOREM 8-2-2 **The median of a trapezoid is parallel to the bases and equal to one-half the sum of their lengths.**

CLASS ACTIVITY

Name the bases, legs, and median.

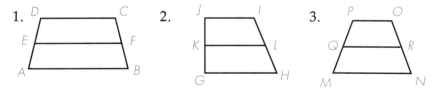

1. 2. 3.

Find the missing lengths.

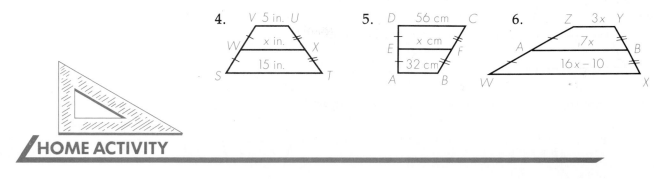

4. V 5 in. U
W x in. X
15 in.
S T

5. D 56 cm C
E x cm F
32 cm
A B

6. Z $3x$ Y
A $7x$ B
$16x - 10$
W X

HOME ACTIVITY

Is the figure a trapezoid? If not, explain why.

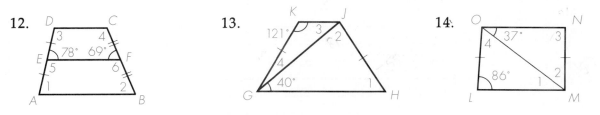

1.   2.   3.   4.   5.

Name the bases, legs, and median for each figure. The points between the vertices are the midpoints of the sides.

6. O N
P Q
L M

7. A B
E F
D C

8. T S
U V
Q R

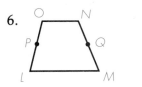

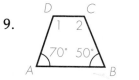

Use pencil and paper to find the measurement of the numbered angles in each trapezoid shown below. Then write Logo commands to draw each trapezoid to check your answers.

9. D C
1   2
70°   50°
A B

10. T S
1   80°
110°
2
Q R

11. P O
1   2
55°   3
M N

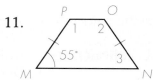

Find the measures of the numbered angles.

12. D C
3   4
E 78°   69° F
5   6
1   2
A B

13. K J
121° 3
2
4
40°   1
G H

14. O N
37°   3
4
86°   2
1
L M

Find the missing lengths.

15.

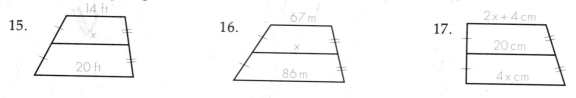

14 ft

x

20 ft

16.

67 m

x

86 m

17.

2x + 4 cm

20 cm

4x cm

18. Write reasons to complete the following proof. Then write a sentence summarizing the result.

Given: *ABCD* is an isosceles trapezoid with $\overline{AB} \parallel \overline{DC}$.
Justify: *AC* = *BD*

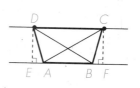

Steps

a. Draw perpendiculars from *D* to $\overleftrightarrow{AB}$ and *C* to $\overleftrightarrow{AB}$, intersecting $\overleftrightarrow{AB}$ at *E* and *F*.
b. *DE* = *CF*
c. *DA* = *CB*
d. △*DEA* ≅ △*CFB*
e. m∠*DAE* = m∠*CBF*
f. m∠*DAB* = m∠*CBA*
g. △*DAB* ≅ △*CBA*
h. *AC* = *BD*

Reasons

a.

b.

c.

d.

e.

f.

g.

h.

Refer to isosceles trapezoid *MNOP*.

19. Name the pairs of similar triangles you can find. Explain how you know.

20. Name the pairs of congruent triangles.

CRITICAL THINKING

Draw each polygon and its diagonals. Copy and complete the following table. Can you predict the number of diagonals in a polygon with 12 sides?

Polygon	Number of Sides	Number of Diagonals
Triangle	3	0
Quadrilateral	4	2
Pentagon		
Hexagon		
Heptagon		
Octagon		
Nonagon		
Decagon		

SECTION 8.3

Parallelograms

After completing this section, you should understand
▲ the properties of a parallelogram.
▲ how to apply the theorems about parallelograms.

Quadrilaterals and trapezoids are all four-sided figures, but trapezoids have the additional condition that one pair of sides is parallel. Parallelograms are another subset of quadrilaterals.

no parallel sides   1 pair of parallel sides   2 pairs of parallel sides

DEFINITION 8-3-1

A PARALLELOGRAM is a quadrilateral with both pairs of opposite sides parallel.

$\overline{AB} \parallel \overline{DC}$
$\overline{AD} \parallel \overline{BC}$

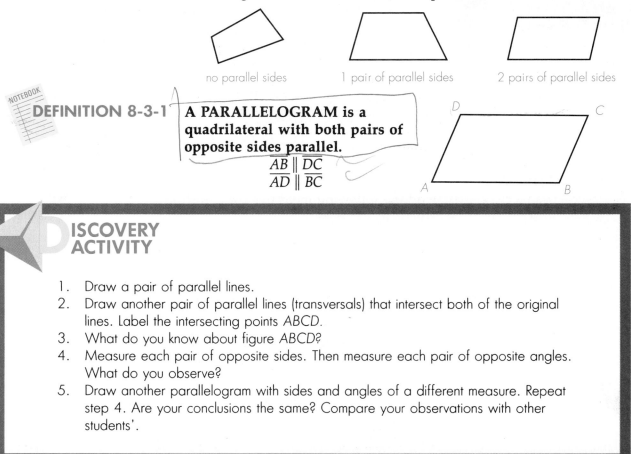

DISCOVERY ACTIVITY

1. Draw a pair of parallel lines.
2. Draw another pair of parallel lines (transversals) that intersect both of the original lines. Label the intersecting points *ABCD*.
3. What do you know about figure *ABCD*?
4. Measure each pair of opposite sides. Then measure each pair of opposite angles. What do you observe?
5. Draw another parallelogram with sides and angles of a different measure. Repeat step 4. Are your conclusions the same? Compare your observations with other students'.

You may have found that the measures of opposite sides and opposite angles are equal for the parallelograms you drew. Do you think that this property will be true for all parallelograms?

Given: $\square ABCD$
Justify: $m\angle A = m\angle C$, $m\angle B = m\angle D$,
$\qquad AB = DC$, $AD = BC$

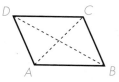

Steps	Reasons
1. Draw diagonals $\overline{AC}$ and $\overline{DB}$.	1. Two points determine a line.
2. $\overline{AB} \parallel \overline{DC}$ and $\overline{AD} \parallel \overline{BC}$	2.
3. $m\angle CAB = m\angle DCA$ and $m\angle DAC = m\angle ACB$	3.
4. $AC = AC$	4. Identity
5. $\triangle ACD \cong \triangle CAB$	5.
6. $AB = DC$ and $AD = BC$	6.
7. $m\angle ADC = m\angle CBA$	7.
8. $\quad m\angle CAB = m\angle DCA$ $\underline{+\ m\angle DAC = m\angle ACB}$ $\quad m\angle DAB = m\angle DCB$	8. Algebra
9. $m\angle A = m\angle C$	9.

THEOREM 8-3-1   **If a quadrilateral is a parallelogram, then the opposite sides are equal in length and the measures of the opposite angles are equal.**

CLASS ACTIVITY

Find the following measures for $\square ABCD$.

1. DC
2. BC
3. $m\angle B$
4. $m\angle C$
5. $m\angle D$

Find the following measures for $\square PQRS$.

6. SP
7. SR
8. $m\angle P$
9. $m\angle Q$
10. $m\angle R$
11. $m\angle S$

Find the angle measures.

12.

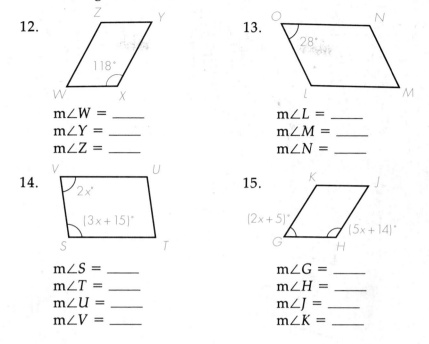

m∠W = _____
m∠Y = _____
m∠Z = _____

13.

m∠L = _____
m∠M = _____
m∠N = _____

14.

m∠S = _____
m∠T = _____
m∠U = _____
m∠V = _____

15.

m∠G = _____
m∠H = _____
m∠J = _____
m∠K = _____

DISCOVERY ACTIVITY

1. Draw a parallelogram and label it $ABCD$, with diagonals $\overline{AC}$ and $\overline{BD}$.
2. Label the point where $\overline{AC}$ and $\overline{BD}$ intersect O.
3. Find the lengths of $\overline{AO}$ and $\overline{OC}$. Then find the lengths of $\overline{DO}$ and $\overline{OB}$. What do you observe?
4. Draw another parallelogram of a different size. Repeat steps 2 and 3. What do you find?
5. Write a conclusion about the diagonals of a parallelogram.

You may have discovered that the diagonals of a parallelogram bisect each other. This conclusion can be justified for all parallelograms.

Given: $\square ABCD$ with diagonals $\overline{AC}$ and $\overline{BD}$ intersecting at O

Justify: $AO = OC$, $DO = OB$

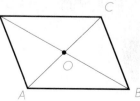

Steps	Reasons
1. $\overline{AB} \parallel \overline{DC}$	1.
2. $m\angle ODC = m\angle OBA$ and $m\angle OCD = m\angle OAB$	2.
3. $AB = DC$	3.
4. $\triangle AOB \cong \triangle COD$	4.
5. $AO = OC$ and $DO = OB$	5.

NOTEBOOK

THEOREM 8-3-2 **If the quadrilateral is a parallelogram, then the diagonals bisect each other.**

CLASS ACTIVITY

Find the missing lengths for each parallelogram.

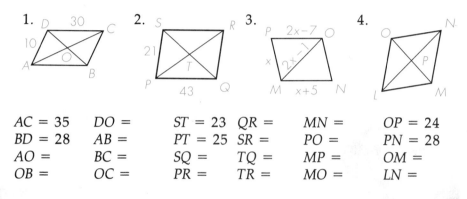

1.
2.
3.
4.

$AC = 35$	$DO =$	$ST = 23$	$QR =$	$MN =$	$OP = 24$
$BD = 28$	$AB =$	$PT = 25$	$SR =$	$PO =$	$PN = 28$
$AO =$	$BC =$	$SQ =$	$TQ =$	$MP =$	$OM =$
$OB =$	$OC =$	$PR =$	$TR =$	$MO =$	$LN =$

Find the missing angle measures for each parallelogram.

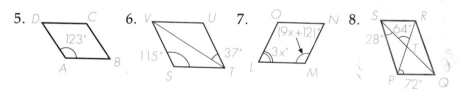

5.
6.
7.
8.

$m\angle B =$	$m\angle U =$	$m\angle L =$	$m\angle SRP =$	$m\angle STP =$
$m\angle C =$	$m\angle UVT =$	$m\angle M =$	$m\angle PTQ =$	$m\angle SPR =$
$m\angle D =$	$m\angle SVT =$	$m\angle N =$	$m\angle PQS =$	$m\angle SQR =$
	$m\angle STV =$	$m\angle O =$	$m\angle QSR =$	$m\angle PRQ =$

Is the quadrilateral a parallelogram? Give a reason for your answer.

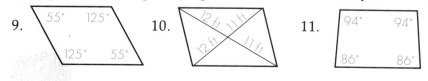

9.
10.
11.

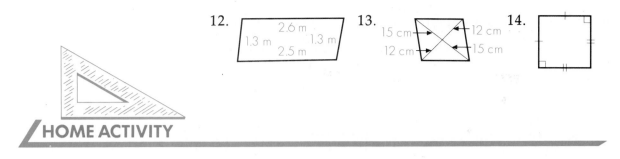

12. 2.6 m / 1.3 m / 1.3 m / 2.5 m

13. 15 cm / 12 cm / 12 cm / 15 cm

14.

HOME ACTIVITY

Find the angle measures for each parallelogram.

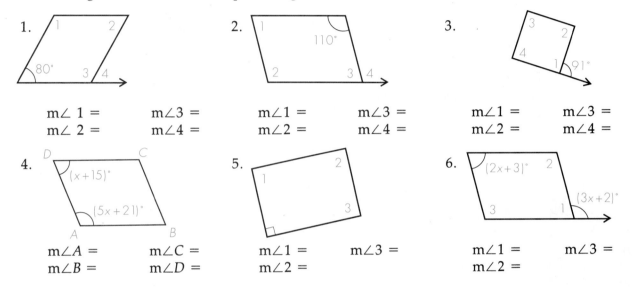

1.
80°

m∠1 =   m∠3 =
m∠2 =   m∠4 =

2.
110°

m∠1 =   m∠3 =
m∠2 =   m∠4 =

3.
91°

m∠1 =   m∠3 =
m∠2 =   m∠4 =

4.
$(x+15)°$
$(5x+21)°$

m∠A =   m∠C =
m∠B =   m∠D =

5.

m∠1 =   m∠3 =
m∠2 =

6.
$(2x+3)°$
$(3x+2)°$

m∠1 =   m∠3 =
m∠2 =

Find the missing measures for each parallelogram.

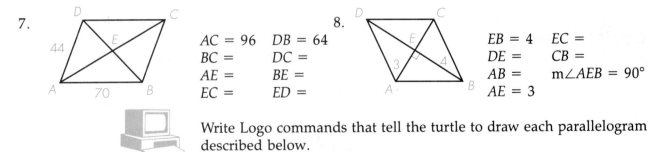

7.
44
70

AC = 96   DB = 64
BC =   DC =
AE =   BE =
EC =   ED =

8.

EB = 4   EC =
DE =   CB =
AB =   m∠AEB = 90°
AE = 3

Write Logo commands that tell the turtle to draw each parallelogram described below.

9. ▱ABCD with m∠ABC = 80°, AB = 75, BC = 50
10. ▱HIJK with m∠IJK = 40°, IJ = 40, JK = 110

11. The seats in a stadium are supported by the structure outlined in the picture.
$\overline{WX} \parallel \overline{ZY}$, $\overline{ZB} \parallel \overline{YX}$, $\overline{WZ} \parallel \overline{AY}$, m∠ZWX = 52°, and m∠YXW = 33°. What is the measure of ∠ZCA?

52°
33°

Is the quadrilateral a parallelogram? Give a reason for your answer.

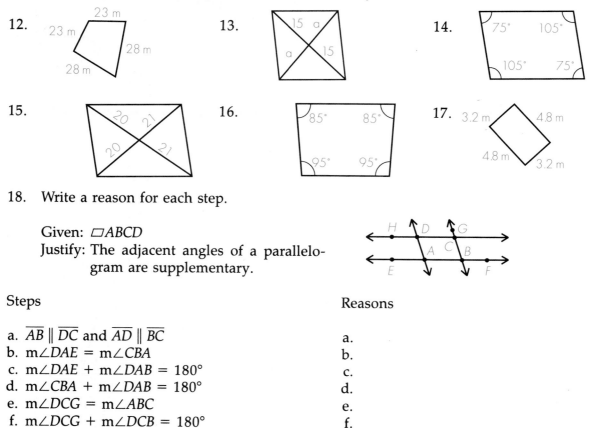

12. 23 m / 23 m / 28 m / 28 m

13. 15 / a / a / 15

14. 75° 105° / 105° 75°

15. 20 21 / 20 21

16. 85° 85° / 95° 95°

17. 3.2 m 4.8 m / 4.8 m 3.2 m

18. Write a reason for each step.

Given: ☐ABCD
Justify: The adjacent angles of a parallelogram are supplementary.

Steps

a. $\overline{AB} \parallel \overline{DC}$ and $\overline{AD} \parallel \overline{BC}$
b. $m\angle DAE = m\angle CBA$
c. $m\angle DAE + m\angle DAB = 180°$
d. $m\angle CBA + m\angle DAB = 180°$
e. $m\angle DCG = m\angle ABC$
f. $m\angle DCG + m\angle DCB = 180°$
g. $m\angle ABC + m\angle DCB = 180°$
h. Do you think this will be true for all four angles of the parallelogram? Why?

Reasons

a.
b.
c.
d.
e.
f.
g.

CRITICAL THINKING

19. Tessellations are patterns made by using congruent figures placed side by side so that there are no gaps or overlapping parts in the pattern. Any two figures have only one side in common.

   a. Construct and cut out an equilateral triangle. Then cut out 11 copies of the triangle.
   b. Use the triangles to make a tessellation. What is the sum of the measures of the angles that meet at any one point?
   c. Draw and cut out a quadrilateral. Label the angles A, B, C, and D. Make 11 copies of the quadrilateral.
   d. Use the quadrilaterals to make a tessellation. Be sure that there are no gaps or overlapping parts in the pattern. What angles meet at any one point?
   e. What is the sum of the measures of the angles that meet at any one point?

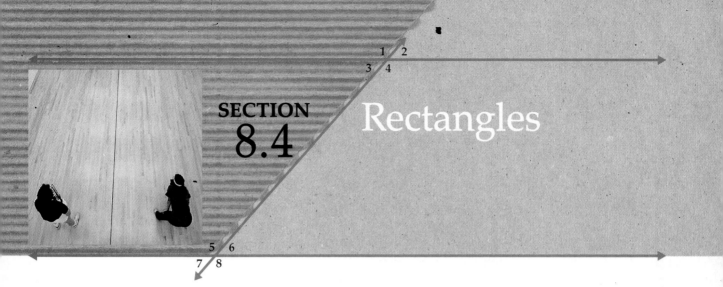

Rectangles

After completing this section, you should understand
▲ the properties of a rectangle.

The diagram below shows the set of quadrilaterals and the subsets of figures you have investigated thus far in this chapter. Also included, as a subset of parallelograms, is another figure, the rectangle.

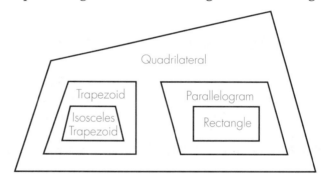

You have discovered that the following properties are true.

A **quadrilateral** has four sides and four angles.
The sum of the measures of the angles is 360°.

A **trapezoid** has one pair of parallel sides.
The median is parallel to both bases and equal to one-half the sum of the lengths of the bases.

A **parallelogram** has two pairs of parallel sides.
Both pairs of opposite angles have equal measures.
The diagonals bisect each other.

Rectangles possess several new properties in addition to the ones listed above.

DISCOVERY ACTIVITY

1. Draw a line and mark on it two points 6 centimeters apart. Label the line segment $\overline{AB}$.

2. Use a protractor to construct 90° angles at A and B. The rays of each angle should be on the same side of $\overline{AB}$.

3. Measure 2 centimeters up on the ray at point A and mark point D. Do the same on the ray at point B and mark point C.

4. Draw DC, then draw diagonals $\overline{AC}$ and $\overline{BD}$.

5. Does quadrilateral ABCD appear to be a parallelogram?

6. Measure the four sides and the four angles. What do you find?

7. Measure the diagonals. What do you find?

You may have discovered two new properties in the quadrilateral ABCD you constructed: the four angles have equal measures, and the diagonals are equal in length.

NOTEBOOK

DEFINITION 8-4-1   A RECTANGLE is a parallelogram with one right angle.

CLASS ACTIVITY

1. Why do you think the definition of a rectangle does not state that all four of the angles are right angles? Use your knowledge of the properties of a parallelogram to answer.

2. Justify that the diagonals of a rectangle are always equal in length. Complete the reasons for the steps.

   Given: Rectangle ABCD with diagonals $\overline{AC}$ and $\overline{BD}$

   Justify: AC = BD

Steps	Reasons
a. $AD = BC$	a.
b. $AB = AB$	b.
c. $m\angle ABC = m\angle BAD$	c.
d. $\triangle DAB \cong \triangle CBA$	d.
e. $AC = BD$	e.

NOTEBOOK

THEOREM 8-4-1 **If a quadrilateral is a rectangle, then the diagonals are equal in length and all four angles are right angles.**

Example: *ABCD* is a rectangle. *AB* = 12, *BC* = 5, and *AC* = 13. Find the following measures.

$AD = ?$ $\qquad$ $DC = ?$ $\qquad$ $BD = ?$ $\qquad$ $m\angle ADC = ?$

Example: *LMNO* is a rectangle. *LM* = 4 and *MN* = 3. Find *LN*.
Hint: What kind of a right triangle is △*LMN*?
LN = ?

CLASS ACTIVITY

JKLM is a rectangle.

1. What kind of triangle is △*JOK*? Why?
2. What kind of a triangle is △*MJK*? Why?
3. If *MO* + *OL* = 30, what is the measure of *MK*?
4. Name all pairs of congruent segments in rectangle *WXYZ*.

Study the information given for figures 5–7. Which of the parallelograms would be rectangles? (Assume that the information is correct even though the figure may not be accurately drawn.)

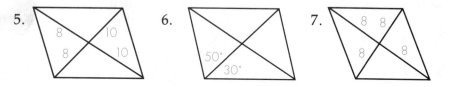

5. 6. 7.

Use a computer and a geometric drawing tool to complete Exercises 8–10.

8. Draw a parallelogram. Measure each angle. Repeat this activity several times, using different angles each time. Is every parallelogram a rectangle?
9. Draw a quadrilateral with one right angle. Measure the other three angles. Repeat this activity several times, using different angles each time. If a quadrilateral has one right angle, is it a rectangle?
10. Draw a parallelogram and its diagonals. Measure each diagonal. Repeat this activity several times, using different angle measures each time. If the diagonals of a parallelogram are equal in length, is it a rectangle?

11. Write steps and reasons to justify the following.

Given: WXYZ is a rectangle.

Justify: △WXO is isosceles.

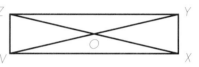

PROJECT 8-4-1 Rectangles and triangles are the geometric building blocks of architecture and construction. As you travel from home to school, take note of these quadrilaterals as you see them used in homes, playgrounds, office buildings, sports fields, bridges, and other structures.

Work with a group to create a display or collage illustrating the use of triangles, rectangles, trapezoids, and parallelograms in the architecture of your community. Use your own photographs and sketches, or pictures cut from local newspapers and magazines.

HOME ACTIVITY

ABCD is a rectangle. Find the length of the segments

1. AC
2. DC
3. DB
4. BC
5. DO
6. CO

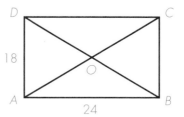

LMNO is a rectangle.

7. What kind of triangle is △OLN?

8. What is the length of LN?

9. What kind of triangle is △LPM?

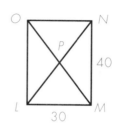

10. List the pairs of congruent triangles in rectangle *PQRS*.

△

11. Study the information given for figures a–d. Which of the parallelograms would be rectangles? (Assume that the information is correct even though the figure may not be drawn accurately.)

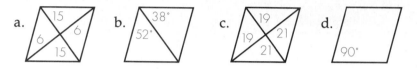

a.   b.   c.   d.

12. Write the steps and reasons to justify the following.

Given: *WXYZ* is a rectangle.
      WVXZ is a parallelogram.
Justify: △*WYV* is isosceles.

13. *ABCD* and *EFGH* are two rectangles. Find the sum of the measures of angles 1, 2, 3, and 4. (Hint: Use what you know about parallel lines and transversals.)

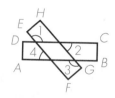

CRITICAL THINKING

The ancient Greeks considered the Golden Rectangle to be one of the most beautifully proportioned geometric forms, and used it in much of their architecture. All of the figures below are Golden Rectangles. What do they have in common?

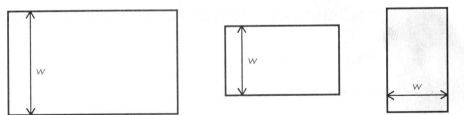

Measure the length and width of each rectangle in millimeters. Write the ratio of width (*w*) to length (*l*) as a decimal rounded to hundredths. What do you observe about the decimal ratio?
The ratio 0.61803 . . . is called the golden ratio.

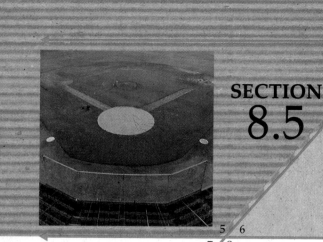

SECTION 8.5

Squares and Rhombuses

After completing this section, you should understand
▲ the properties of a rhombus.
▲ the properties of a square.

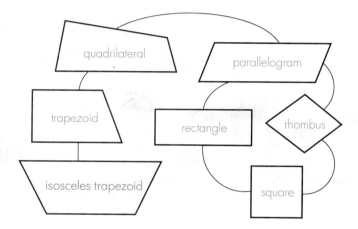

The diagram shows two new figures to investigate—rhombus and square. You can tell from the diagram that a rhombus will have all of the properties of a parallelogram, and a square will have properties common to both the rectangle and rhombus.

DISCOVERY ACTIVITY

1. Use four strips of cardboard equal in length and fasten them at the endpoints with brads to form figure *ABCD*. Be sure the figure is flexible.

2. Place the figure on a piece of paper, holding it down on $\overline{AB}$. Move point *C* back and forth as far as it will go in each direction.

3. Notice the position of the opposite sides as you move C back and forth. In every position, except when $\overline{DC}$ is on $\overline{AB}$, the figure forms a rhombus. Why is a rhombus a parallelogram?

4. Now hold a pencil point at C, still holding the figure firmly on AB. As you move point C, use the pencil to trace the path of point C. What do you observe about the line you traced?

This activity may have shown you that the sides of a rhombus are always parallel as well as equal in length.

DEFINITION 8-5-1   **A RHOMBUS is a parallelogram with sides of equal length.**

CLASS ACTIVITY

ABCD is a rhombus.

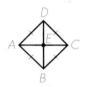

1. Does $DE = EB$? Does $AE = EC$? Why?

2. Which triangles in ABCD are congruent? How do you know?

From the definition you know that a rhombus is a parallelogram with four sides of equal length. There is another property that distinguishes the rhombus.

DISCOVERY ACTIVITY

1. Draw three different rhombuses and label each one ABCD.
2. Draw diagonals $\overline{AC}$ and $\overline{BD}$. Label the intersection O.
3. With a protractor, measure the four angles formed at the intersection of the diagonals. Do this for each figure and record your measurements.
4. Write a conclusion to summarize your results.

You may have discovered that, since all four angles at the intersection of the diagonals are right angles, the diagonals of a rhombus are perpendicular to each other.

THEOREM 8-5-1

If a parallelogram is a rhombus, then the diagonals are perpendicular.

Given: $ABCD$ is a rhombus with diagonals $\overline{AC}$ and $\overline{BD}$ intersecting at O.

Justify: $\overline{AC} \perp \overline{BD}$ and $m\angle AOD = 90°$

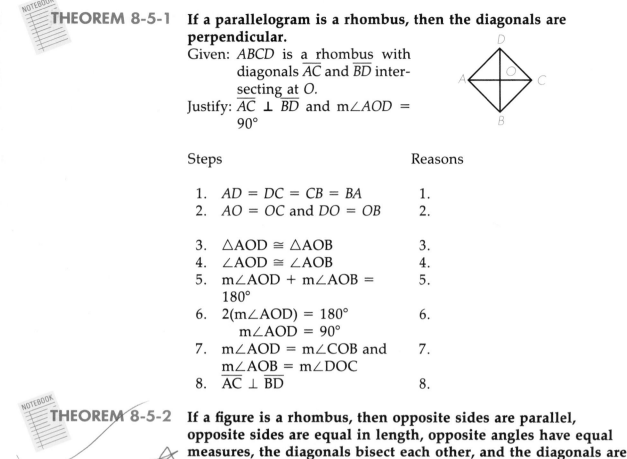

Steps	Reasons
1. $AD = DC = CB = BA$	1.
2. $AO = OC$ and $DO = OB$	2.
3. $\triangle AOD \cong \triangle AOB$	3.
4. $\angle AOD \cong \angle AOB$	4.
5. $m\angle AOD + m\angle AOB = 180°$	5.
6. $2(m\angle AOD) = 180°$ $\quad m\angle AOD = 90°$	6.
7. $m\angle AOD = m\angle COB$ and $m\angle AOB = m\angle DOC$	7.
8. $\overline{AC} \perp \overline{BD}$	8.

THEOREM 8-5-2

If a figure is a rhombus, then opposite sides are parallel, opposite sides are equal in length, opposite angles have equal measures, the diagonals bisect each other, and the diagonals are perpendicular.

SQUARES

From the tree diagram at the beginning of this section you can see that a square is both a rectangle and a rhombus.

DEFINITION 8-5-2

A SQUARE is a rectangle with sides that are equal in length.

DISCOVERY ACTIVITY

1. Construct two squares, each with sides of a different length. Recall that since a square is a rectangle, each of its angles measures 90°.

2. Draw the diagonals for each square.

3. With a protractor, measure the angles formed by the intersection of the diagonals. What do you observe?

You may have found that the diagonals of a square are perpendicular.

THEOREM 8-5-3 **If a figure is a square, then opposite sides are parallel, opposite sides are equal in length, there are four right angles, the diagonals are equal in length, the diagonals bisect each other, and the diagonals are perpendicular.**

CLASS ACTIVITY

Given: Square $ABCD$ with diagonals $\overline{AC}$ and $\overline{BD}$ intersecting at O; $AB = 5$ cm

Find the following measures. Use a calculator when necessary.

1. $AD = ?$ 2. $BC = ?$ 3. $CD = ?$ 4. $AC = ?$

5. $BD = ?$ 6. $OA = ?$ 7. $OD = ?$ 8. $OC = ?$

Name the quadrilateral described by each set of diagonals.

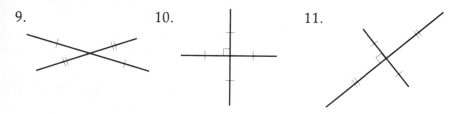

9.   10.   11.

12. Use a computer and a geometric drawing tool to draw a quadrilateral that is not a parallelogram. Find and connect the midpoints of adjacent sides. What kind of quadrilateral has been formed?

HOME ACTIVITY

1. Copy and complete the following table. Mark an X in the column under each figure if the property is true for that figure.

Properties of Quadrilaterals					
Property	Parallelogram	Rectangle	Square	Rhombus	Trapezoid
Both pairs of opposite sides equal.	X	X	X	X	
All sides are equal.					
Both pairs of opposite sides parallel.					
Each pair of opposite angles equal.					
All angles right.					
Diagonals bisect each other.					
Diagonals are equal.					
Diagonals are perpendicular.					
Adjacent angles are supplementary.					
Two sides are parallel.					
Sum of measures of angles = 360°.					
Diagonals bisect opposite angles.					

Name the quadrilateral described by each set of diagonals.

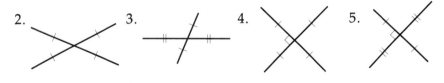

2. 3. 4. 5.

CRITICAL THINKING

6. The game of checkers is played on a square game board that is divided into 64 small squares, 8 on a side. How many squares are there on a checkerboard?

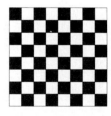

The total cannot be 64, since you can immediately see 65 squares—64 small squares and 1 large square—the checkerboard itself.

On a board with only one square, the number of squares is 1.

On a board that is a 2 by 2 array of squares, the total number of squares is 5: 1 + 4.

On a 3 by 3 board, the total number of squares is 13: 1 + 4 + 9. What is being added in each case?

Use inductive reasoning to predict the total number of squares for the 8 by 8 checkerboard. Show your work.

8.1 A quadrilateral is a closed, plane, four-sided figure. It may be convex or concave. The sum of the measures of the interior angles of a quadrilateral as 360°. What is the measure of $\angle D$?

$$50 + 108 + 92 + x = 360$$
$$250 + x = 360$$
$$x = 110 \qquad m\angle D = 110°$$

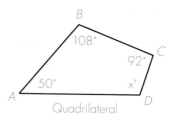

Quadrilateral

8.2 A trapezoid is a quadrilateral with one pair of parallel sides. An isosceles trapezoid has equal non-parallel sides and equal base angles. The median of a trapezoid joins the midpoints of the non-parallel sides, is parallel to the bases, and has a length that is half the sum of the lengths of the two bases. What is the length of JG?

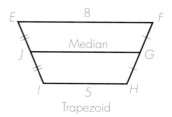

$$JG = \tfrac{1}{2}(8 + 5)$$
$$JG = 6.5$$

Trapezoid

8.3 A parallelogram is a quadrilateral with two pairs of opposite sides parallel. Opposite sides and opposite angles of a parallelogram are equal. What is the measure of $\angle L$?

$$m\angle L = m\angle N = 122°$$
$$m\angle L = 122°$$

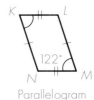

Parallelogram

8.4 A rectangle is a parallelogram with one right angle. A rectangle has diagonals equal in length. The diagonals bisect each other. $SQ = 13$. What is the length of TR?

$$PR = SQ = 13$$
$$TR = \tfrac{1}{2}PR = \tfrac{1}{2}(13) = 6.5$$

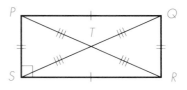

Rectangle

8.5 A rhombus is a parallelogram with all four sides equal in length. In a rhombus the diagonals intersect at right angles and bisect each other. A square is a rectangle with all four sides equal in length. The diagonals of a square are equal and bisect each other at right angles. If $UV = 5$ and $AC = 8$, what are the lengths of VW and ED?

$$VW = UV = 5 \qquad\qquad BD = AC = 8$$
$$VW = 5 \qquad\qquad ED = \tfrac{1}{2}(BD) = \tfrac{1}{2}(8) = 4$$

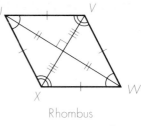

Rhombus

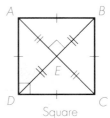

Square

COMPUTER ACTIVITY

You can use Logo to draw a parallelogram on the x–y plane when given just the coordinates of three of the vertices.

Example: The coordinates of three of the vertices of $\square ABCD$ are $A(0,0)$, $B(2,2)$, and $C(5,2)$. Find the coordinates of vertex D. Then write a procedure using the procedure XYPLANE from page 118 and the command SETXY to draw the parallelogram.

First, graph the coordinates for the given vertices on the x–y plane. Then, use the fact that the opposite sides of a parallelogram are equal and the distance formula to find the coordinates of D. $\overline{BC}$ and $\overline{AD}$ are opposite sides of the parallelogram, so $BC = AD$. $BC = |5 - 2| = 3$ so $AD = 3$. Thus, the coordinates of D must be $(3,0)$.

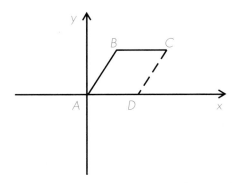

The Logo procedure below tells the turtle to draw this parallelogram.

```
TO PARA
    XYPLANE
    SETXY 2 2
    SETXY 5 2
    SETXY 3 0
    SETXY 0 0
END
```

Modify the procedure PARA from the example above to tell the turtle to draw each parallelogram.

1.

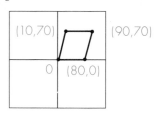

2.

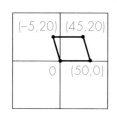

3.

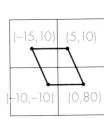

4.

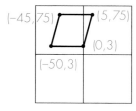

The coordinates of three of the vertices of ▱ABCD are given in each exercise below. Find the coordinates of vertex D. Then modify the procedure PARA to tell the turtle to draw each parallelogram.

5. A(0,0), B(6,12), C(15,12)

6. A(2,4), B(7,8), C(10,8)

7. A(5,−1), B(7,12), C(13,12)

8. A(−4,2), B(−6,−3), C(1,−3)

Use the procedure XYPLANE in procedures that tell the turtle to draw each figure. Give each procedure the indicated name.

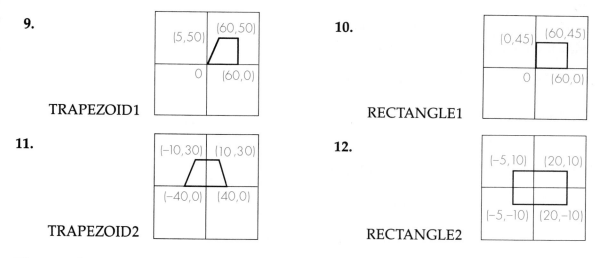

9. TRAPEZOID1

10. RECTANGLE1

11. TRAPEZOID2

12. RECTANGLE2

The coordinates of three of the vertices of the specified quadrilateral are given below. Find the coordinates of the fourth vertex. Then modify the indicated procedure from Exercises 9–12 to tell the turtle to draw each quadrilateral.

13. (0,4), (6,2), (2,2); TRAPEZOID1

14. (3,5), (8,5), (8,2); RECTANGLE1

15. (−3,2), (4,7), (6,7); TRAPEZOID2

16. (−4,−1), (−4,2), (3,−1); RECTANGLE2

Refer to quadrilateral *PQRS*.

   1. Name one pair of opposite sides.
   2. Name one pair of adjacent angles.
   3. m∠*P* + m∠*Q* + m∠*R* + m∠*S* = ?

Is the figure a parallelogram? Give a reason for your answer.

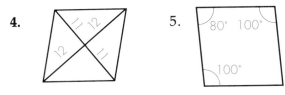

4.   **5.**   **6.**   **7.**

$\overline{EF}$ is the median of trapezoid *ABCD*. *BC* = 18.
Find the following measures.

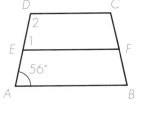

   8. *BF* = ?   **9.** m∠1 = ?   **10.** m∠2 = ?

11. Do you know that *AD* = 18? Why or why not?

JKLM is a parallelogram. Find the following measures.

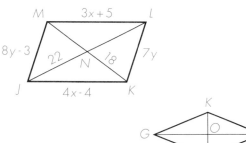

12. *JK*   **13.** *ML*   **14.** *KL*

15. *MJ*   **16.** *NL*   **17.** *NM*

18. Name four pairs of congruent triangles in rhombus *GHJK*.

Find the indicated angle measures.

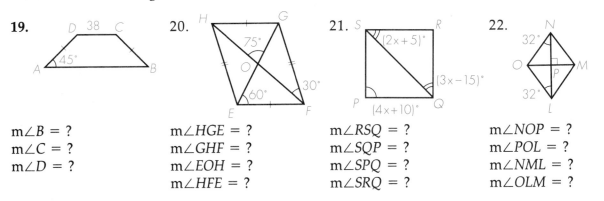

19.   **20.**   **21.**   **22.**

m∠*B* = ?   m∠*HGE* = ?   m∠*RSQ* = ?   m∠*NOP* = ?
m∠*C* = ?   m∠*GHF* = ?   m∠*SQP* = ?   m∠*POL* = ?
m∠*D* = ?   m∠*EOH* = ?   m∠*SPQ* = ?   m∠*NML* = ?
        m∠*HFE* = ?   m∠*SRQ* = ?   m∠*OLM* = ?

23. A rectangular garden measures 30 feet by 40 feet. How many feet of garden hose are needed to reach from one corner to the opposite corner?

Which of the following figures are

1. quadrilaterals?
2. parallelograms?
3. rectangles?
4. trapezoids?
5. squares?
6. none of these?

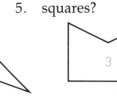
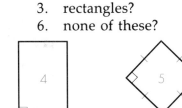

ABCD is a trapezoid.

7. Name one pair of opposite sides and one pair of adjacent sides.
8. Name one pair of opposite angles and one pair of adjacent angles.
9. If *ABCD* is an isosceles trapezoid, what is true for m∠A and m∠B?
10. m∠A + m∠D = ?
11. If $\overline{EF}$ is the median, what is true for $\overline{AE}$ and $\overline{ED}$?
12. If *AB* = 15 and *DC* = 12, what is the length of $\overline{EF}$?

Is the figure a parallelogram? Give a reason for your answer.

13. 14. 15. 16.

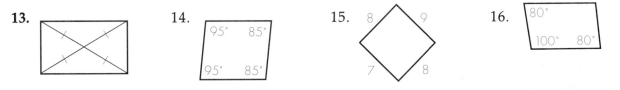

Find the indicated lengths or angle measures for each figure.

17. Rectangle

18. Isosceles trapezoid

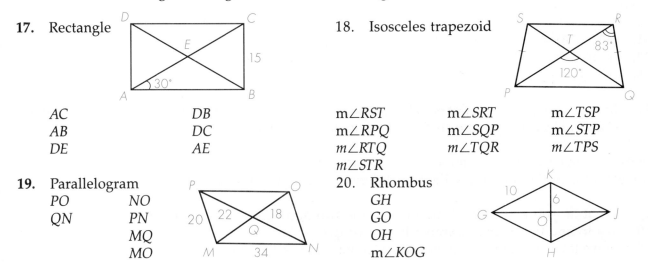

AC	DB
AB	DC
DE	AE

m∠RST	m∠SRT	m∠TSP
m∠RPQ	m∠SQP	m∠STP
m∠RTQ	m∠TQR	m∠TPS
m∠STR		

19. Parallelogram

PO	NO
QN	PN
	MQ
	MO

20. Rhombus

GH
GO
OH
m∠KOG

ALGEBRA SKILLS

To add or subtract expressions containing radicals, first simplify the expression, then combine like terms.

Example: Find $x + y$ where $x = 4 + \sqrt{12}$ and $y = 2 - \sqrt{27}$.
First simplify terms. $x = 4 + \sqrt{12} = 4 + \sqrt{4 \cdot 3} = 4 + 2\sqrt{3}$
$$y = 2 - \sqrt{27} = 2 - \sqrt{9 \cdot 3} = 2 - 3\sqrt{3}$$
So, $x + y = 4 + 2\sqrt{3} + 2 - 3\sqrt{3} = 6 - \sqrt{3}$

Multiply radical expressions the same way you multiply binomials.

Example:
$$
\begin{aligned}
xy &= (4 + 2\sqrt{3})(2 - 3\sqrt{3}) \\
&= 4(2 - 3\sqrt{3}) + 2\sqrt{3}(2 - 3\sqrt{3}) \\
&= 8 - 12\sqrt{3} + 4\sqrt{3} - 6 \cdot 3 \\
&= 8 - 8\sqrt{3} - 18 \\
&= -10 - 8\sqrt{3}
\end{aligned}
$$

When you divide radical expressions, you may have to rationalize the denominator—that is, rewrite the denominator without the radical. Use the idea that $(a - b)(a + b) = a^2 - b^2$.

Example: $x/y = \dfrac{4 + 2\sqrt{3}}{2 - 3\sqrt{3}} \cdot \dfrac{2 + 3\sqrt{3}}{2 + 3\sqrt{3}} = \dfrac{8 + 4\sqrt{3} + 12\sqrt{3} + 18}{4 - 27}$

$x/y = \dfrac{26 + 16\sqrt{3}}{-23}$

Let $x = 1 + \sqrt{8}$ and $y = 3 - \sqrt{18}$. Evaluate the following expressions.

1. $x + y$
2. $x - y$
3. xy
4. $y - x$
5. x/y
6. y/x
7. $(x - y)^2$
8. $(x + y)^2$

The following expressions represent geometric figures when they are graphed on the number line. Draw the graph and name the figure.

9. $x \geq 2$
10. $3x \leq 12$
11. $3(2x - 6) = -4(3x - 6)$
12. $x \leq 5$ and $x \geq -5$
13. $|x| \leq 5$
14. $|x - 5| = 0$
15. $|x - 10| \leq 5$
16. $|x| \leq |x|$

17. In a school of 850 students, 62% registered for math classes and 175 signed up for music classes. How many more students took math than music?

Suppose you begin with a number. You subtract 2 from your number. Then you subtract 8 from the original number. You multiply the two differences and get a product of -9.

18. What do you know about the signs of the two numbers you multiplied? Why?
19. Try to find the original number by guessing.
20. Solve for the original number using algebra. Let $n =$ the number.

314

9

PERIMETER AND AREA

To measure is to know.

Johannes Kepler

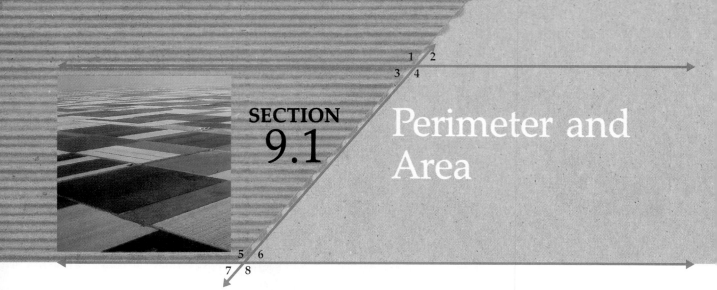

SECTION 9.1
Perimeter and Area

After completing this section, you should understand
▲ the definitions of perimeter and area.
▲ how to find the perimeter and area of a rectangle.

In Chapter 8 you investigated the properties that define different geometric figures. Since all of these figures are part of the structure of the natural and artificial world we live in, it is important to understand how to find the amount of space they occupy.

DISCOVERY ACTIVITY

Below are some different geometric figures.

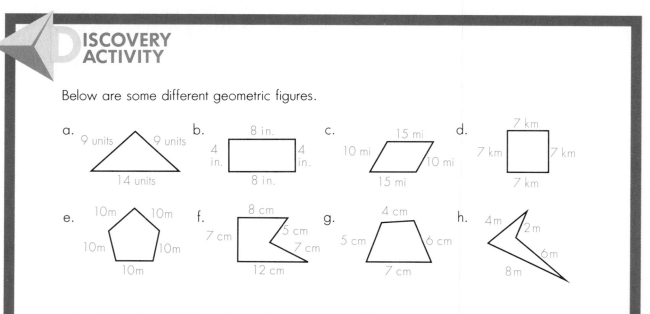

1. What do all of these figures have in common?
2. Suppose figure a is a fenced-in piece of land. Without the use of measuring tools, how could you find the distance around it?
3. How can you find the distance around each figure using the measurements given? Find this distance for figures a–h.

You may have discovered that you can find the distance around any polygon by adding the lengths of the sides.

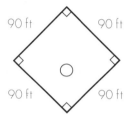

DEFINITION 9-1-1   **The PERIMETER of a figure is the distance around the figure or the sum of the lengths of all the sides.**

The word perimeter comes from the Greek words for distance, *meter*, and around, *peri*.

Example: How far do you have to run when you hit a home run in a baseball game?

Perimeter = 90 ft + 90 ft + 90 ft + 90 ft = 360 ft

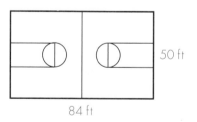

Example: What is the perimeter (*P*) of a basketball court?

P = 84 ft + 50 ft + 84 ft + 50 ft = 268 ft

50 ft

84 ft

AREA

Any closed plane figure has an interior region. It is possible to assign a number called *area* to describe this interior region. The shaded regions are the interiors of these polygons.

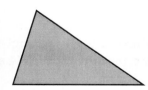

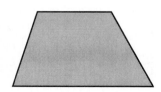

DISCOVERY ACTIVITY

1. Count the number of squares in the interior of each figure.

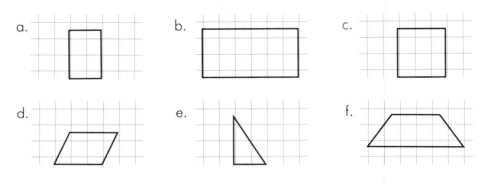

a.   b.   c.

d.   e.   f.

2. In figures a–c, it is easy to count squares. How did you count squares for figures d–f?

3. There is a pattern to the number of squares you counted in figures a–c. Copy and complete the following table.

Figure	Number of Interior Squares	Measure of Sides
a		
b		
c		

4. What do you observe?

You may have found that multiplying the length times the width of a rectangular figure gives the number of square units found in the interior. This number is the *area* of the figure.

DEFINITION 9-1-2

The AREA of a plane figure is the number of square units contained in the interior.

The figures a–c above are all rectangles. The table you completed shows a pattern that can be stated as a formula.

POSTULATE 9-1-1

If the figure is a rectangle, then the number assigned for the area is the length times the width.

Area is always expressed in square units. If the unit of measure is feet, the area is expressed as feet times feet, or square feet.

Example: What is the area of a basketball court?

The court is a rectangle.

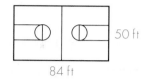

50 ft

84 ft

So, $A = lw$
$A = 84 \text{ ft} \times 50 \text{ ft} = 4200 \text{ sq ft}$

CLASS ACTIVITY

Find the area of the rectangle.

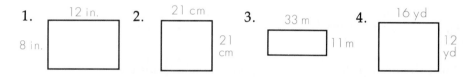

1. 12 in. / 8 in.
2. 21 cm / 21 cm
3. 33 m / 11 m
4. 16 yd / 12 yd

Find the length and width of each rectangle.

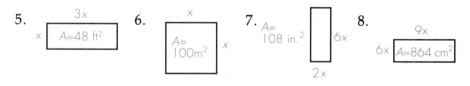

5. 3x / x / A=48 ft²
6. x / x / A=100m²
7. A=108 in.² / 6x / 2x
8. 9x / 6x / A=864 cm²

Find the area of the shaded region for each rectangle.

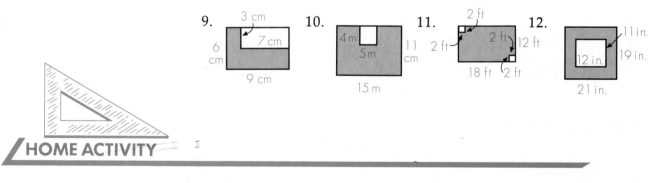

9. 3 cm / 7 cm / 6 cm / 9 cm
10. 4m / 5m / 11 cm / 15 m
11. 2 ft / 2 ft / 12 ft / 18 ft / 2 ft
12. 2 ft / 11 in. / 12 in. / 19 in. / 21 in.

HOME ACTIVITY

Find the perimeter.

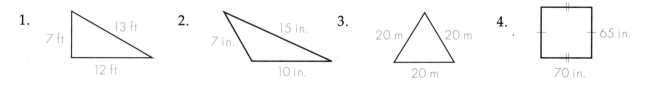

1. 7 ft / 13 ft / 12 ft
2. 7 in. / 15 in. / 10 in.
3. 20 m / 20 m / 20 m
4. 65 in. / 70 in.

Write a LOGO procedure to tell the turtle to draw a rectangle that has each perimeter. Then modify the procedure to draw another rectangle with the same perimeter.

5.   20   6.   100   7.   14

Find the area of the rectangle.

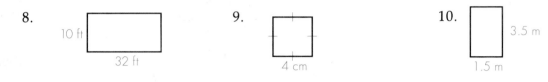

8. 10 ft, 32 ft

9. 4 cm

10. 3.5 m, 1.5 m

Find the length and width of each rectangle.

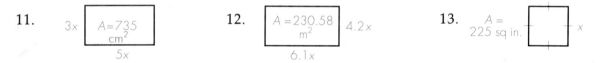

11. $3x$, $A = 735$ cm^2, $5x$

12. $A = 230.58$ m^2, $4.2x$, $6.1x$

13. $A = 225$ sq in., x

Find the area of the shaded region of each rectangle.

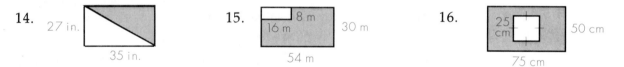

14. 27 in., 35 in.

15. 8 m, 16 m, 30 m, 54 m

16. 25 cm, 50 cm, 75 cm

A rectangular backyard measures 150 ft by 75 ft.

The owner of the property wants to sod the yard and then fence it.

17.   Sod is sold for $1.55 per sq yd. How much will it cost to sod the yard?

18.   Fencing costs $3.75 per foot. How much will it cost to fence the yard?

CRITICAL THINKING

A rectangle has a length of 20 units and a width of 15 units.

19.   What is the area? What is the perimeter?
20.   What is the area if you double the length and width? What is the perimeter?
21.   What happens to the area and perimeter when the dimensions of the rectangle are doubled?
22.   What do you think will happen to the area and perimeter if the length and width are halved?

SECTION 9.2

Perimeter and Area of a Triangle

Precal
Book
Pg 25
+ Pg 352

After completing this section, you should understand
▲ how to find the perimeter and area of a triangle.
▲ how to solve problems involving the area of a triangle.

Very often, the shape of a piece of land or of a building is made up of a group of triangles. In Section 9.1, you found the area of some triangular figures by counting the number of squares contained in the figure. This section will investigate an easier way to find the area of a triangle.

DISCOVERY ACTIVITY

1. On graph paper, draw a triangle with sides of 3 and 4 units.

2. What is the length of the third side? How do you know?

3. Notice that you cannot make an exact count of the number of squares in the interior of the triangle. Instead, make the triangle into a rectangle.

4. What is the area of the rectangle?

5. How many congruent triangles does the rectangle contain?

6. What is the area of one of the triangles?

7. What is the relationship between the area of the rectangle and the area of a right triangle?

8. Now suppose that you want to find the area of a triangle that is not a right triangle. On graph paper, draw an acute triangle and an obtuse triangle, each with sides having lengths a, b, and c.

9. Draw congruent triangles with sides a, b, and c next to the original triangles so that they form parallelograms.

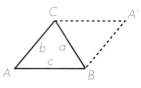
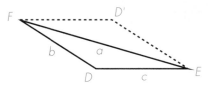

10. In each figure, draw an altitude to base, forming triangles (△CHA and △D'HE). Slide the triangles to the left or right, so that they transform the parallelograms into rectangles.

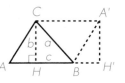
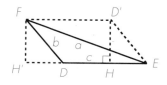

11. What is the area of the rectangle?

12. What can you conclude about the area of the triangle?

The rectangles you formed are made up of the two original triangles. Since the area of a rectangle is length times width or base times height, you may have concluded that the area of the triangle is $\frac{1}{2}bh$.

THEOREM 9-2-1 If the figure is a triangle, then the number assigned to the area is $\frac{1}{2}$ the base times the height.

Example: What is the area of △NMP?

$A = \frac{1}{2} \cdot b \cdot h$
$\overline{PO}$ is the altitude or height. So,
$A = \frac{1}{2}(10 \cdot 4) = 20$ cm²

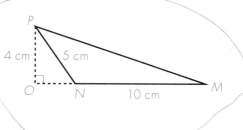

Example: $(\triangle PQR) = 272$ cm². $PQ = 34$ cm. What is the length of the altitude $\overline{RS}$?

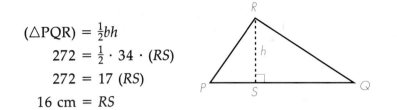

$(\triangle PQR) = \frac{1}{2}bh$

$272 = \frac{1}{2} \cdot 34 \cdot (RS)$

$272 = 17\ (RS)$

$16\text{ cm} = RS$

CLASS ACTIVITY

Find the area.

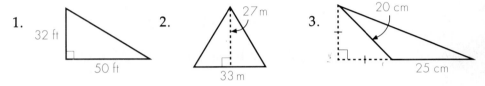

1. 32 ft 50 ft

2. 27 m 33 m

3. 20 cm 25 cm

4. The diagram shows a plan for a large garden. The triangular area is to be planted with roses. What is the area of the rest of the garden?

35 ft 7 ft 36 ft 12 ft 72 ft

Find the missing altitude or base for each triangle.

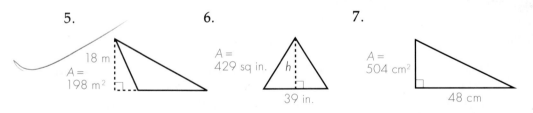

5. 18 m $A = 198$ m²

6. $A = 429$ sq in. h 39 in.

7. $A = 504$ cm² 48 cm

AREAS OF SPECIAL TRIANGLES

In Chapter 7, you studied several triangles with special properties.

You can use the properties of these triangles to find the area or a missing base or altitude measurement.

Example: What is the area of $\triangle ABC$?

You need to find the lengths of the base and altitude. Since $m\angle B = m\angle C = 45°$, you know that $\triangle ABC$ is an isosceles right triangle. The hypotenuse $BC = 5\sqrt{2}$, so $AB = 5$ cm and $AC = 5$ cm.
Area $\triangle ABC = \frac{1}{2}(5 \cdot 5) = \frac{1}{2} \cdot 25 = 12.5$ cm²

Example: What is the area of $\triangle LMN$?

Altitude $\overline{NL} = 3$ m. What is the length of the base?
$\triangle LMN$ is a 30°–60° right triangle, so $LM = NL\sqrt{3}$ or $3\sqrt{3}$ m.
Area $\triangle LMN = \frac{1}{2}(3\sqrt{3} \cdot 3)$ or about 7.79 m²

CLASS ACTIVITY

Find the area.

1. 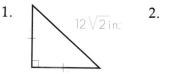 12√2 in.

2. 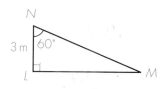 15 m

3. 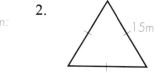 60° 14 ft. 7√3 ft

4. A farmer's field is fenced off in the shape of an isosceles right triangle. The lengths of the sides are shown. How much fence did it take to enclose the field? What is the area of the field?

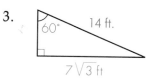
38 yd 38 yd 54 yd

5. Place a meter stick or a long piece of string diagonally across your desktop. Estimate the area in square inches of one of the triangles formed. Then measure the base and altitude and find the area. Was your estimate reasonable?

THE PERIMETER OF A TRIANGLE

You learned in Section 9.1 how to find the distance around an area. The perimeter of an area is the sum of the lengths of the sides.

The perimeter of a figure should be expressed in an appropriate unit of measure.

Example: The perimeter of a field is 300 yards.
Since 1 yd = 3 ft, it is also true that $P = 900$ ft.
Since 1 ft = 12 in., the perimeter can also be expressed as 10,800 in., but this measurement would probably not be useful.

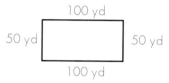

CLASS ACTIVITY

Find the perimeter.

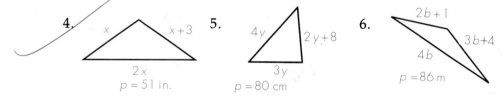

1.

2.

3.

Find the missing measurements.

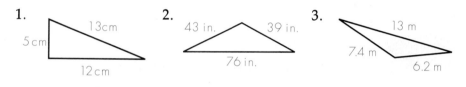

4.

5.

6.

7. A pyramid-shaped sculpture at the left has three triangular sides. All of the edges, including the base, are to be outlined with reflective tape to illuminate the sculpture at night. How many feet of tape will be needed?

8. The perimeter of a triangle is 150 meters. What is the perimeter in centimeters?

Refer to $\triangle ABC$.

9. What is the length of $\overline{BC}$?

10. What is the perimeter of $\triangle ABC$?

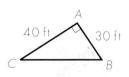

HOME ACTIVITY

Find the area.

1.
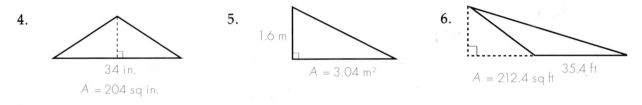

12 ft
16 ft

2.
35 m
48.5 m

3.
30 cm

Find the missing altitude or base.

4.
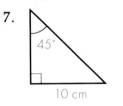

34 in.
A = 204 sq in.

5.
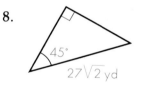

1.6 m
A = 3.04 m²

6.
A = 212.4 sq ft   35.4 ft

Find the area. The triangles are isosceles right triangles.

7.
45°
10 cm

8.
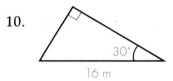

45°
27√2 yd

Find the area. The triangles are 30°–60° right triangles.

9.
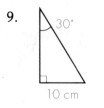

30°
10 cm

10.
30°
16 m

Find the perimeter.

11.
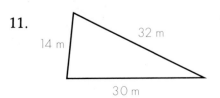

14 m   32 m
30 m

12.
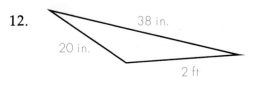

38 in.
20 in.
2 ft

Write a LOGO procedure to tell the turtle to draw a triangle that has each area. Then modify the procedure to draw another triangle with the same area.

13. 24   14. 72   15. 144

16. Find the perimeter and area of *PQRS*.

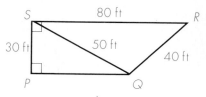

The diagram shows a floor plan for a log cabin.

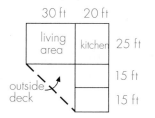

17. What is the total indoor and outdoor deck area of the cabin?

18. What is the area of the deck?

19. The drawing shows a driveway leading to a 3-car garage. What are the area and perimeter of the driveway?

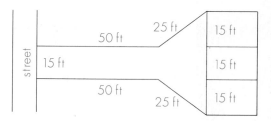

CRITICAL THINKING

Lauren said that her room had a perimeter of 528 inches.

20. Do you think this a reasonable size for a room?
21. What would be a more meaningful expression of perimeter?

22. Can you tell the shape of the room from knowing only the perimeter?
23. Assume that the room is rectangular. Give two possible sets of measurements for the length and width.

24. What are the measurements if the room is a square?

SECTION 9.3

Perimeter and Area of a Parallelogram

After completing this section, you should understand
▲ how to find the perimeter and area of a parallelogram.

In learning how to find the area of a triangle, you saw that two cases were considered: the right triangle, in which the altitude was one of the sides of the triangle, and the non-right triangle, in which you had to determine the altitude.

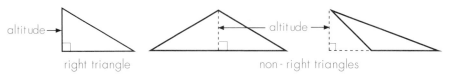

altitude →   ← altitude →   ← altitude →

right triangle   non-right triangles

Now you will investigate finding the area of any parallelogram. Recall that the set of parallelograms contains several different figures.

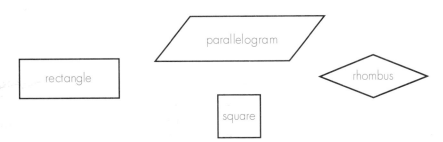

parallelogram

rectangle

rhombus

square

So a method for finding the area of a parallelogram will also give a method for finding the area of a rectangle, rhombus, or square.

DISCOVERY ACTIVITY

1. On graph paper, draw a large parallelogram. Label it *ABCD*.

2. Using a compass or protractor, construct a line segment perpendicular to $\overline{AB}$ from point D. Label the point of intersection E.

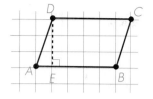

3. Cut off $\triangle ADE$ and place it at the other side of the parallelogram, so that $\overline{AD}$ lies on $\overline{BC}$.

4. What kind of figure have you formed? How can you find the area?

5. Write a conclusion telling what you have learned about the area of a parallelogram.

You may have discovered from the activity that the area of a parallelogram is the same as the area of a rectangle with congruent base and altitude.

THEOREM 9-3-1 **If the figure is a parallelogram, then the area is the length of the base times the height.**

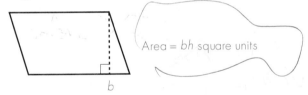

Area = bh square units

Notice that when you refer to the area of parallelograms, you use the terms _base_ and _height_ instead of _length_ and _width_. A parallelogram has two pairs of parallel bases. Once you have chosen a base, the corresponding height or altitude is a line segment perpendicular to that base, with its endpoint on the opposite side.

Example: What is the area of *GHJK*?

$A = bh$
$A = 18 \text{ cm} \cdot 9 \text{ cm} = 162 \text{ cm}^2$

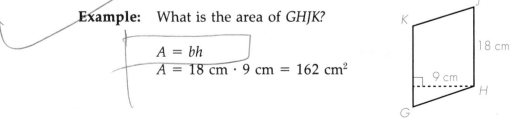

CLASS ACTIVITY

Find the area of each parallelogram.

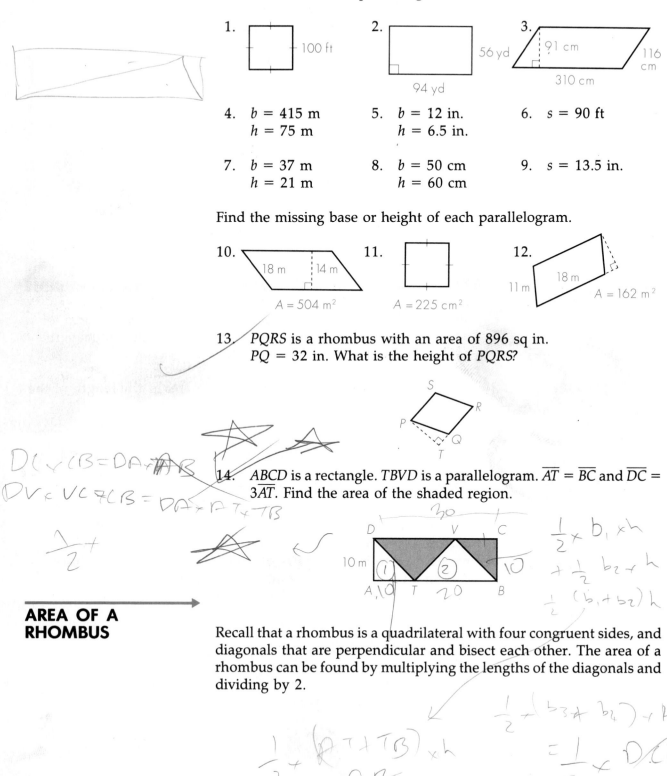

1. 100 ft

2. 56 yd, 94 yd

3. 91 cm, 116 cm, 310 cm

4. $b = 415$ m
   $h = 75$ m

5. $b = 12$ in.
   $h = 6.5$ in.

6. $s = 90$ ft

7. $b = 37$ m
   $h = 21$ m

8. $b = 50$ cm
   $h = 60$ cm

9. $s = 13.5$ in.

Find the missing base or height of each parallelogram.

10. 18 m, 14 m, $A = 504$ m²

11. $A = 225$ cm²

12. 18 m, 11 m, $A = 162$ m²

13. PQRS is a rhombus with an area of 896 sq in. $PQ = 32$ in. What is the height of PQRS?

14. ABCD is a rectangle. TBVD is a parallelogram. $\overline{AT} = \overline{BC}$ and $\overline{DC} = 3\overline{AT}$. Find the area of the shaded region.

10 m

AREA OF A RHOMBUS

Recall that a rhombus is a quadrilateral with four congruent sides, and diagonals that are perpendicular and bisect each other. The area of a rhombus can be found by multiplying the lengths of the diagonals and dividing by 2.

To show that this method works, first try it for a square with sides equal to 1 unit.

From the Pythagorean Theorem, $c^2 = a^2 + b^2$,

$$\text{so } AC^2 = 1^2 + 1^2 = 2,$$
$$AC = \sqrt{2}.$$
Since $AC = BD$, $DB = \sqrt{2}$.

$$\tfrac{1}{2}(AC \cdot BD) = \tfrac{1}{2} \cdot \sqrt{2} \cdot \sqrt{2} = 1$$

From the theorem of the area of a parallelogram, you know that $A = bh$, so for square $ABCD$, $= 1 \cdot 1 = 1$.

Both methods give the same answer for the area of a square. Now we justify the method for any rhombus.

> Given: Rhombus $ABCD$ with diagonals d_1 and d_2
> $DB = d_1$, $AC = d_2$
> Justify: Area $ABCD = \tfrac{1}{2}d_1 d_2$

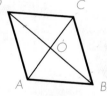

Supply the missing reasons.

Steps	Reasons
1. $A(ABCD) = A(\triangle ADC) + A(\triangle ABC)$	1.
2. $DO = OB = \tfrac{1}{2}d_1$	2.
3. $d_1 \perp d_2$, so $\tfrac{1}{2}d_1$ is an altitude of both $\triangle ADC$ and $\triangle ABC$.	3.
4. $A(\triangle ADC) = \tfrac{1}{2}d_2 \cdot \tfrac{1}{2}d_1 = \tfrac{1}{4}d_2 d_1$ $A(\triangle ABC) = \tfrac{1}{2}d_2 \cdot \tfrac{1}{2}d_1 = \tfrac{1}{4}d_2 d_1$	4.
5. $A(ABCD) = \tfrac{1}{4}d_1 d_2 + \tfrac{1}{4}d_1 d_2$ $= \tfrac{1}{2}d_1 d_2$	5.

PERIMETER OF A PARALLELOGRAM

If quadrilateral *ABCD* is a parallelogram with *AB* = 16 m and *BC* = 12 m, you know that $\overline{DC}$ and $\overline{AD}$ are equal to the lengths of the opposite sides.

Then from the definition of perimeter,

$P = \ell + w + \ell + w$, or $P = 2\ell + 2w$.

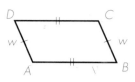

THEOREM 9-3-2 **If the figure is a parallelogram, then the perimeter is twice the sum of two adjacent sides.**

Example: What is the perimeter of *STUV*?

$P = 2\ell + 2w$

$P = 2(47 \text{ yd}) + 2(20 \text{ yd}) = 134 \text{ yd}$

Example: *ABCD* is a rhombus with *AB* = 80 in.

What is the perimeter?

$P = 4s = 4(80 \text{ in.}) = 320 \text{ in.}$

CLASS ACTIVITY

Find the perimeter of the parallelogram.

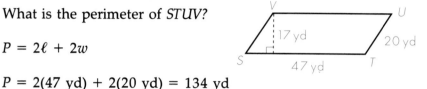

1. 100 ft

2. 101 m   75 m

3. 75 yd   100 yd   165 yd

4. 50 cm   75 cm

5. 90 ft

6. 12 in.   24 in.

Find the missing length or width of the parallelogram.

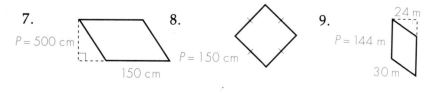

7. $P = 500 \text{ cm}$ 150 cm

8. $P = 150 \text{ cm}$

9. $P = 144 \text{ m}$ 24 m   30 m

PROJECT 9-3-1

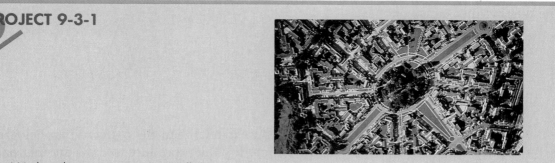

Work with a partner or a small group.
1. Decide on a subject for your map. Choose several blocks in an area of your community that you know well, one level of your school building, a small park, or another accessible location.
2. Decide on an appropriate unit of measure.
3. Measure the area by pacing it off or by using a measuring tool. Be sure to find the perimeters of all structures or objects you plan to include in the map.
4. Decide on a size for your finished map. Choose a scale.
5. Transfer the data to the map. Check the accuracy of measurements on the map by using the scale. For example, a scale might be 1 cm = 10 m or $\frac{1}{2}$ in. = 10 yd.
6. Title the map. Identify what you have included (label streets, buildings, room, and so on), and make a key.
7. Use the map to write several problems involving perimeter and area.
8. Display your map and exchange sets of problems with other groups. Solve the problems using the maps.

HOME ACTIVITY

Find the area of each parallelogram.

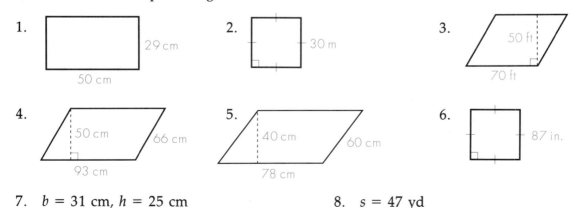

1. 29 cm
    50 cm

2. 30 m

3. 50 ft
    70 ft

4. 50 cm
    66 cm
    93 cm

5. 40 cm
    60 cm
    78 cm

6. 87 in.

7. $b = 31$ cm, $h = 25$ cm

8. $s = 47$ yd

Find the missing measures.

9.

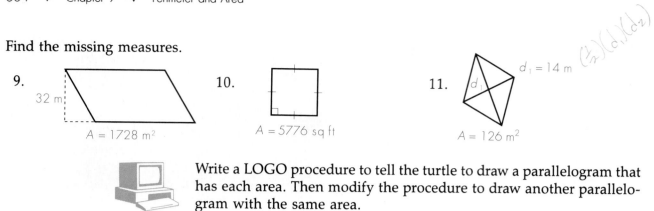

32 m

$A = 1728 \text{ m}^2$

10.

$A = 5776 \text{ sq ft}$

11.

$d_1 = 14 \text{ m}$

$A = 126 \text{ m}^2$

Write a LOGO procedure to tell the turtle to draw a parallelogram that has each area. Then modify the procedure to draw another parallelogram with the same area.

12. 48

13. 36

14. 160

Find the area of the parallelogram.

15.

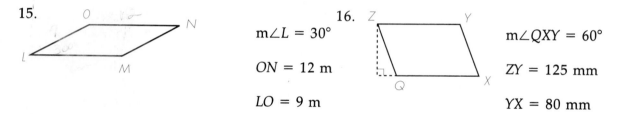

16.

$m\angle L = 30°$

$ON = 12 \text{ m}$

$LO = 9 \text{ m}$

$m\angle QXY = 60°$

$ZY = 125 \text{ mm}$

$YX = 80 \text{ mm}$

Find the area of the shaded region.

17. *DEFG* and *HJKL* are parallelograms.

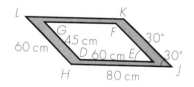

60 cm   45 cm   30°

D 60 cm E   30°

80 cm

Use the diagonals of the rhombus to find the area.

18.

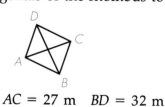

$AC = 27 \text{ m}$ $BD = 32 \text{ m}$

19.

$GJ = 15 \text{ ft}$ $KH = 28 \text{ ft}$

Write a LOGO procedure to tell the turtle to draw a rhombus for each of the following areas. Then modify the procedure to draw another rhombus with the same area.

20. 16

21. 28

22. 99

Find the perimeter of each parallelogram.

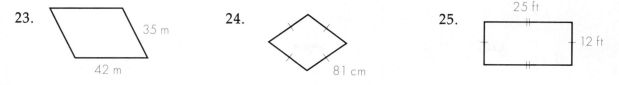

23.

35 m

42 m

24.

81 cm

25.

25 ft

12 ft

26. The plan shows a classroom 30 feet wide by 40 feet long. If carpet costs $6.45 per square foot, what would be the cost of carpeting the whole room?

40 ft | Plan of classroom

30 ft

27. The drawing shows a room in a house.
Find the area of the ceiling and the area of the four walls.

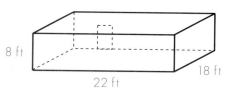

8 ft

22 ft

18 ft

28. One gallon of paint covers 425 square feet. If you paint the walls and ceiling in the same color with two coats of paint, how many gallons will you need to buy?

29. If the paint costs $18.75 per gallon, how much will you spend?

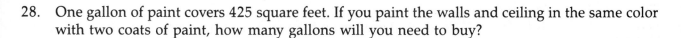

CRITICAL THINKING

30. Terry has 120 feet of fencing that he wants to put up around his garden. Copy and complete the table.

Length (ft)	Width (ft)	Perimeter (ft)	Area
50		120	
45		120	
40		120	
35		120	
30		120	

31. What will be the shape and dimensions of the greatest area Terry can enclose?

Perimeter and Area of a Trapezoid

After completing this section, you should understand
▲ how to find the perimeter and area of a trapezoid.

In Chapter 8, you learned that there are three general types of quadrilaterals.

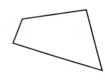
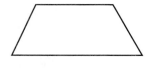
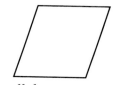

quadrilateral—
no sides parallel

trapezoid—one pair
of sides parallel

parallelogram—two
pairs of sides parallel

The method used to find the area of a trapezoid takes advantage of the area theorems you have already investigated, and uses algebra to combine them.

DISCOVERY ACTIVITY

1. Draw a trapezoid and label it *ABCD*.
2. From points *D* and *C*, draw or construct perpendiculars to $\overline{AB}$. Label the intersections of the perpendiculars with the base *E* and *F*.
3. What familiar figures have you formed?
4. Describe how you would find the area of *ABCD*.
5. Draw another trapezoid and label it *GHJK*. How could you divide the trapezoid into two triangles?
6. How would you find the area of trapezoid *GHJK*?

You may have found that the area of a trapezoid is equal to the sum of the areas of the two triangles it contains.

Given: Trapezoid $GHJK$
with diagonal $\overline{GJ}$,
$\overline{GH} \parallel \overline{KJ}$

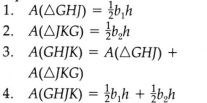

Justify: Area $GHJK = \frac{1}{2}h(b_1 + b_2)$

Supply the missing reasons.

Steps	Reasons
1. $A(\triangle GHJ) = \frac{1}{2}b_1 h$	1.
2. $A(\triangle JKG) = \frac{1}{2}b_2 h$	2.
3. $A(GHJK) = A(\triangle GHJ) +$ $A(\triangle JKG)$	3.
4. $A(GHJK) = \frac{1}{2}b_1 h + \frac{1}{2}b_2 h$	4.
5. $A(GHJK) = \frac{1}{2}h(b_1 + b_2)$	5.

THEOREM 9-4-1 **If the figure is a trapezoid, then the area is $\frac{1}{2}$ the height times the sum of the bases.**

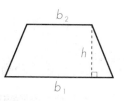

$A = \frac{1}{2}h(b_1 + b_2)$ square units

Recall that the height is the length of the altitude, or perpendicular, between the two parallel bases or sides.

Example: A city park is in the shape of a trapezoid. What is its area?

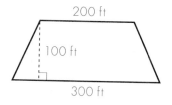

$A = \frac{1}{2}h(b_1 + b_2)$

$h = 100$ ft, $b_1 = 300$ ft, $b_2 = 200$ ft

$A = \frac{1}{2} \cdot 100(300 + 200)$

$= 50 \cdot 500 = 25{,}000$ sq ft

Example: *ABCD* is an isosceles trapezoid. What is the area?

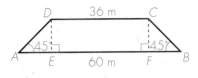

1. What information is missing?

2. Use what you know about 45° right triangles to find the height. Draw perpendiculars $\overline{DE}$ and $\overline{CF}$ from points *D* and *C* to $\overline{AB}$. What is figure *EFCD*?

3. What is the length of $\overline{EF}$? Find the length of $\overline{AE}$ and $\overline{FB}$.

4. What do you know about $\overline{AE}$ and $\overline{DE}$ in △*ADE* and $\overline{BF}$ and $\overline{CF}$ in △*BCE*? Why?

5. So, if *AE* = *BF* and *AE* + *BF* = 24 m, then *AE* = 12 m and *BF* = 12 m. Since *AE* = *DE* and *BF* = *CF*, *DE* = 12 m and *CF* = 12 m. Thus, the height of the trapezoid is 12 m.

6. Find the area.

CLASS ACTIVITY

Find the area of the trapezoid.

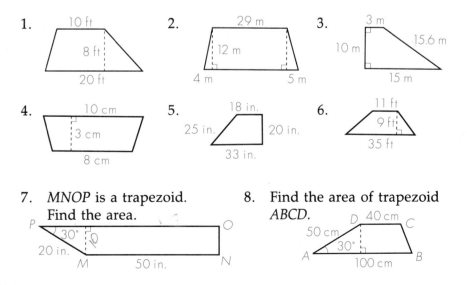

7. *MNOP* is a trapezoid. Find the area.

8. Find the area of trapezoid *ABCD*.

9. Find the area of isosceles trapezoid *STWX*.

10. Find the area of trapezoid *GHJK*.

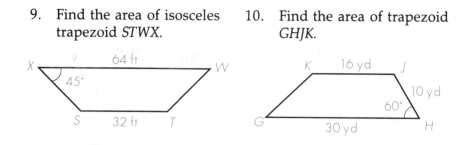

Find the missing base or height for each trapezoid.

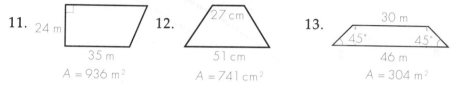

11. 24 m
35 m
A = 936 m²

12. 27 cm
51 cm
A = 741 cm²

13. 30 m
45° 45°
46 m
A = 304 m²

There is another way to derive the formula for the area of a trapezoid. Trapezoid *ABCD* can be seen as one-half of parallelogram *AGHD*.

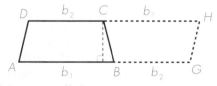

14. Since *AGHD* is a parallelogram, $AG = DH = ? + ?$
15. Area of *AGHD* = ?
16. Since $ABCD = \frac{1}{2}AGHD$, Area *ABCD* = ?

PERIMETER OF A TRAPEZOID

You have seen that to find the perimeter of any figure, you add the lengths of the sides. For a trapezoid,

Perimeter $(P) = a + b + c + d$.

Example: Find the perimeter of isosceles trapezoid *ABCD*.

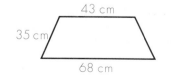

$P = 68$ cm $+ 35$ cm $+ 43$ cm $+ 35$ cm

$P = 181$ cm

CLASS ACTIVITY

Find the perimeter and the area of each trapezoid.

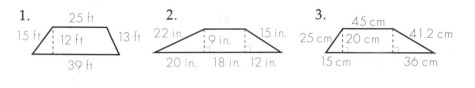

1. 25 ft, 15 ft, 12 ft, 13 ft, 39 ft

2. 22 in., 9 in., 15 in., 20 in., 18 in., 12 in.

3. 45 cm, 25 cm, 20 cm, 41.2 cm, 15 cm, 36 cm

4. *ABCD* is an isosceles trape-zoid with $DC = \frac{2}{3}AB$. What is the perimeter? the area?

5. *LMNO* is a trapezoid with $LM = 2ON$. What is the pe-rimeter? the area?

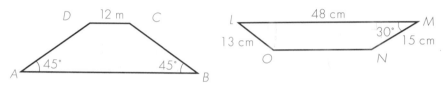

D 12 m C
45° 45°
A B

L 48 cm M
13 cm 30° 15 cm
O N

CONSTRUCTION AND AREA

On dot paper, draw a square with an area of 1 square unit, and a triangle and a trapezoid, both with areas of $1\frac{1}{2}$ square units. Draw the figures below. Find the area of each figure without using any of the area formulas.

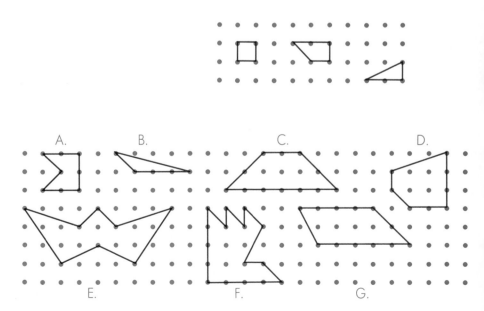

A.   B.   C.   D.

E.   F.   G.

HOME ACTIVITY

Find the perimeter and the area of the trapezoid.

1.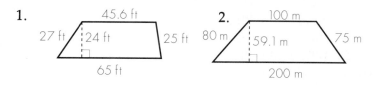
   45.6 ft
   27 ft   24 ft   25 ft
   65 ft

2. 100 m
   80 m   59.1 m   75 m
   200 m

3. 20 ft   120 ft   20 ft
   15 ft
   25 ft

Find the missing measure of the trapezoid.

4.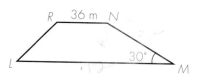
   16 m
   40 m
   A = 568 m²

5. 84 cm
   65 cm
   A = 3948.5 cm²

6.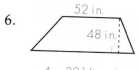
   52 in.
   48 in.
   A = 3216 sq in.

Find the area of each trapezoid.

7.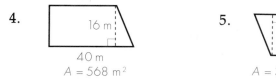
   D   27 m   C
   45°
   A
   65 m   B

8.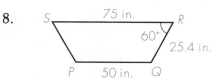
   S   75 in.   R
   60°
   25.4 in.
   P   50 in.   Q

9. $RN = \frac{1}{2}LM$, $NM = 32$ m

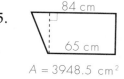
   R   36 m   N
   30°
   L   M

10. $KJ = \frac{3}{5}GH$

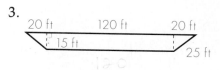
   K   J
   45°
   G   100 cm   H

11. $\overline{EF}$ is the median of trapezoid *ABCD*. *DG* = 10 m, *AB* = 36 m, and *DC* = 25 m. What is the area of trapezoid *EFCD*?

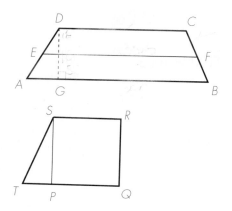
   D   C
   E   F
   A   G   B

12. *TQRS* is a trapezoid with an area of 180 cm². *PQRS* is a square with *PQ* = 12 cm. Find the length of $\overline{SR}$ and $\overline{TQ}$.

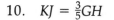
   S   R
   T   P   Q

Write a LOGO procedure to tell the turtle to draw a trapezoid that has each area. Then modify the procedure to draw another trapezoid with the same area.

13. 64   14. 210   15. 150

A marching band planned a new formation to use during the football season. They will march in two lines from points A and B at 45° angles. When the first person in each line reaches the center line in the field, the lines will turn and march toward each other. The band will form the outline of a trapezoid.

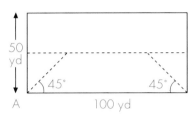

16. What is the area of the trapezoid?

17. If the band members stand 1 yard apart and there is a person on points A and B, how many players are in the band?

The diagram shows a city park.
18. What is the perimeter of trapezoid ADFG?
19. What is the perimeter of the entire park?
20. What is the area of the trapezoid?
21. What is the area of the entire park?
22. If the scale for the drawing were 1 inch = 200 feet, then how many inches long would $\overline{AD}$ be on the drawing?

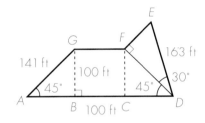

CRITICAL THINKING

23. In the Discovery Activity on page 336, you found that a trapezoid can be divided into two right triangles and a rectangle. Show that the area of trapezoid ABCD is equal to the combined areas of triangles I and II and rectangle III. Write a two-column proof.

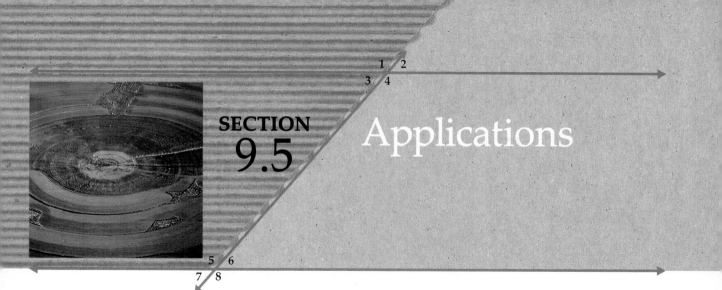

Applications

After completing this section, you should understand
▲ how to find the area of a quadrilateral with no parallel sides.
▲ the relationship between area and perimeter of similar figures.

In this chapter, you have investigated methods of finding area for several different kinds of quadrilaterals. Often it is necessary to find the areas of quadrilaterals that do not have a specific formula.

Look at figures 1–3.

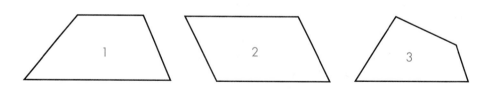

The areas of figures 1 and 2 can be found by formula; figure 3 has no parallel sides, so it is not a parallelogram and has no specific area formula. To find the area of quadrilaterals with no parallel sides, partition the figure into other figures for which you know a formula for the area.

AREA OF QUADRILATERALS WITH NO PARALLEL SIDES

Example: Find the area of quadrilateral *ABCD*. You can partition *ABCD* into a triangle and a trapezoid by drawing a line segment from *C*, parallel to $\overline{AB}$, and intersecting $\overline{AD}$ at *E*.

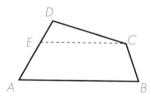

343

The area of quadrilateral *ABCD* thus becomes the sum of the areas of trapezoid *ABCE* and triangle *DEC*.

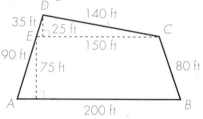

To find the areas of trapezoid *ABCE* and △*DEC* you need to know the altitudes of triangle and trapezoid and the lengths of $\overline{AB}$ and $\overline{EC}$ (bases). In actual situations, these lengths are usually obtained by measurement. Here, the measurements have been given to you.

$$A(ABCD) = A(ABCE) + A(\triangle DEC)$$
$$= \tfrac{1}{2}h(b_1 + b_2) + \tfrac{1}{2}bh$$
$$= \tfrac{1}{2} \cdot 75 \text{ ft } (150 \text{ ft} + 200 \text{ ft}) + \tfrac{1}{2}(150 \text{ ft} \cdot 25 \text{ ft})$$
$$= 13,125 \text{ sq ft} + 1875 \text{ sq ft}$$
$$= 15,000 \text{ sq ft}$$

Since there is no formula for finding the area of quadrilaterals with no parallel sides, you must partition the figure into triangles, trapezoids, or parallelograms, and find the sum of the combined areas.

To find the perimeter of quadrilateral *ABCD*, find the sum of the lengths of the sides.
$P(ABCD)$ = 200 ft + 125 ft + 140 ft + 80 ft = 545 ft

CLASS ACTIVITY

Find the area and perimeter of each quadrilateral.

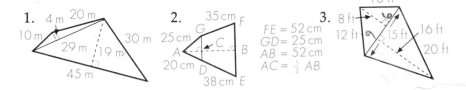

1. 4 m 20 m 10 m 30 m 29 m 19 m² 45 m

2. 35 cm F 25 cm G C A B 20 cm D 38 cm E

3. 10 ft 8 ft 12 ft 15 ft 16 ft 20 ft
   FE = 52 cm
   GD = 25 cm
   AB = 52 cm
   AC = $\tfrac{1}{3}$ AB

4. The garden shown at the right is made up of flower beds; a reflecting pond, and lawn. To buy grass seed and fertilizer for the lawn, you need to know its area. Partition the space and find the area of the lawn.

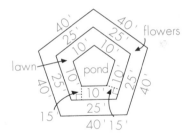

RELATIONSHIPS BETWEEN SIDES, AREA, AND PERIMETER

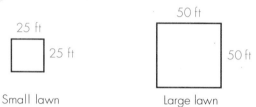

Small lawn   Large lawn

Suppose that you have a job mowing the smaller lawn. You charge $4 for the job. What would you charge to mow the larger lawn?

a. $8   b. $10   c. $12   d. $16

Before you decide on an amount, compare the areas of the two lawns. Both yards are squares, so the area is ℓw or s^2.

A(small lawn) $= 25 \cdot 25 = 625$ square feet
A(large lawn) $= 50 \cdot 50 = 2500$ square feet

The area of the large lawn is four times that of the small lawn, so you should charge $16.

You can use the concept of ratio and proportion (from Chapter 4) to solve problems like this one.

$$\frac{\text{Area}_1}{\text{Area}_2} = \frac{\text{Cost}_1}{\text{Cost}_2} = \frac{625}{2500} = \frac{\$4}{C_2}$$

$$625 \cdot C_2 = 2500 \cdot 4$$
$$625 C_2 = 10{,}000$$
$$C_2 = \$16$$

DEFINITION 9-5-1   Two figures are SIMILAR if the corresponding angles are equal in measure and the corresponding sides are in equal ratio.

Notice that this definition of similar figures is valid for all similar figures, not just for triangles.

DISCOVERY ACTIVITY

1. Draw two squares, one with sides of 15 units and one with sides of 30 units.
2. Are the two figures similar?
3. Are the corresponding angles equal?
4. Are corresponding sides in equal ratio? What is the ratio?
5. Map the figures.
6. What is the area of each square?
7. Write the ratio of the areas, $\frac{A_1}{A_2}$.
8. In place of A_1 and A_2, write the formula for the area of any parallelogram, b_1h_1 and b_2h_2.

   So, $\frac{A_1}{A_2} = \frac{b_1h_1}{b_2h_2} = \frac{225}{900}$

9. Rewrite the proportion.

   $$\frac{b_1}{b_2} \cdot \frac{h_1}{h_2} = \frac{225}{900}$$

   $$\frac{1}{2} \cdot \frac{1}{2} = \frac{225}{900}$$

   $$\left(\frac{1}{2}\right)^2 = \frac{1}{4}$$

10. What does the proportion tell you about the ratios of corresponding sides and areas of similar figures? Write a sentence stating your conclusion.

You may have discovered that the square of the ratio of the corresponding sides of similar figures is the same as the ratio of their areas.

POSTULATE 9-5-2 **If two similar figures have a ratio of a/b for their corresponding sides, then**
a. the ratio of their perimeters is a/b, and
b. the ratio of their areas is $(a/b)^2$.

Example: Figure A and figure B are similar.

$\dfrac{\text{Perimeter A}}{\text{Perimeter B}} = \dfrac{6}{18} = \dfrac{1}{3}$ Fig. A

$\dfrac{\text{Area A}}{\text{Area B}} = \dfrac{1}{9} = \left(\dfrac{1}{3}\right)^2$ Fig. B

CLASS ACTIVITY

Each pair of figures is similar. Find the ratios of the perimeters and areas.

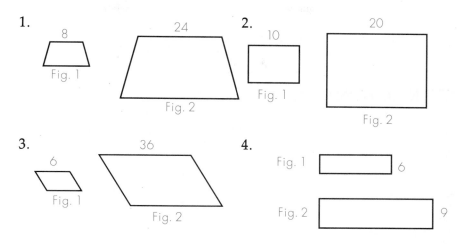

1.
8
Fig. 1
24
Fig. 2

2.
10
Fig. 1
20
Fig. 2

3.
6
Fig. 1
36
Fig. 2

4.
Fig. 1 6
Fig. 2 9

5. A garage is 34 feet wide. It has an area of 1234 square feet. A similar garage is 40 feet wide. What is its area?

HOME ACTIVITY

Find the area and perimeter of each figure.

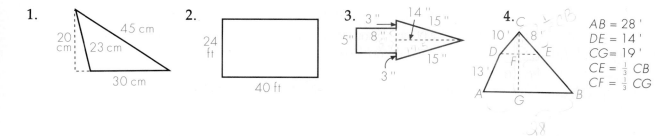

1.
20 cm
45 cm
23 cm
30 cm

2.
24 ft
40 ft

3.
3" 14" 15"
5" 8"
3"
15"

4. C
$AB = 28'$
$DE = 14'$
$CG = 19'$
$CE = \frac{1}{3} CB$
$CF = \frac{1}{3} CG$
10' 8'
D E
F
13
A G B

The ratio of the area of the figure shown to the area of a similar figure is given. Write a Logo procedure to tell the turtle to draw the figure given and the similar figure.

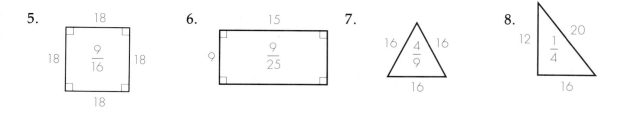

5.
18
18 $\frac{9}{16}$ 18
18

6.
15
9 $\frac{9}{25}$
16

7.
16 $\frac{4}{9}$ 16
16

8.
12 $\frac{1}{4}$ 20
16

9. X, Y, and Z are the midpoints of the sides of $\triangle QRS$. What is the ratio of the area of $\triangle XYZ$ to the area of $\triangle QRS$? Explain your answer.

CRITICAL THINKING

The figures below are made up of arrays of dots and represent geometric numbers. Find each number and record it.

1   4   ____   ____   ____   ____

10. What geometric number is represented by each of these numbers?

1   3   ____   ____   ____   ____

11. What geometric number is represented by each of these numbers?

The numbers associated with the square arrays of dots are called square numbers. The ones found in the triangular arrays are called triangular numbers. You can use an abbreviation to represent a square or triangular number. For example, S_3 means the third square number, a square with three dots on each side. T_6 means the sixth triangular number, a triangle with six dots on a side.

There are several interesting patterns involving these geometric numbers. Copy and complete the table.

n	1	2	3	4	5	6	7	8	9	10	11	12	13	14
S_n	1	4	9	16	25	36								
T_n	1	3	6	10	15	21								

12. Find the sum of any two consecutive triangular numbers. What do you notice about the sum?

9.1 Find the perimeter of figure *ABCD*.
$P = 3.5\ m + 3.5\ m + 2.5\ m + 2\ m$
$P = 11.5\ m$

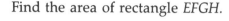

Find the area of rectangle *EFGH*.

$A = 45\ ft \cdot 20\ ft$
$A = 900\ sq\ ft$

9.2 Find the area of △*JKL*.

$A = \frac{1}{2}(3\ cm \cdot 5\ cm)$
$A = 7.5\ cm^2$

9.3 Find the perimeter and the area of parallelogram *MNOP*.

$P = 2(9\ mi + 10\ mi)$
$P = 38\ mi$
$A = 6.5\ mi \cdot 10\ mi$
$A = 65\ sq\ mi$

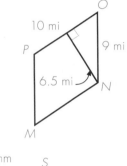

9.4 Find the area of trapezoid *QRST*.

$A = \frac{1}{2}(22\ mm)(52\ mm + 70\ mm)$

$A = 1342\ mm^2$

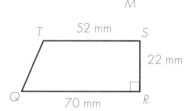

9.5 What is the ratio between the perimeter of △*UVW* and that of △*XYZ*?

The ratio between corresponding sides of △*UVW* and △*XYZ* is 1:2, so the two triangles are similar. By Postulate 9-5-1, the ratio between the perimeters is also 1:2.

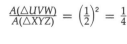

What is the ratio of the area of △*UVW* to that of △*XYZ*?

By Postulate 9-5-1, square the ratio of the sides of similar figures to get the ratio of the areas.

$\frac{A(\triangle UVW)}{A(\triangle XYZ)} = \left(\frac{1}{2}\right)^2 = \frac{1}{4}$

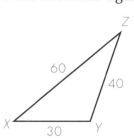

CALCULATOR ACTIVITY

Carla wants to plant a garden in a triangular plot. She knows that the lengths of the three sides of the plot are 9 feet, 10 feet, and 13 feet. Carla wants to find the area of her plot.

To find the area, Carla could make a scale drawing of her plot, find the height of the triangle, and then find the area. Or she could use a formula developed in the first century A.D. by the Greek mathematician Heron of Alexandria.

Heron's formula states that if a triangle has sides with lengths a, b, and c, then the area of the triangle is

$$\sqrt{s(s-a)(s-b)(s-c)}$$

where $s = \frac{1}{2}(a + b + c)$.

A calculator can be very useful when using Heron's formula to find an area. The Parentheses keys, which are labeled (and), can be used to group numbers together just as in the formula. For example, to calculate s for Carla's garden plot, press the keys shown below.

Press: | (| 9 | + | 1 | 0 | + | 1 | 3 |) | ÷ | 2 | = |

Display: 9   19   32   16

Another key that would be very useful in calculating area using Heron's formula is the Store key, which is labeled STO. The value for s that was calculated above can be stored in the memory of the calculator by pressing the Store key after pressing the = key. The value of s can then be retrieved from memory each time it is needed by pressing the Recall key, which is labeled RCL.

To calculate the radicand to find the area of Carla's garden plot using Heron's formula, press the keys shown below. Notice that the RCL key is pressed in place of s, and the (and) keys are pressed in place of the parentheses.

Formula: s (s − a)   (s − b)   (s − c)

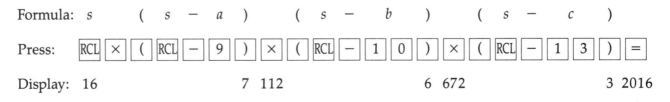

Press: | RCL | × | (| RCL | − | 9 |) | × | (| RCL | − | 1 | 0 |) | × | (| RCL | − | 1 | 3 |) | = |

Display: 16   7 112   6 672   3 2016

To find the area of Carla's garden plot, we must now find the square root of 2016. To do this, press the Square Root key, which is labeled $\sqrt{}$ or $\sqrt{x}$, after pressing the $\boxed{=}$ key.

Press: $\boxed{\sqrt{}}$

Display: 44.899889

Rounded to the nearest hundredth, the area of Carla's triangular garden plot is 44.90 square feet.

Use a calculator and Heron's formula to calculate the area of each triangle with side lengths as given below. Round each answer to the nearest hundredth.

1. 3 in., 7 in., 8 in.

2. 4 m, 11 m, 12 m

3. 15 ft, 15 ft, 15 ft

4. 8 cm, 9 cm, 10 cm

5. 6 yd, 10 yd, 12 yd

6. 7 in., 11 in., 16 in.

7. 5 m, 9 m, 13 m

8. 6 mi, 8 mi, 10 mi

9. 9 ft, 17 ft, 21 ft

10. 5 cm, 12 cm, 13 cm

11. Mark wants to paint the front of his triangular house including the door. To do this, he needs to find the area of the front of the house. Use a calculator and Heron's formula to find this area.

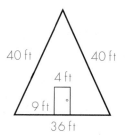

12. Suppose Mark decides not to paint the door of his house. Find the area of the surface he wants to paint.

CHAPTER REVIEW

Find the perimeter and area.

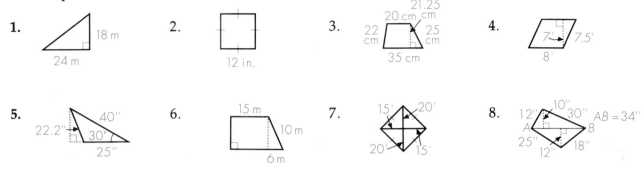

1.

2.

3.

4.

5.

6.

7.

8.

9. Find the area and the perimeter of trapezoid *PQRS*.

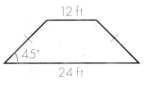

The area of each figure is given. Find the missing base and/or height.

10.

11.

12.

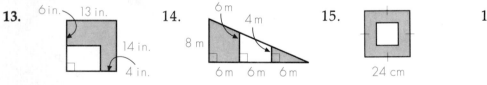
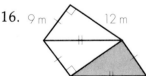

Find the area of the shaded region.

13.

14.

15.

16.

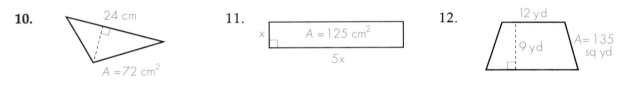

Find the ratios of the perimeters and area for each pair of similar figures.

17.

18.

19.

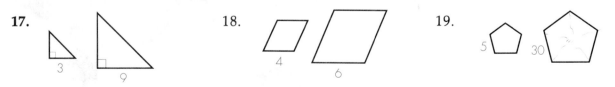

20. Ellen wants to rope off a section of a gym and use it as an exercise area. She has a 100-meter length of rope. What are the dimensions of the largest area she can enclose?

21. The bottom of a swimming pool measures 15 meters by 20 meters. It is to be tiled with rectangular tiles that are 50 centimeters by 25 centimeters. How many tiles are needed?

Find the area and perimeter.

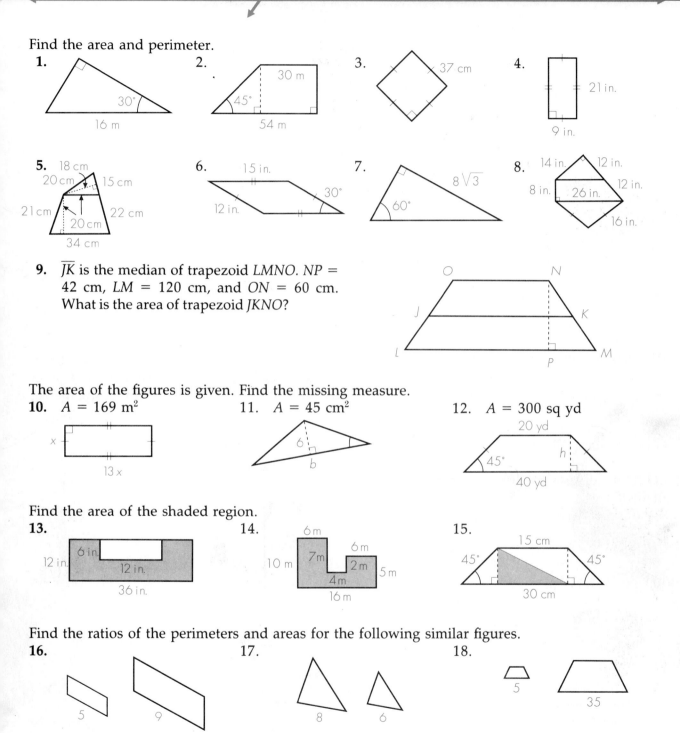

1. 30° 16 m

2. 30 m 45° 54 m

3. 37 cm

4. 21 in. 9 in.

5. 18 cm 20 cm 15 cm 21 cm 20 cm 22 cm 34 cm

6. 15 in. 12 in. 30°

7. $8\sqrt{3}$ 60°

8. 14 in. 12 in. 8 in. 26 in. 12 in. 16 in.

9. $\overline{JK}$ is the median of trapezoid *LMNO*. *NP* = 42 cm, *LM* = 120 cm, and *ON* = 60 cm. What is the area of trapezoid *JKNO*?

The area of the figures is given. Find the missing measure.

10. $A = 169$ m² x $13x$

11. $A = 45$ cm² 6 b

12. $A = 300$ sq yd 20 yd 45° h 40 yd

Find the area of the shaded region.

13. 6 in. 12 in. 12 in. 36 in.

14. 6 m 6 m 10 m 7 m 2 m 4 m 5 m 16 m

15. 15 cm 45° 45° 30 cm

Find the ratios of the perimeters and areas for the following similar figures.

16. 5 9

17. 8 6

18. 5 35

The area of a yard is 10,800 square feet. A similar yard has a side of 50 feet that corresponds to a side of 60 feet in the larger yard.

19. What is the area of the smaller yard?

20. What is the ratio of the perimeters, smaller to larger?

ALGEBRA SKILLS

Several properties of equations are useful in solving geometry problems.

Addition Property: If $x = y$, then $x + a = y + a$.
You can add the same number to each side of an equation.

Multiplication Property: If $x = y$, then $x(a) = y(a)$.
You can multiply both sides of an equation by the same number.

Division Property: If $x = y$, then $x/a = y/a$, $a \neq 0$.
You can divide both sides of an equation by the same number.

Solve the following equations. Round decimals to the nearest tenth.

1. $\frac{13}{x} = \frac{144}{69}$
2. $37(12x - 14) = -7(15x - 40)$
3. $3 - 4(3x - 6) = 4x - 11$
4. $3 - 4(\frac{3}{8}x - 6) = \frac{3}{4}x - 9$
5. $2x^2 - 5x + 3 = 2(x - 1)(x - 2)$
6. $5 - (2(x + 5) + 1) = x - 3$
7. $(2x - 7)(3x + 15) = 0$
8. $x^2 + 5x = -6$

9. The side of an equilateral triangle is 2 units longer than the side of a square. The perimeter of the square is greater than the perimeter of the triangle. The length of the side of the square is an integer. The side of the square must be at least how long? Are there an infinite number of solutions?

Given: Square $ABCD$. $\overline{HA} = \overline{EB} = \overline{FC} = \overline{GD}$.

$\triangle AEH \cong \triangle BFE \cong \triangle CGF \cong \triangle DHG$.
$A(ABCD) - A(EFGH) = A(\triangle AEH) +$
$A(\triangle BFE) + A(\triangle CGF) + A(\triangle DHG)$.

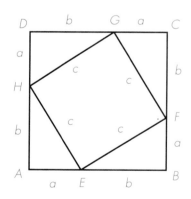

Complete the following.

10. $A(ABCD) = (a + b)^2 =$

11. Area of each triangle =

12. $A(EFGH) =$

13. Substitute the values for the areas that you obtained in Exercises 10–12 in the equation for Area $ABCD$ above. Solve the equation so that c^2 is on one side and all other terms are on the other side.

14. What theorem does this equation represent?

10

REGULAR POLYGONS AND CIRCLES

In brief, the flight into abstract generality must start from and return again to the concrete and specific.

Richard Courant

Regular Polygons

After completing this section, you should understand
▲ the definition of a regular polygon.
▲ how to find the measure of the interior angles of a polygon.
▲ how to find the perimeter of a polygon.

In Chapters 8 and 9 you investigated the properties of polygons with three and four sides—triangles and quadrilaterals. A polygon can have any number of sides, it can be convex or concave, and it can be regular or nonregular.

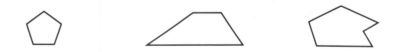

     convex, regular   convex, nonregular   concave, nonregular

You can classify a polygon by the number of its sides.

CLASS ACTIVITY

The number of sides of a polygon is given under *n*. Match each figure with its name.

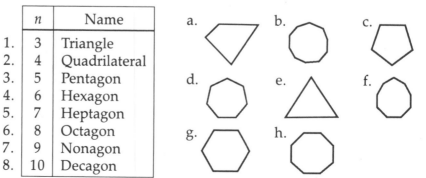

	n	Name
1.	3	Triangle
2.	4	Quadrilateral
3.	5	Pentagon
4.	6	Hexagon
5.	7	Heptagon
6.	8	Octagon
7.	9	Nonagon
8.	10	Decagon

In general, the term *n*-gon is used to describe a polygon with *n* sides.

DEFINITION 10-1-1 **A POLYGON is a closed plane figure consisting of line segments.**

Recall that the points making up a polygon are coplanar, and that the line segments (or sides) determined by these points do not intersect.

polygons   not polygons

You have already worked with several kinds of polygons that have sides of equal length and angles of equal measure.

equilateral triangle   square

DEFINITION 10-1-2 **A REGULAR POLYGON is a polygon that is equilateral and equiangular.**

Example:

 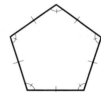

regular hexagon   regular pentagon

CLASS ACTIVITY

Identify each polygon by the number of sides. Tell whether it is convex or concave, regular or nonregular.

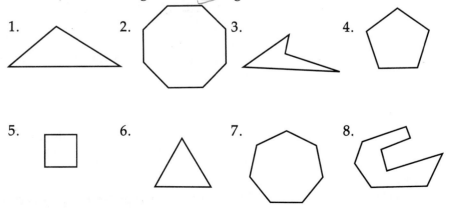

1. 2. 3. 4.

5. 6. 7. 8.

DISCOVERY ACTIVITY

1. Trace this regular pentagon. Notice that all of the sides are of equal length and all of the angles have equal measures.

2. Draw all the diagonals from point A. How many triangles did you form?

3. What is the sum of the measures of the angles of the triangles you formed?

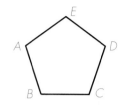

4. Do you think that the sum of the measures of the angles in the triangles is equal to the sum of the measures of the angles of the pentagon? Why?

5. How can you find the measure of each angle of the pentagon? How do you know?

6. Repeat steps 1–5 for a regular hexagon. What measure do you get for each angle of the hexagon?

You may have discovered that it is possible to calculate the measures of the interior angles of a regular polygon if you know the number of sides.

For a regular pentagon, $n = 5$
Sum of angle measures =

$3 \cdot 180° = 540°$

For a regular hexagon, $n = 6$
Sum of angle measures =

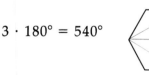

$4 \cdot 180° = 720°$

Measure of each angle =

$$\frac{3 \cdot 180°}{5} = \frac{540°}{5} = 108°$$

Measure of each angle =

$$\frac{4 \cdot 180°}{6} = \frac{720°}{6} = 120°$$

Or in general, for a regular polygon with n sides, the measure of each angle of the polygon is $\frac{(n-2)180°}{n}$ _Fan ✗._

THEOREM 10-1-1   Each angle of a regular n-gon is equal to $(n - 2)$ times $180°$ divided by n, where n is the number of sides in the n-gon.

Example: Find the measure of an interior angle of a regular 15-gon.

$$\frac{(n-2)180°}{n} = \frac{(15-2)180°}{15} = \frac{13 \cdot 180°}{15} = 156°$$

Example: What is the sum of the angle measures of a 30-gon?

$$(n-2)180° = (30-2)180° = 28 \cdot 180° = 5040°$$

CLASS ACTIVITY

Find the sum of the angle measures for each polygon.

1. Hexagon   2. 12-gon   3. 20-gon   4. 42-gon

Find the measure of an interior angle and an exterior angle for each regular polygon.

5. Heptagon) 6. 18-gon   7. Nonagon   8. 24-gon

9. The measure of an interior angle of a regular polygon is 156°. How many sides does the polygon have?

PERIMETER OF A POLYGON

In Chapter 9 you learned that the perimeter of any quadrilateral is the distance around it or the sum of the lengths of the sides.

The same is true for any polygon.

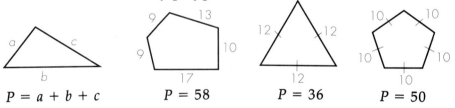

$P = a + b + c$ $P = 58$ $P = 36$ $P = 50$

CLASS ACTIVITY

Use a calculator to find the perimeter.

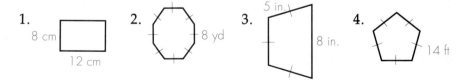

1. 8 cm, 12 cm   2. 8 yd   3. 5 in., 8 in.   4. 14 ft

5. Use a calculator to find the perimeter of a regular 15-gon with a side measuring 13 meters.

6. The perimeter of a regular decagon is 325.8 ft. Use a calculator to find the measure of one side.

HOME ACTIVITY

Identify each polygon by the number of its sides. Tell whether the polygon is convex or concave, regular or nonregular.

1.   2.   3.   4.

Find the sum of the angle measures for each polygon.

5. Square   6. Triangle   7. 12-gon   8. 100-gon   9. Octagon

 Write a LOGO procedure to tell the turtle to draw each regular polygon.

10. Pentagon   11. Hexagon   12. Octagon

Find the measure of an exterior angle for each regular polygon.

13. Square   14. Hexagon   15. 28-gon   16. Nonagon

The measure of an angle of a regular polygon is given. Find the number of sides.

17. 144°   18. 156°   19. 165.6°

20. Find the perimeter of polygon ABCD.

21. What is the area of △BDC?

22. What is the area of ABCD?

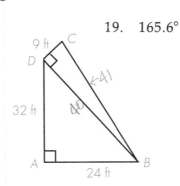

CRITICAL THINKING

Suppose you could program the LOGO turtle to move one foot and then turn right one second, repeating the routine until the turtle arrived at its starting point.

23. What regular polygon would it draw?
24. How many miles would it travel to form the polygon?

SECTION 10.2

Area of Regular Polygons

After completing this section, you should understand
▲ how to find the area of a regular polygon.

How could you find the area of the side of the Ferris wheel shown in the photograph? What geometric figures does it contain? In the last chapter, you investigated finding the area of quadrilaterals with no parallel sides and no specific area formula. To find the area you could

1. count the number of square units of measure contained in the figure; or

2. divide the figure into triangles, parallelograms, or trapezoids, find those areas, and combine them.

For many-sided polygons such as the Ferris wheel, these methods may be inexact or impractical.

DISCOVERY ACTIVITY

1. Trace this regular pentagon.
2. Use a protractor or compass to construct the bisector of each angle. Extend the bisectors until they intersect in the center of the pentagon. Label the point of intersection O.
3. You have divided the pentagon into five triangles. Are the triangles congruent? How do you know?
4. How could you write a formula for finding the area of a regular pentagon?

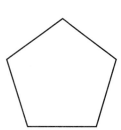

5. Trace this regular hexagon and repeat steps 2 and 3. How could you write a formula for finding the area for a regular hexagon?

You may have discovered that the area of a regular pentagon or hexagon is equal to the sum of the areas of the congruent triangles contained in them.

When you drew the angle bisectors from the vertices of the hexagon, congruent triangles were formed. Can you explain why the six triangles are congruent? Where do the bisectors intersect?

The line segments connecting the vertices of a regular polygon with the center bisect the angles and form congruent triangles. You can see that the area of a regular polygon is a multiple of the area of one of these triangles.

The regular heptagon has perpendicular bisectors of the seven sides. Note that the bisectors intersect at point O. If you put the point of a compass on point O and the pencil on a vertex of the heptagon and draw a circle, the vertices of the heptagon will all be on the circle.

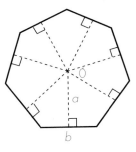

The center of any regular polygon can be located by constructing the perpendicular bisectors of two sides. The point where the bisectors intersect is the center. The length of the bisector, from a base to the center, is the *apothem* of the polygon.

For a regular heptagon, $A = 7 \cdot \frac{1}{2}ba$. $P = 7b$
You can express the area of the heptagon in terms of its perimeter.

$A = \frac{1}{2}(7ba)$

$A = \frac{1}{2}Pa$ (by substitution)

THEOREM 10-2-1   The area of a regular polygon is $\frac{1}{2}Pa$, where P is the perimeter of the polygon and a is the length of the apothem.

Example: What is the area of a regular pentagon with a side of 4 meters and an apothem of 2.75 meters?

$P = 5 \times 4 = 20$ m, $a = 2.75$ m $A = \frac{1}{2}Pa = \frac{1}{2} \times 20 \times 2.75 = 27.5$ m^2

Example: What is the area of this regular polygon?

Since the polygon is a hexagon, m∠DEF is 120° and m∠DEO is 60°.

From the relationships of the sides of a 30°-60°-right triangle:

$DE = 5$ in., $OD = 5\sqrt{3}$ in. and
$CD = 10$ in.
$P = 6 \cdot 10 = 60$

$A = \frac{1}{2}Pa = \frac{1}{2} \cdot 60 \cdot 5\sqrt{3} = 150\sqrt{3}$ sq in.

Example: What is the area of a regular hexagon with a side of 8 centimeters?

Since a side of the hexagon is 8 cm, you can find the perimeter: $6 \cdot 8 = 48$ cm.

$m∠ABC = 120°$, so $m∠ABO = 60°$.

$AB = 4$ cm, so $AO = 4\sqrt{3}$ cm (height).

$A = \frac{1}{2}Pa = \frac{1}{2} \cdot 48 \cdot 4\sqrt{3} = 96\sqrt{3}$ cm²

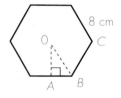

CLASS ACTIVITY

Use a calculator to find the area of each regular polygon.

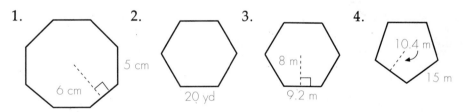

1.   6 cm

2.   5 cm   20 yd

3.   8 m   9.2 m

4.   10.4 m   15 m

Use a calculator to solve the problems.

5. The top of a circus tent is made up of a regular hexagon with sides of 50 feet. What is its area?

6. A regular polygon has three sides, each measuring 10 feet. The apothem is 8.7 feet. What is the area?

7. Find the area of a regular hexagon with a side of 10 meters and an apothem of 8.7 meters.

Show that the angle bisectors of a regular hexagon divide the hexagon into six equilateral triangles.

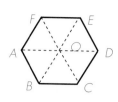

8. Find the measure of an angle of a regular hexagon.

9. What is the measure of ∠OAB? of ∠OBA? Why?

10. What is the measure of each of the six angles around point O? How do you know?

11. Are the six triangles congruent? Why?

12. Are the six triangles equilateral? Why?

HOME ACTIVITY

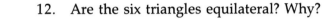

Find the area of each regular polygon.

1.

$8\sqrt{3}$ cm

16 cm

2.

18 cm

15 cm

3.

9.3 in.

18.6 in.

4.
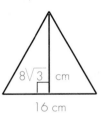

$8\sqrt{3}$

5.

5.5 in.

8 in.

6.

22 ft   32 ft

7. Copy and complete the table.

Figure	Number of Sides	Length of Side	Apothem	Area
Hexagon		1 m	0.87 m	
Heptagon		1 m	1.04 m	
Octagon		1 m	1.21 m	
Nonagon		1 m	1.37 m	
Decagon		1 m	1.54 m	

8. What do you observe about the areas of the polygons as the number of sides increases?

9. What is the perimeter of each polygon in Exercise 7?

One of the most famous buildings in the United States is the Pentagon in Arlington, Virginia, where the Defense Department is located. This tremendous office complex has 17.5 miles of corridors. Each side of the building is 921 feet long, and the apothem of the pentagon is 633.8 feet.

10. What is the perimeter of the Pentagon?

11. What is the area (in square feet) of the Pentagon? What is the area in square yards?

12. How many square yards are there in a football field?

13. How many football fields can fit into the area of the Pentagon?

A band shell is to be built on a 40-foot by 50-foot piece of land in an amusement park. The shell will have either 10 sides, each 10 feet long, or 12 sides, each 12 feet long, depending on how it fits on the piece of land.

14. What is the area of a 10-sided band shell with 10-foot sides and a height of 15.4 feet?

15. What is the area of a 12-sided band shell with 12-foot sides and a height of 22.4 feet?

16. Which shell will fit on a 40-foot by 50-foot piece of land?

17. Write a LOGO procedure using the procedure XYPLANE from page 118 and the command SETXY to plot and connect the points $A(2, 0)$, $B(4, 0)$, $C(6, 2)$, $D(4, 4)$, $E(2, 4)$, and $F(0, 2)$. Show the output on a piece of graph paper. What is the name of the polygon that is drawn? Is the polygon regular? Explain.

Use a ruler, compass, and protractor to do the following construction.

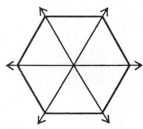

18. Construct a regular hexagon. Draw a circle and mark the center.
Since a hexagon has 6 sides, divide 360° by 6.
Draw 6 angles, each 60°, with the vertices around the point at the center of the circle.
Connect the points where the rays of the angles intersect the circle.

19. Construct a regular nonagon. Follow the steps above. What angle measure did you use?

20. Construct a regular octagon. What angle measure did you use?

21. Construct a regular pentagon. What angle measure did you use?

CRITICAL THINKING

Compare the areas of a square and a hexagon with the same perimeter. Give reasons for the following steps.

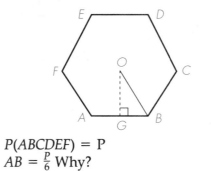
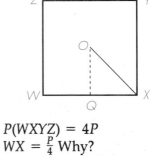

$P(ABCDEF) = P$

$AB = \frac{P}{6}$ Why?

$OG = \frac{P\sqrt{3}}{12}$

$A = \frac{1}{2}Ph$

$A = \frac{1}{2}\left(P\frac{\sqrt{3}}{12}\right) \cdot P$

$A = \frac{\sqrt{3}}{24} \cdot P \cdot P$

$P(WXYZ) = 4P$

$WX = \frac{P}{4}$ Why?

$OQ = \frac{P}{8}$

$A = \frac{1}{2}Ph$

$A = \frac{1}{2}\left(\frac{P}{8}\right) \cdot P$

$A = \frac{1}{16} \cdot P \cdot P$

22. Which area is greater? Why?

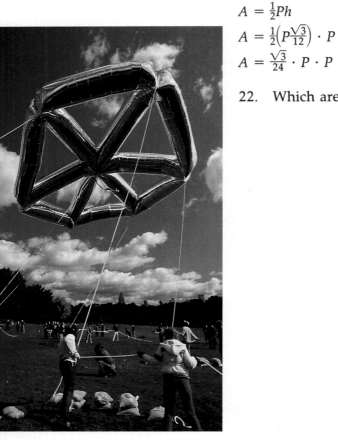

SECTION 10.3

Circumference and Area of a Circle

After completing this section, you should understand
▲ the definition of a circle and its parts.
▲ how to find the circumference of a circle.
▲ how to find the area of a circle.

Pg 394
+ 395

The problem of finding an accurate method for determining the circumference and area of a circle has occupied mathematicians, engineers, and other scientists for thousands of years. Recall that a circle is the set of all points on a plane that are equidistant from a point called the center.

NOTEBOOK

DEFINITION 10-3-1 **The RADIUS of a circle is a line segment whose endpoints are the center of a circle and a point on the circle.**

The *diameter* is the distance across a circle through the center. It is twice the length of the radius. The *circumference* of a circle is the distance around the circle.

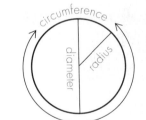

Use a compass to draw three circles, each one with a greater radius than the last.

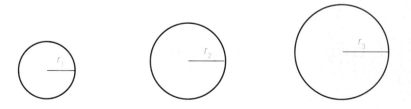

You can see that the circles get larger as you increase the length of the radius. And as the circles get larger, the circumference, or distance around the circle, also gets larger.

DISCOVERY ACTIVITY

1. Collect a set of different-sized circular objects, such as bicycle or wagon wheels, bangle bracelets, plastic or metal rings. Use a small piece of tape to mark a point on the circumference of each circle.

2. Determine the circumference of each circle by rolling it through one revolution, marking the points where the taped point begins and ends.

start   end

3. Measure and record the distance between the points.

4. Measure the diameter, or distance across the circle through the center.

5. Find the ratio of the circumference to the diameter. (distance from step 3 ÷ diameter)

6. Repeat the activity for several other circular objects. Record the ratios you find.

7. What do you observe about the ratio of circumference to diameter (C/d)?

You may have discovered that the circumference of a circle is approximately three times the diameter. The more accurate your measurements, the closer the value of the ratio will be to 3.14 . . . . The ratio of circumference to diameter, or circumference to two times the radius, is a *constant*. This means that this ratio will be the same for any circle.

This constant is an irrational number that has been given a symbol, π, **pi.** The value of π is only approximate—to 10 decimal places, $\pi \approx$ 3.1415926535.

POSTULATE 10-3-1   **The circumference of a circle is π times the diameter (d).**
$$C = \pi d \approx 3.14d$$
Since the diameter is twice the radius, $C = \pi(2r) = 2\pi r$.

Example: What is the circumference of a 26-inch bicycle wheel?

$C = \pi d$ $d = 26$ in.

$C = 26\pi \approx 26(3.14) \approx 81.64$ in.

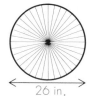

26 in.

Example: What is the circumference of a circle with a radius of 12 meters?

$C = 2\pi r$ $C = 2\pi \cdot 12 = 24\pi \approx 24 \cdot 3.14 \approx 75.36$ m

CLASS ACTIVITY

For each diameter find the length of the radius.

1. 32.4 cm   2. 12 in.   3. 15.32 cm   4. 8.5 m

For each radius find the length of the diameter.

5. 3.75 ft   6. 7 yd   7. 32.1 mm   8. $18\frac{1}{2}$ ft

9. You can use the $\boxed{\pi}$ key on a calculator to help you find the circumference of a circle. Press the $\boxed{\pi}$ key on a calculator. What number is shown on the display?

Use a calculator to find the circumference. Round your answer to the nearest hundredth.

10. $d = 2$ in.   11. $r = 2$ ft   12. $d = 8$ cm   13. $r = 50$ yd

14. π is the ratio of the _____ to the _____.

THE AREA OF A CIRCLE

It is possible to determine the area of a circle by counting the number of square units in the interior, but this is a difficult and inaccurate method. The area of a circle is also related to its diameter and radius and to the constant, π. You can see that, as the length of the diameter increases, so does the area.

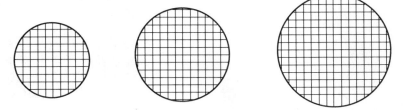

1. Construct a circle with a 1-inch radius. Mark the center O. Label point A on the circle.

2. Place the point of a compass on A and the pencil tip on O and mark off equal segments around the circle. Label the points on the circle.

3. Connect the points. What polygon is inscribed in the circle? Is it a regular polygon?

4. What is the length of $\overline{OA}$? of $\overline{AB}$? Why?

5. What is the perimeter of the hexagon? What is the area?

6. How is the circumference of the circle related to the perimeter of the hexagon? Use the formula for circumference, $C = \pi d$, to explain.

7. How do you think the area of the circle is related to the area of the hexagon?

8. How might you get a closer estimate of the circle's area?

You may have discovered that the perimeter and area of the hexagon are close to, but less than, the circumference and area of the circle.

Now find the area of a regular 12-gon inscribed in a circle with a 1-inch radius.

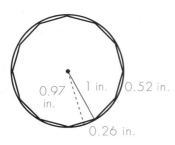

0.97 in.   1 in.   0.52 in.

0.26 in.

$$A = \tfrac{1}{2}Ph = \tfrac{1}{2}(0.97)6.24 = 3.03 \text{ sq in.}$$

Notice that as the number of sides increases,
• the height of the polygon approaches the length of the radius;
• the perimeter of the polygon approaches the circumference of the circle;
• the area of the polygon approaches the area of the circle.

$$h \to r$$
$$P \to C = 2\pi r$$
$$A = \tfrac{1}{2}Ph \to \tfrac{1}{2} \cdot 2\pi r \cdot r \text{ or } \pi r^2$$

THEOREM 10-3-1 — **The area of a circle is the radius squared times the number π.**

Example: A circle has a diameter of 4 feet. What is the area?

$$A = \pi r^2 \approx 3.14(2)^2 \approx 12.56 \text{ sq ft}$$

Example: A circle has an area of 625π. What is the radius?

$$A = \pi r^2$$
$$625\pi = \pi r^2$$
$$625 = r^2$$
$$25 = r$$

REPORT 10-3-1   Prepare and present an oral report on one of the following mathematicians of ancient Greece. In your report, pay particular attention to the mathematician's contributions to geometry and to the history of π.

Plato   Euclid   Archimedes
Pythagoras   Hippocrates of Chios

CLASS ACTIVITY

Use a calculator to find the area. Round your answer to the nearest hundredth.

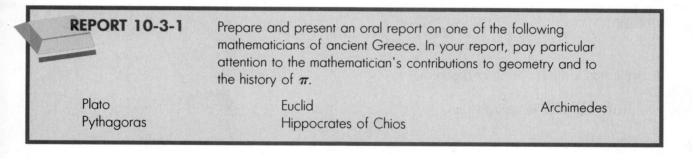

1. 15 ft
2. 1.4 in.
3. 5 yd
4. 10 cm

Find the area of each circle with the given measure. Use $\pi \approx 3.14$.

5. $d = 4$ ft   6. $C = 6\pi$ cm   7. $r = 4$ m   8. $C = 10\pi$ in.

9. $d = 100$ m   10. $r = 12$ yd   11. $d = 1.5$ km   12. $C = 12\pi$ m

Find the diameter of each circle.

13. $A = 11{,}304$ cm^2 14. $A = 628$ sq ft   15. $A = 3140$ m^2

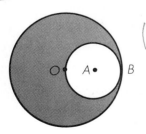

16. Use a calculator to find the area of the shaded region. The centers of the circles are O and A, the length of radius $\overline{AB}$ is 8 cm, and $\overline{OB} = 2\overline{AB}$.

HOME ACTIVITY

Copy and complete the table. Use $\pi \approx 3.14$.

	Diameter	Radius	Circumference	Area
1.		8 in.		
2.		30 cm		
3.			131.88 yd	
4.				109.9 m²
5.		10 in.		
6.				78.5 sq ft

A circle has a radius of 2 centimeters.

7. Find the area in terms of π.

8. Find the circumference in terms of π.

An equilateral triangle circumscribes a circle with a diameter of 5.8 feet. The side of the triangle is 10 feet.

9. What is the area of the triangle?

10. What is the area of the circle? Use $\pi \approx 3.14$.

11. What is the area of the shaded region?

CRITICAL THINKING

Two circles have radii of 3 centimeters and 5 centimeters. Use $\pi \approx 3.14$.

12. What is the circumference of the smaller circle?
13. What is the circumference of the larger circle?
14. What is the ratio of the radii, smaller to larger?
15. What is the ratio of the circumferences, smaller to larger?
16. Find the area of the smaller circle.
17. Find the area of the larger circle.
18. What do you think the ratio of the areas, smaller to larger, will be? Why?
19. Write a conclusion about what you found.

SECTION 10.4

Angles and Circles

After completing this section, you should understand

▲ how to identify central and inscribed angles and their intersected arcs.
▲ how to find the measures of angles intersecting circles.
▲ how to identify a tangent to a circle.

Before beginning an investigation of angles and circles, it will be helpful to define some more terms associated with a circle.

CHORD: a line segment whose endpoints are two points on the circle

SECANT: a line that intersects the circle at two points

TANGENT: a line that intersects the circle at one point

ARC: a part of the circumference of the circle

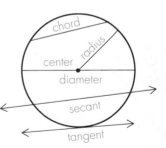

DISCOVERY ACTIVITY

1. Use a compass to draw six circles on a sheet of paper.

2. Look at the pair of intersecting lines shown at the right. How many ways can you find for these lines to intersect a circle? Think of placing point X at different places in, on, or outside of the circle.

3. Draw as many cases as you can find.

You may have discovered that there are five ways for these lines to intersect a circle.

INTERSECTION AT THE CENTER

DEFINITION 10-4-1 A CENTRAL ANGLE is an angle whose vertex is at the center of the circle.

You can measure a central angle with a protractor. Since the rays of the angle intersect the circle and the complete circle contains 360°, the arc intersected by the angle contains the same number of degrees as the central angle.

mADB=130°

DEFINITION 10-4-2 The ARC of the circle has the same measure as the central angle that intersects the arc. The symbol for arc is $\overset{\frown}{AB}$ or $\overset{\frown}{ADB}$.

INTERSECTION ON THE CIRCLE

Angle ABC intersects $\overset{\frown}{AC}$ of the circle, with the vertex B on the circle. This angle is called an inscribed angle.

DEFINITION 10-4-3 An INSCRIBED ANGLE is an angle whose vertex is on the circle and whose rays intersect the circle in two other points.

It is possible to find the measure of $\angle ABC$ if you know the measure of the intersected arc AC.

Steps

1. Draw $\overleftrightarrow{BO}$ and mark the point where $\overleftrightarrow{BO}$ intersects the circle O.

2. Draw $\overline{OA}$ to form $\triangle ABO$. $\triangle ABO$ is isosceles.

3. $m\angle ABO = m\angle BAO$

Reasons

1.

2.

3.

Let $\angle y = \angle AOD$ and $\angle x = \angle ABO$.

4. $m\angle y = m\angle x + m\angle BAO$ 4.
5. $m\angle y = 2(m\angle x)$ and   5.
    $\frac{1}{2}(m\angle y) = m\angle x$
6. $m\angle y = m\widehat{AD}$ 6.
7. $m\angle x = \frac{1}{2}(m\widehat{AD})$ 7.

Repeat steps 2–7 to show that $m\angle CBO = \frac{1}{2}(m\widehat{CD})$.

8. $m\angle ABD + m\angle CBD =$ 8.
    $\frac{1}{2}(m\widehat{AD}) + \frac{1}{2}(m\widehat{CD})$ and
    $m\angle ABC = \frac{1}{2}(m\widehat{ADC})$

THEOREM 10-4-1   **The measure of an inscribed angle is $\frac{1}{2}$ the number of degrees in the intersected arc.**

INTERSECTION INSIDE THE CIRCLE

To find the measure of $\angle AED$, follow these steps.

1. Draw chord $\overline{CA}$.
2. Does $m\angle x = m\angle CAB + m\angle ACD$? Why?
3. $m\angle ACD = \frac{1}{2}(m\widehat{AD})$ and
    $m\angle CAB = \frac{1}{2}(m\widehat{BC})$. Why?
4. $m\angle x = \frac{1}{2}(m\widehat{AD} + m\widehat{BC})$

THEOREM 10-4-2   **If the vertex of the angle is within the circle, then the measure of the angle is $\frac{1}{2}$ the sum of the degrees in the intersected arcs.**
This theorem is true even for the case where the vertex of the angle is at the center of the circle.

Example: Find $m\angle x$.

$m\angle x = \frac{1}{2}(m\widehat{AD} + m\widehat{BC})$

$m\angle x = \frac{1}{2}(60° + 80°) = 70°$

Example: Find $m\angle PQR$.
The measure of all the arcs is 360°.
Solve for x to find the measure of each arc.

$x + (2x + 5) + (3x + 20) + (5x + 5) = 360°$
$11x + 30 = 360°$
$11x = 330°$
$x = 30°$

So, $m\widehat{PR} = 155°$ and $m\widehat{ST} = 110°$.
$m\angle PQR = \frac{1}{2}(155° + 110°) = 132.5°$

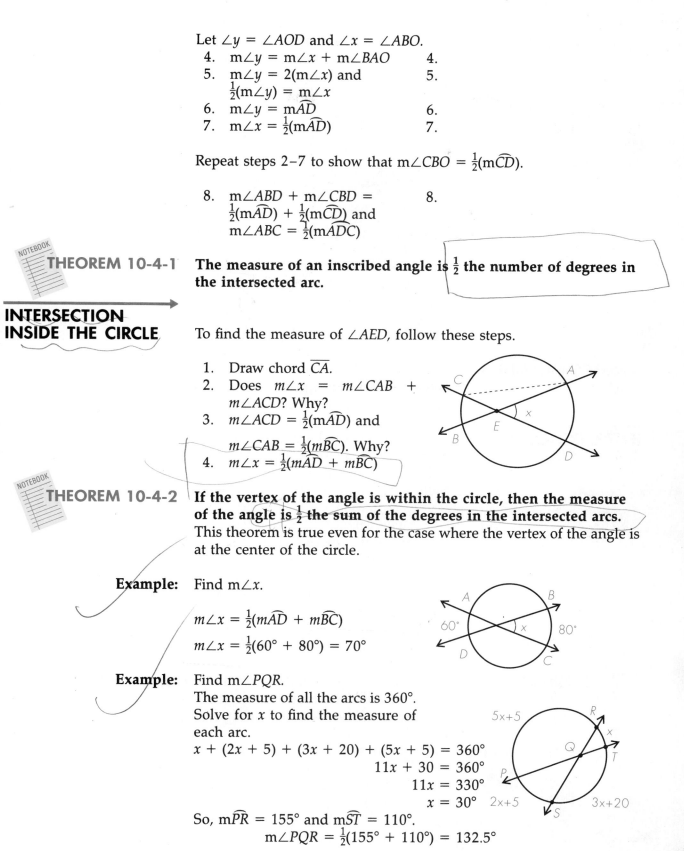

CLASS ACTIVITY

For each circle, name the radii, diameters, chords, secants, tangents, central angles, and inscribed angles.

1.

2.

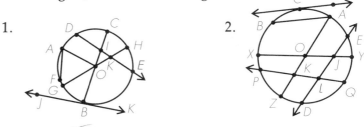

In circle O, $m\widehat{KL} = 65°$, $m\widehat{CL} = 48°$, and $\overline{JL}$ is a diameter. Find the measures of

3. $\angle KOL$
4. $\angle KJL$
5. $\angle LOC$
6. $\angle JOK$

ANGLES OUTSIDE A CIRCLE

In the circle to the right, the two lines intersect so that the vertex of the angle is outside the circle and the rays intersect the circle. Find $m\angle x$.

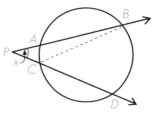

Steps	Reasons
1. Draw $\overline{CB}$.	1.
2. $m\angle DCB = m\angle ABC + m\angle CPB$	2.
3. $m\angle CPB = m\angle DCB - m\angle ABC$	3.
4. $m\angle DCB = \frac{1}{2}(m\widehat{BD})$ and $m\angle ABC = \frac{1}{2}(m\widehat{AC})$	4.
5. $m\angle CPB = \frac{1}{2}(m\widehat{BD} - m\widehat{AC})$	5.

NOTEBOOK

THEOREM 10-4-3 If the vertex of the angle is outside the circle and the rays intersect the circle, then the measure of the angle is $\frac{1}{2}$ the difference of the measure in degrees of the two intersected arcs.

Example: $m\widehat{RT} = 100°$ and $m\widehat{QS} = 30°$. Find $m\angle P$.

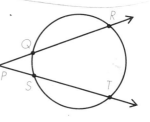

$$m\angle P = \frac{1}{2}(m\widehat{RT} - m\widehat{QS})$$

$$m\angle P = \frac{1}{2}(100° - 30°) = 35°$$

TANGENTS TO A CIRCLE

Two lines intersect so that the vertex of an angle is outside the circle; $\overleftrightarrow{PC}$ passes through the circle, and $\overleftrightarrow{PA}$ is tangent to the circle at point A.

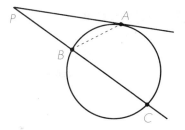

This case is another case of an exterior angle. $m\angle APB = m\angle ABC - m\angle PAB$, so $m\angle APB = \frac{1}{2}(m\widehat{AC} - m\widehat{AB})$.

Example: $m\widehat{TQ} = 40°$ and $m\widehat{TR} = 130°$. Find $m\angle P$.

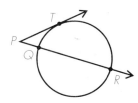

$$m\angle P = \frac{1}{2}(m\widehat{TR} - m\widehat{TQ})$$

$$m\angle P = \frac{1}{2}(130° - 40°) = 45°$$

Notice that $\overleftrightarrow{PT}$ intersects, or is tangent to, the circle at only one point.

DEFINITION 10-4-4 **If a line intersects a circle at one point, then the line is TANGENT to the circle.**

What is the measure of the angle formed by a tangent to a circle and a line through the center of the circle? Given circle O with tangent $\overleftrightarrow{PT}$, find the measure of $\angle PTO$.

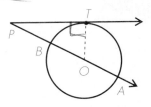

Steps	Reasons
1. $m\angle TOA = m\angle OPT + m\angle PTO$	1.
2. $m\angle PTO = m\angle TOA - m\angle OPT$	2.
3. $m\angle TOA = m\widehat{TA}$	3.
4. $m\angle OPT = \frac{1}{2}(m\widehat{TA} - m\widehat{TB})$	4.
5. $m\angle PTO = m\widehat{TA} - \frac{1}{2}(m\widehat{TA} - m\widehat{TB})$ $\quad = \frac{1}{2}(m\widehat{TA}) + \frac{1}{2}(m\widehat{TB})$ $\quad = \frac{1}{2}(m\widehat{TA} + m\widehat{TB})$	5.
6. $m\widehat{TA} + m\widehat{TB} = 180°$	6.
7. $m\angle PTO = \frac{1}{2}(180°) = 90°$	7.

THEOREM 10-4-4 **A tangent to a circle forms a 90° angle with the radius of the circle at the point of intersection.**
This theorem tells you that the tangent and the radius are perpendicular.

CLASS ACTIVITY

1. Use a computer and a geometric drawing tool to draw any circle. Label a point on the circle and draw a radius to the point. Then draw a line that is perpendicular to the radius at the point. Repeat this activity several times, using a circle with a different radius each time. What seems to be true about the line? Do you think it is possible to draw such a line through a point on a circle so that the line also passes through the interior of the circle? Do your results show that Theorem 10-4-4 is true?

2. Draw an angle inscribed in a semicircle. Find the measure of the angle. Repeat this activity several times, using a circle with a different radius each time. If an angle is inscribed in a semicircle, is the measure of the angle 90°? Justify your answer.

Find the measure of each angle.
$m\widehat{AB} = 100°$, $m\widehat{AD} = 80°$

3. m∠1   4. m∠2
5. m∠3   6. m∠4

$m\widehat{BD} = 170°$, $m\widehat{DF} = 10°$,
$m\widehat{CF} = 100°$

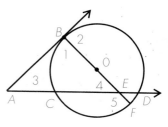

7. m∠1   8. m∠2
9. m∠3   10. m∠4
11. m∠5

HOME ACTIVITY

Identify the following for circle O.

1. Radius   2. Diameter   3. Center   4. Central angle

5. Tangent   6. Inscribed angle   7. Chord   8. Arc

In circle O, m∠TA = 140° and $m\widehat{AB}$ = 120°. Find the measure of each angle.

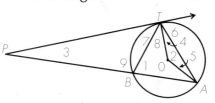

9. m∠1   10. m∠2   11. m∠3
12. m∠4   13. m∠5   14. m∠6
15. m∠7   16. m∠8   17. m∠9

CRITICAL THINKING

18. From point P outside a circle, draw $\overrightarrow{PA}$ and $\overrightarrow{PB}$ tangent to the circle. Show that the ray from P through the center of the circle bisects ∠APB.

SECTION 10.5

Chords and Secants

After completing this section, you should understand
▲ the definitions of a chord and a secant.
▲ how to use the theorems about chords and secants to solve problems.

In Section 10.4, chords were briefly discussed as you investigated finding the measures of angles that intersect circles. Several properties associated with these lines are useful in solving problems about circles.

DISCOVERY ACTIVITY

1. Use a compass to draw a large circle. Label the center O.

2. Mark four points on the circle and label them A, B, C, and D.

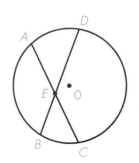

3. Draw $\overline{AC}$ and $\overline{BD}$. Your figure should resemble this one. Label the intersection of $\overline{AC}$ and $\overline{BD}$, E. E is not the center of the circle.

4. Measure (in centimeters to the nearest tenth) $\overline{AE}$, $\overline{EC}$, $\overline{BE}$, and $\overline{ED}$. Copy and complete the following.
   AE = _____ EC = _____ Product of AE and EC = _____
   BE = _____ ED = _____ Product of BE and ED = _____

5. What do you notice about the products?

6. Repeat steps 1–4 with a different circle and different segments. What can you conclude? Do you think you will always get this result?

You may have discovered that when two chords intersect, the product of the segment lengths of one chord is equal to the product of the segment lengths of the other chord.

DEFINITION 10-5-1 **A CHORD of a circle is a line segment whose endpoints are two points on the circle.**

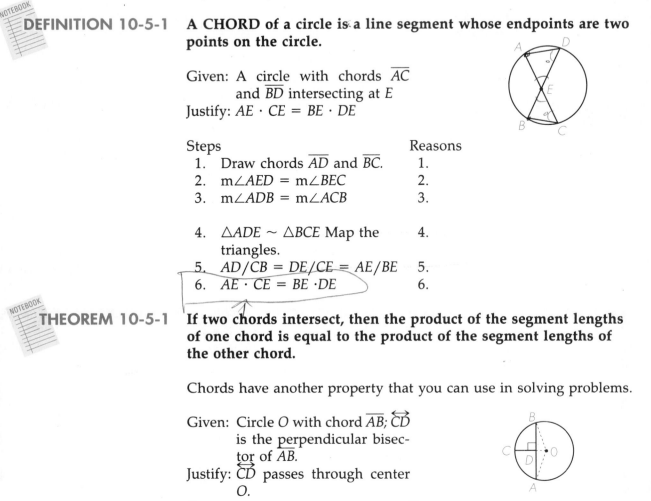

Given: A circle with chords $\overline{AC}$
   and $\overline{BD}$ intersecting at E
Justify: $AE \cdot CE = BE \cdot DE$

Steps	Reasons
1. Draw chords $\overline{AD}$ and $\overline{BC}$.	1.
2. $m\angle AED = m\angle BEC$	2.
3. $m\angle ADB = m\angle ACB$	3.
4. $\triangle ADE \sim \triangle BCE$ Map the triangles.	4.
5. $AD/CB = DE/CE = AE/BE$	5.
6. $AE \cdot CE = BE \cdot DE$	6.

THEOREM 10-5-1 **If two chords intersect, then the product of the segment lengths of one chord is equal to the product of the segment lengths of the other chord.**

Chords have another property that you can use in solving problems.

Given: Circle O with chord $\overline{AB}$; $\overleftrightarrow{CD}$
   is the perpendicular bisec-
   tor of $\overline{AB}$.
Justify: $\overleftrightarrow{CD}$ passes through center O.

Steps	Reasons
1. $AD = DB$	1. Definition of bisector
2. Draw $\overline{OA}$, $\overline{OD}$, and $\overline{OB}$.	2.
3. $OA = OB$	3.
4. $m\angle OAD = m\angle OBD$	4.
5. $\triangle AOD \cong \triangle BOD$	5.
6. $m\angle ADO = m\angle BDO$	6.
7. $m\angle ADO + m\angle BDO = 180°$	7.
8. $m\angle ADO = m\angle BDO = 90°$	8.
9. $\overline{DO}$ is perpendicular bisector of $\overline{AB}$.	9.
10. $\overleftrightarrow{DO}$ is the same line as $\overleftrightarrow{CD}$.	10.
11. $\overleftrightarrow{CD}$ passes through center O.	11.

THEOREM 10-5-2 | **The perpendicular bisector of a chord of a circle passes through the center of the circle.**

Example: In circle O, $\overline{AB}$ is a diameter, $\overline{CD} \perp \overline{AB}$, $AE = 5$, and $EB = 15$. Find DE and CE.

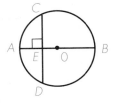

Since diameter $\overline{AB}$ is perpendicular to $\overline{CD}$, $DE = CE$.

$$DE \cdot CE = AE \cdot EB$$
$$(DE)^2 = 5 \cdot 15 = 75$$
$$DE = 5\sqrt{3} = CE$$

▶ DISCOVERY ACTIVITY

1. Use a compass to draw a large circle. Label the center O.

2. Label a point on the circle T and draw $\overline{OT}$.

3. At point T, construct a perpendicular to $\overline{OT}$.

4. Label a point P on the perpendicular, about 6 centimeters from T.

5. Label another point on the circle B, so that $\overline{BP}$ intersects the circle at point A.

6. Measure in centimeters the lengths of $\overline{PT}$, $\overline{PA}$, and $\overline{PB}$.

   $PT =$ _____ $PA =$ _____ $PB =$ _____
   $(PT)^2 =$ _____ $PA \cdot PB =$ _____

7. What do you observe? Do you think this will always be true? Try steps 1–6 for a different-size circle and tangent.

You may have discovered that the square of the length of the tangent equals the product of the lengths of the secant and its external segment.

DEFINITION 10-5-2 **A SECANT is a line segment that intersects a circle in two points and has one endpoint on the circle and the other endpoint outside the circle.**

DEFINITION 10-5-3 **The LENGTH OF A TANGENT is the length of a segment from a point outside a circle to the point of intersection of the tangent with the circle.**

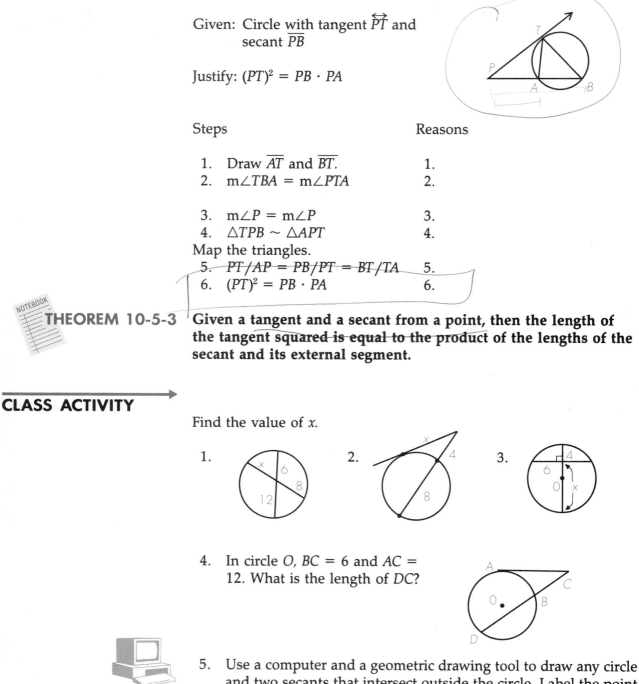

Given: Circle with tangent $\overleftrightarrow{PT}$ and
secant $\overline{PB}$

Justify: $(PT)^2 = PB \cdot PA$

Steps	Reasons
1. Draw $\overline{AT}$ and $\overline{BT}$.	1.
2. $m\angle TBA = m\angle PTA$	2.
3. $m\angle P = m\angle P$	3.
4. $\triangle TPB \sim \triangle APT$	4.
Map the triangles.	
5. $PT/AP = PB/PT = BT/TA$	5.
6. $(PT)^2 = PB \cdot PA$	6.

NOTEBOOK

THEOREM 10-5-3 **Given a tangent and a secant from a point, then the length of the tangent squared is equal to the product of the lengths of the secant and its external segment.**

CLASS ACTIVITY

Find the value of x.

1.

2.

3.

4. In circle O, $BC = 6$ and $AC = 12$. What is the length of DC?

5. Use a computer and a geometric drawing tool to draw any circle and two secants that intersect outside the circle. Label the point at which the two secants intersect A. Label the points at which one of the secants intersects the circle B and C and the points at which the other secant intersects the circle D and E. Find the lengths of $\overline{AC}$, $\overline{AB}$, $\overline{AE}$, and $\overline{AD}$. Repeat this activity several times, using a circle with a different radius and secants of different lengths each time. What do you notice about the product of AC and AB and the product of AE and AD in each case?

6. Use the figure you drew in Exercise 5 and draw $\overline{DC}$ and $\overline{BE}$. Write reasons to show that $AC \cdot AB = AE \cdot AD$.

Steps	Reasons
a. Draw $\overline{DC}$ and $\overline{BE}$.	a.
b. $m\angle BCD = m\angle DEB$	b.
c. $m\angle A = m\angle A$	c.
d. $\triangle AEB \sim \triangle ACD$	d.

Map the triangles.

e. $AE/AC = AB/AD = BE/DC$	e.
f. $AC \cdot AB = AE \cdot AD$	f.

Write a conclusion to summarize the result of the proof.

HOME ACTIVITY

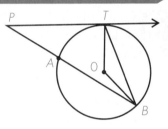

For circle O, classify the following.

1. $\overleftrightarrow{PT}$ 2. $\overline{AB}$ 3. $\overline{PB}$ 4. $\angle PBT$

5. O 6. $\widehat{AB}$ 7. $\overline{TO}$ 8. $\angle TOB$

9. Given: Circle O with $AC = 24$, $DX = 14$, and $BX = 10$. Find CX.

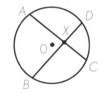

10. Given: Circle O with $AE = 15$ $BE = 7$, and $DE = 12$. Find CE and CD.

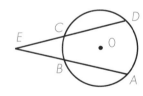

11. Given: Circle O with $BC = 4$ and $DC = 8$. Find AC.

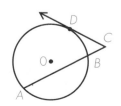

12. Given: Circle O with radius $\overline{OC}$, $AB = 12$, and $OD = 8$. Find the length of the radius.

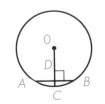

The diagram shows a section of a broken wheel. $\overline{AB} \perp \overline{DE}$, $AF = BF = 6$ cm, $EF = 2$ cm, and $\overline{ED}$ is a diameter.

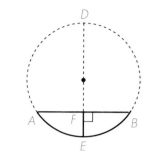

13. What would have been the length of $\overline{ED}$?

14. What was the original circumference of the wheel?

15. What was the area of the wheel?

Given: Circle O with $PT = 12$, $PF = 18$, m$\angle TOA = 90°$, m$\widehat{TF} = 168°$. $\overleftrightarrow{PT}$ is an tangent.

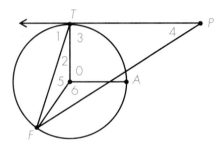

16. Find the measures of the numbered angles.

17. Find PA.

18. Given: Circle O with tangents $\overleftrightarrow{PQ}$ and $\overleftrightarrow{PR}$.
$PQ = 6$ cm, m$\angle RPO = 20°$.
Find m$\angle QRP$ and the length of $\overline{PR}$.

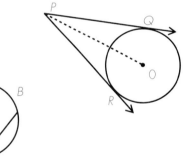

Given: Circle O with m$\widehat{CB} = 60°$ and m$\widehat{AXB} = 220°$. Find the measure.

19. m$\angle AXC$ 20. m$\angle CXB$ 21. m$\widehat{AB}$

CRITICAL THINKING

22. I'm tired of mowing the lawn outside my house, so I bought a sheep to eat the grass. The sheep is tied to one corner of the house. If it takes about 4000 square feet of grass to feed an adult sheep, will mine get enough to eat where she is?

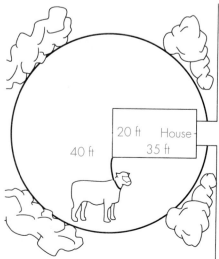

20 ft House

40 ft 35 ft

SECTION 10.6

Applications Involving Circles

After completing this section, you should understand
▲   how to solve problems using theorems about regular polygons and circles.

From your investigation of the properties of regular polygons and circles, you can see that there is a special relationship between these figures.

DISCOVERY ACTIVITY

1. Draw two circles, each with a radius of 8 centimeters. Cut them out.

2. Fold one circle in half three times.

3. Open the circle. Draw a line connecting the adjacent endpoints of the fold lines. What figure have you formed?

4. Fold the other circle in half, then fold the sides in so that they meet in a cone shape.

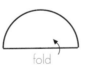
fold

folds

5. Open the circle and draw chords to connect adjacent endpoints. What figure have you formed?

6. Cut out several more circles and try to form a hexagon, an equilateral triangle, and a pentagon.

7. What do these figures all have in common?

You may have found that each of the polygons you drew inside a circle has its vertices on the circumference of the circle. The polygon is **inscribed** in the circle. The circle is **circumscribed** about the polygon.

When a circle is inscribed in a polygon, each side of the polygon is tangent to the circle.

Example: △ABC is equilateral and inscribed in circle O. AB = 8 in. What is the radius of the circle?

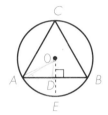

Radius $\overline{OE}$ is perpendicular to chord $\overline{AB}$ at D. AD = DB. Why?

Since AB = 8 in., AD = 4 in. What is OD?

$$(AO)^2 = (AD)^2 + (OD)^2$$
$$(AO)^2 = 4^2 + \left(\frac{4}{\sqrt{3}}\right)^2$$
$$(AO)^2 = 16 + 5\tfrac{1}{3} = 21\tfrac{1}{3}$$
$$AO = 4.6 \text{ in.} \quad \text{The radius of the circle is 4.6 in.}$$

CLASS ACTIVITY

1. Use a computer and a geometric drawing tool to draw any circle. Draw two diameters of the circle and label them $\overline{AC}$ and $\overline{BD}$. Draw $\overline{AB}$, $\overline{BC}$, $\overline{CD}$, and $\overline{DA}$. Repeat this activity several times, using a circle with a different diameter each time. Describe figure ABCD.

2. △ABC is equilateral and inscribed in a circle. Show that the angles of △ABC divide the circle into three arcs with the same measure. Give a reason for your answer.

3. A circular clock face is inscribed in a square frame. The side of the square measures 8 feet. What is the area of the part of the square that is not used for the face of the clock?

AREA OF A SECTOR

You can use the formula for the area of a circle to find the area of a part of the circle called a sector.

$\overparen{POQ}$ and $\angle POQ$ are the boundaries of sector POQ. You know that $m\angle POQ = m\overparen{POQ}$, because $\angle POQ$ is a central angle. You can use this ratio to find the area of sector POQ.

$$\frac{\text{Area(sector } POQ)}{\text{Area(circle } O)} = \frac{m\angle POQ}{360°}$$

Example: $\overline{AB}$ is a chord of circle O. Radius OA = 10 cm and $m\angle AOB = 90°$. What is the area of the shaded region?

Area(circle O) = $\pi r^2 = 100\pi$ cm^2
Area($\triangle AOB$) = $\frac{1}{2}bh = \frac{1}{2}(10 \cdot 10) =$ 50 cm^2

$$\frac{\text{Area(sector } AOB)}{100\pi} = \frac{90°}{360°}$$

Area(sector AOB) = $\frac{1}{4} \cdot 100\pi = 25\pi \approx 78.5$ cm^2
Area(shaded region) = Area(sector AOB) − Area($\triangle AOB$)
= $78.5 - 50 = 28.5$ cm^2

CLASS ACTIVITY

1. A circular theater is divided into six congruent sectors. The diameter of the theater is 60 feet. Two of the sectors are used for audience seating. What is the area of the space taken up by seats?

Find the area of the shaded region. Use $\pi \approx 3.14$.

2.

$r = 6$ ft
$m\angle AOB = 60°$

3.

$r = 12$ in.
$m\angle XOY = 90°$

4.

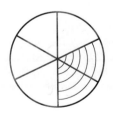

$AB = 36$ cm
$m\angle AOD = 60°$

CIRCLES AND SIMILARITY

In Chapter 9, you learned that for two similar polygons with sides S_1 and S_2, perimeters P_1 and P_2, and areas A_1 and A_2,

$$\frac{S_1}{S_2} = \frac{P_1}{P_2} \text{ and } \left(\frac{S_1}{S_2}\right)^2 = \frac{A_1}{A_2}.$$

S_1 ▱ S_2 ▱

Do you think that similar ratios will also be true for circles? Try it for two circles, one with a radius of 4 feet and one with a radius of 8 feet.

1. Find the circumference of each circle.

   Use $\pi \approx 3.14$.

2. What is the ratio of C_1/C_2? Is this ratio equal to the ratio of the radii of the two circles?

3. Predict the ratio of the areas of $circle_1$ and $circle_2$.

4. Find the area of each circle.

5. What is the ratio $\frac{A_1}{A_2}$? Does your answer match your prediction?

So, for two circles with radii r and R,

$$\frac{\text{Circumference (circle } r)}{\text{Circumference(circle } R)} = \frac{r}{R} \text{ and } \frac{\text{Area(circle } r)}{\text{Area(circle } R)} = \left(\frac{r}{R}\right)^2$$

HOME ACTIVITY

Find the area of the shaded region. Use $\pi = 3.14$.

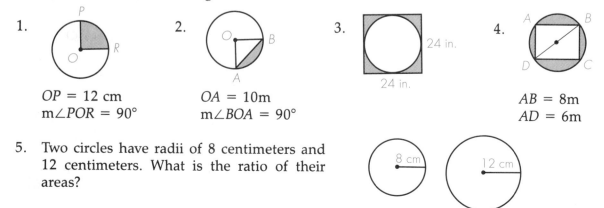

1.
   $OP = 12$ cm
   $m\angle POR = 90°$

2.
   $OA = 10$m
   $m\angle BOA = 90°$

3. 24 in. / 24 in.

4.
   $AB = 8$m
   $AD = 6$m

5. Two circles have radii of 8 centimeters and 12 centimeters. What is the ratio of their areas?

Find the value of *x* for each circle.

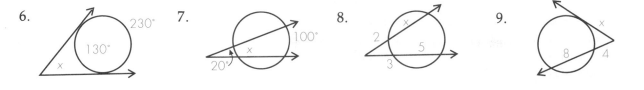

6.   230° 130° x

7.   100° x 20°

8.   x 2 5 3

9.   x 8 4

10.   A circular theater building has a stage that is 4 meters from the center of the circle. The radius of the building is 20 meters. How wide is the front edge of the stage?

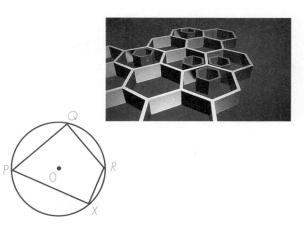

STAGE
4 m
20 m

11.   A Ferris wheel is made up of a regular hexagon divided in triangles and surrounded by a circular frame. The wheel has a radius of 50 feet. How many feet of lights would it take to outline the circumference and the spokes of the wheel?

12.   Quadrilateral *PQRX* is inscribed in circle *O*. Show that ∠*P* is supplementary to ∠*R* and ∠*Q* is supplementary to ∠*X*.

Q P O R X

CRITICAL THINKING

In Montana there is an unusual natural stone formation in the shape of a tall pillar, called Pompey's Pillar. Assume that the formation has a circular base. A surveyor stands 1000 feet from the base of the pillar and finds a tangent to the pillar is 1500 feet.

13.   How can you use this information to find the diameter of the pillar?

14.   What is the diameter?

10.1 *ABCDE* is a regular polygon. Its sides have equal lengths and its angles have equal measures. Where n is the number of sides, the measure of $\angle A$ is

$$m\angle A = \frac{(n-2)180°}{n}$$
$$m\angle A = \frac{3 \cdot 180°}{5} = 108°$$

10.2 The hexagon is a regular polygon with a side of 6 centimeters and a height of $3\sqrt{3}$ centimeters. The perimeter of the hexagon is 36 centimeters. The area is

$$A = \tfrac{1}{2}Ph$$
$$A = \tfrac{1}{2} \cdot 36 \cdot 3\sqrt{3} = 54\sqrt{3} \text{ cm}^2$$

10.3 Circle *O* has a diameter of 20 inches and a radius of 10 inches. The circumference of the circle is
$$C = \pi d \text{ or } C = 2\pi r \qquad \pi \approx 3.14 \ldots$$
$$C \approx 3.14 \cdot 20 \approx 62.8 \text{ in.}$$
The area of the circle is
$$A = \pi r^2$$
$$A = \pi(10)^2 = \pi \cdot 100 \approx 314 \text{ sq in.}$$

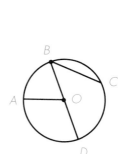

10.4 A central angle has its vertex at the center of a circle. For circle *O*, $m\widehat{AB} = 70°$ and $m\widehat{DC} = 110°$.

$\angle AOB$ is a central angle, so
$m\angle AOB = m\widehat{AB} = 70°$.

An inscribed angle has its vertex on the circle.

$\angle DBC$ is an inscribed angle, so
$m\angle DBC = \tfrac{1}{2}(m\widehat{DC}) = \tfrac{1}{2} \cdot 110° = 55°$.

For circle *O*, $m\widehat{QR} = 85°$ and $m\widehat{PS} = 75°$. $\overline{PR}$ and $\overline{QS}$ are chords that intersect inside the circle.

$m\angle PTS = \tfrac{1}{2}(m\widehat{QR} + m\widehat{PS})$
$m\angle PTS = \tfrac{1}{2}(85° + 75°) = 80°$

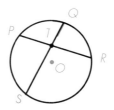

A secant is a line segment that intersects a circle in two points. $\overline{XZ}$ and $\overline{XT}$ are secants. $m\widehat{ZT} = 130°$ and $m\widehat{YS} = 40°$.

$$m\angle X = \tfrac{1}{2}(m\widehat{ZT} - m\widehat{YS})$$
$$m\angle X = \tfrac{1}{2}(130° - 40°) = 45°$$

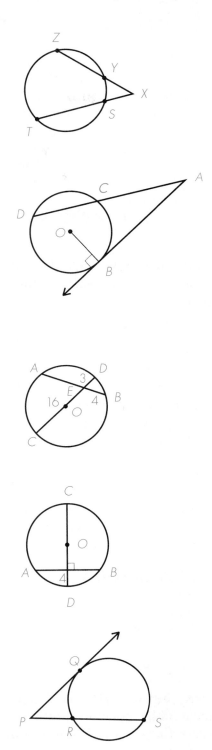

A tangent is a line that intersects a circle in one point. Line $\overleftrightarrow{AB}$ is tangent to circle O at point B. Secant $\overline{AD}$ intersects circle O at C and D. Radius $\overline{OB}$ is perpendicular to tangent $\overleftrightarrow{AB}$.

$m\widehat{BD} = 140°$ and $m\widehat{BC} = 80°$.
$$m\angle A = \tfrac{1}{2}(m\widehat{BD} - m\widehat{BC})$$
$$m\angle A = \tfrac{1}{2}(140° - 80°) = 30°$$

10.5 A chord of a circle has its endpoints on the circle. Chords $\overline{AB}$ and $\overline{CD}$ intersect at point E. Find AE.

$$AE \cdot EB = CE \cdot ED$$
$$AE \cdot 4 = 16 \cdot 3$$
$$AE = 12$$

The perpendicular bisector of chord $\overline{AB}$ passes through the center of the circle. $\overline{OD} = 10$. Find AB.

$$(AO)^2 = (OE)^2 + (AE)^2$$
$$10^2 = 6^2 + (AE)^2$$
$$8 = AE$$
$$AB = 2(AE) = 16$$

$\overleftrightarrow{PQ}$ is a tangent and $\overline{PS}$ is a secant of the circle. $PS = 25$ and $PR = 10$. Find PQ.

$$(PQ)^2 = PS \cdot PR$$
$$(PQ)^2 = 25 \cdot 10$$
$$PQ = 15.8$$

To draw a regular polygon using Logo, the turtle must turn through the exterior angles of the polygon. Since you know that the sum of the exterior angles for any polygon is 360°, the measure of each exterior angle of a regular n-gon is 360 ÷ n. So, to draw a regular n-gon, the turtle must make n turns of (360 ÷ n)° each. In Logo, this is called the Total Turtle Trip Theorem, or the TTT Theorem.

Example: Use the TTT Theorem to write a procedure to draw a regular pentagon.

The turtle must make five turns to draw a pentagon. So the measure of the angle at each turn is (360 ÷ 5)° or 72°. The measure of the angle at each turn can be computed in the procedure by the computer.

```
TO PENTAGON
  RT 90
  REPEAT 5 [FD 50 LT 360/5]
END
```

The output is shown below.

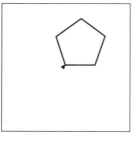

Use the TTT Theorem to write a procedure to tell the turtle to draw each polygon.

1. Octagon
2. Nonagon
3. Decagon
4. 15-gon
5. 20-gon
6. 30-gon

7. What figure does the 30-gon look like?

You have already discovered that as the number of sides of a regular polygon increases, the perimeter of the polygon approaches the circumference of the circle. This information along with the results of Exercises 1–7 indicates that a circle can be drawn in Logo.

The TTT Theorem tells us that to draw a regular polygon, the turtle turns through 360° to return to its starting position. This is also true when the turtle draws a circle. The basic command used to draw a circle using Logo is REPEAT X[FD Y RT Z], where X · Z = 360. When using this command, be sure that as X gets larger, Y gets smaller so that the circle will fit on the screen.

Example: Find X so that the command REPEAT X[FD 2 RT 5] can be used in a procedure to tell the turtle to draw a circle. Then write the procedure.

We know that $X \cdot Z = 360$. In the command REPEAT X[FD 2 RT 5], $Z = 5$. So, to find X, substitute 5 for Z and solve.

$$X \cdot 5 = 360$$
$$X = 72$$

So the command is REPEAT 72[FD 2 RT 5]. The procedure to draw the circle is given below.

```
TO CIRCLE
  RT 90
  REPEAT 72[FD 2 RT 5]
END
```

Find the unknown value so that each command can be used in a procedure to tell the turtle to draw a circle. Then write the procedure.

8. REPEAT X[FD 5 RT 12]   9. REPEAT X[FD 8 RT 36]   10. REPEAT X[FD 7 RT 18]

11. REPEAT 90[FD 2 RT Z]   12. REPEAT 24[FD 2 RT Z]   13. REPEAT 180[FD 1 RT Z]

Write a Logo procedure to draw a circle in each position on the screen.

14. Upper right-hand corner   15. Upper left-hand corner

16. Lower right-hand corner   17. Lower left-hand corner

Write procedures that tell the turtle to draw each figure.

18.

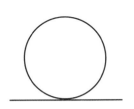

19.

Find the measure of an interior angle for each regular polygon.
1. Hexagon
2. Decagon
3. 21-gon
4. 18-gon
5. The measure of an interior angle of a regular polygon is 150°. How many sides does the polygon have?
6. A regular pentagon has a side of 20 centimeters and an apothem of 14 centimeters. What is its area?

Find the circumference and area. Use $\pi = 3.14$.

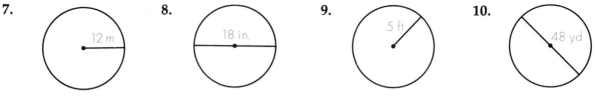

7. 12 m
8. 18 in.
9. 5 ft
10. 48 yd

11. How many revolutions will a 26-inch bicycle wheel make in traveling a distance of one mile? (1 mi = 5280 ft)

In circle O, m$\angle A = 40°$, m$\angle BDA = 20°$, and $\overline{CE}$ is a diameter. Find the measure.

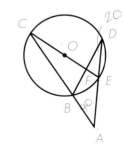

12. m$\angle ACE$ 13. m$\angle CBD$ 14. m$\angle DBA$

15. m$\overarc{BE}$ 16. m$\overarc{CB}$ 17. m$\overarc{CD}$

18. m$\angle CFD$ 19. m$\overarc{DE}$ 20. m$\angle DFE$

Find the value of x for each circle.

21.

22.
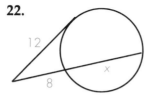

23.

24. $ABCD$ is a square inscribed in circle O. $AB = 10$ centimeters. What is the area of the shaded part of the circle?

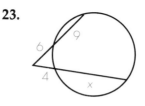

CHAPTER TEST

Find the measure of an interior angle for each regular polygon.
1. Octagon   2. Nonagon   3. 28-gon   4. 40-gon

Find the measure of an exterior angle for each regular polygon.
5. Hexagon   6. Pentagon   7. Decagon   8. 20-gon
9. The sum of the angle measures for a regular polygon is 7200°. How many sides does the polygon have?

Find the circumference and area of each circle. Use $\pi \approx 3.14$.

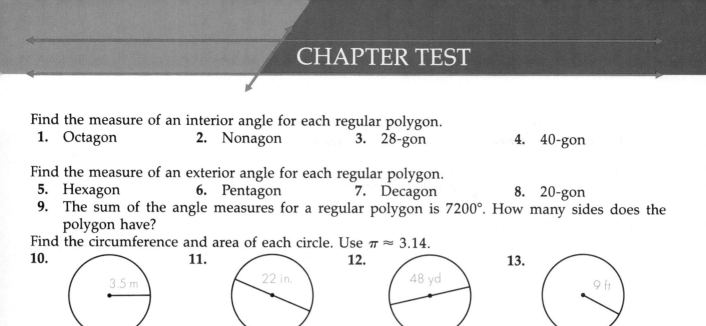

10. 3.5 m   11. 22 in.   12. 48 yd   13. 9 ft

A circular jogging path has a circumference of one mile.
14. What is the area in square miles of the space enclosed by the jogging path? Use $\pi \approx$ 3.14.
15. What is the length in feet of a path from one side of the circle to the other, through the center? (1 mi = 5280 ft)

In circle O, $m\widehat{TS} = 110°$, $m\angle TVU = 50°$, $m\angle QOR = 60°$, and $\overline{TR}$ is a diameter.

Find these measures.
16. $m\widehat{SR}$ 17. $m\widehat{TU}$ 18. $m\angle TPQ$
19. $m\widehat{QR}$ 20. $m\widehat{UQ}$ 21. $m\angle TVS$
22. $m\angle OQP$ 23. $m\angle TOQ$ 24. $m\angle SVR$

25. $ABCD$ is a rectangle inscribed in circle O. AD = 18 cm and AB = 24 cm. What is the area of the shaded region?

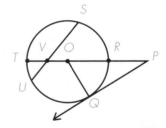

Find the value of x.
26. 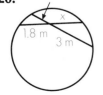 1.2 m, x, 1.8 m, 3 m

27. 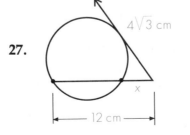 $4\sqrt{3}$ cm, x, 12 cm

28. 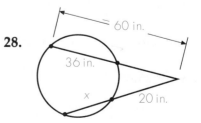 60 in., 36 in., x, 20 in.

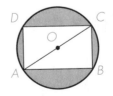

395

ALGEBRA SKILLS

The formulas you use to find the perimeter, area, and circumference of regular polygons and circles are called functions. These functions show the relationships between measures.

A **relation** is a set of ordered pairs.

$$R = \{(2, 4), (4, 8), (8, 16), (16, 32)\}$$

A **function** is a relation in which the first term in each ordered pair is different.

Example: $P = \{s, 4s\} = \{(1, 4), (2, 8), (3, 12), (4, 16)\}$

Function P shows the relation between the first term in the ordered pair (the length of a side of a square) and the second term in the ordered pair (the perimeter of the square).

Example: $Q = \{(1, 3), (2, 6), (1, 9), (3, 12)\}$ Is Q a function?

Two of the ordered pairs in relation Q have the same first term, so Q is not a function.

To evaluate a function, substitute values for the variables and perform the operations.

Function $S = \{4\pi r^2\}$, where $\pi \approx 3.14$.

1. Copy and complete the table.

Value of r		Value of S
1	$S = 4(3.14)1^2$	
2	$S = 4(3.14)2^2$	
4		
8		
16		
0.5		
0.001		

2. If the value of function S is 1234, what is the positive value of r to hundredths?

3. Function $x = \{x^2 + 1\}$. Find the value of f(x) for $x = \{-1, 0, 2, 3, 5\}$.

4. Function $\left\{(x, y)\ y = \frac{-5}{3} x + 2\right\}$. Find the value of y for $x = \{-2, -1, 0, 1, 2\}$.

Solve the equations.

5. $(y + 4) - (y - 6) = 5(y - 8)$

6. $16 - 4(2x + 1) = 6x + (1 - 3x)$

7. $\frac{3}{16} = \frac{x}{(x - 13)}$

8. $\frac{x^2}{3} + \frac{7x}{3} + 4 = 0$

In the diagram, $l \parallel m$ and t is a transversal. Give the degree measure of each angle.

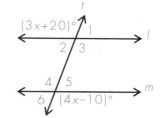

1. $\angle 1$ **2.** $\angle 2$ **3.** $\angle 3$

4. $\angle 4$ **5.** $\angle 5$ **6.** $\angle 6$

7. Measure the angle to the nearest degree. State the error in the measurement.

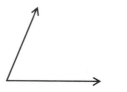

Use the information in the diagram. Find the measure of each angle.

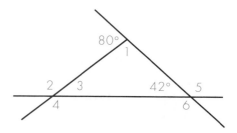

8. $\angle 1$ **9.** $\angle 2$

10. $\angle 3$ **11.** $\angle 4$

12. $\angle 5$ **13.** $\angle 6$

Which of the pairs of triangles are similar and why?

14.   **15.**

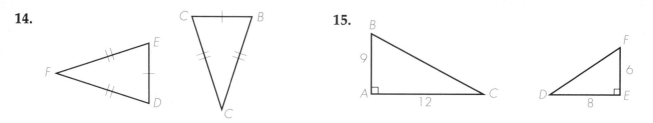

In the diagram, $m\angle BAC = m\angle DEC$. Find each length.

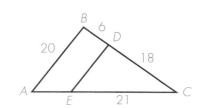

16. AE

17. DE

Refer to figures a–f.
18. Which triangles are congruent?
19. Which triangles are similar?

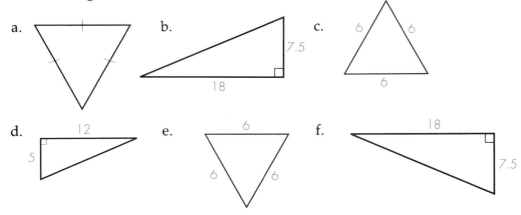

Write the converse of each statement.
20. If you passed the driving test, then you will get a driver's license.

21. If you have been to Paris, then you have been to the capital of France.

22. Draw and label an acute triangle *ABC*. Construct the median from *B* to $\overline{AC}$.

23. In △*DEF*, *G* and *H* are midpoints. What is the length of $\overline{DF}$?

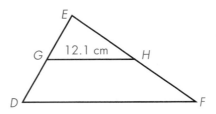

24. Draw and label a triangle KLM. Construct the incenter of the triangle.

Find the missing lengths to the nearest tenth. Use a calculator or the table of squares and square roots on page 562.

25.

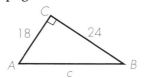

26.

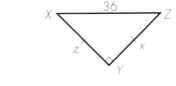

27.

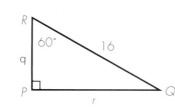

28.

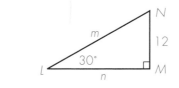

29. List the lengths of the sides of △PQR from least to greatest.

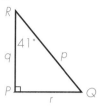

30. Write $m\angle A < m\angle B$, $m\angle A = m\angle B$, $m\angle A > m\angle B$, or "insufficient information."

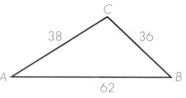

Find the length of the line segment that connects the two points.

31.

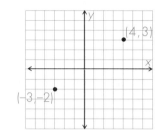

32.

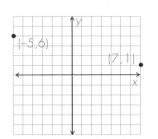

Refer to quadrilateral *ABCD*.

33. Name one pair of opposite sides.

34. Name one pair of adjacent angles.

35. $m\angle A + m\angle B + m\angle C + m\angle D = ?$

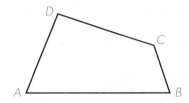

Is the figure a parallelogram? Give a reason for your answer.

36.

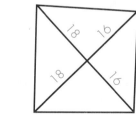

37.

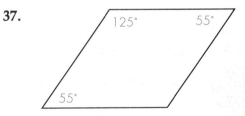

Figure *MNOP* is an isosceles trapezoid. Find the measures of the angles.

38. ∠OMN   **39.** ∠PQM   **40.** ∠MPO

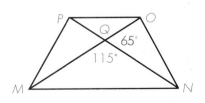

Figure LMNO is a rhombus. Find the lengths of the line segments.

41. $\overline{LM}$ **42.** $\overline{PM}$ **43.** $\overline{LP}$

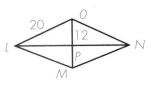

Find the height of the figure.

44.

315 cm²

21 cm

h

45.

20 in

240 in.²

30°

24 in.

h

46. A rectangular plot of land has a length of 300 feet and an area of 45,000 square feet. The length of a similar plot is 500 feet. What is the area?

Find the measure of an interior angle for each regular polygon.

47. hexagon   **48.** 20-gon   **49.** decagon   **50.** 50-gon

51. A regular octagon has a side of 5 cm and an apothem of 6 cm. What is the area?

Find the circumference and area. Use $\pi \approx 3.14$. Express answers to the nearest tenth.

52.

18 m

53.

25 yd

In circle O, $m\widehat{CF} = 115°$, $m\angle CGE = 60°$, $m\angle DOA = 70°$, and $\overline{CB}$ is a diameter. Find each measure.

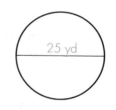

54. $m\widehat{FB}$ **55.** $m\widehat{CE}$ **56.** $m\angle CAD$
57. $m\widehat{ED}$ **58.** $m\angle ODA$ **59.** $m\angle FGB$

Find the value of x.

60.

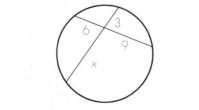

6   3
    9
   x

61.

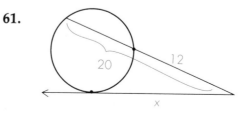

20   12

x

11

FIGURES IN SPACE

The great architect of the universe now begins to appear as a pure mathematician.

James H. Jeans

Area of Prisms and Cylinders

After completing this section, you should understand
▲ what a prism and a cylinder are.
▲ how to find the area of prisms and cylinders.

The figures at the right all have the shape of a *prism*.

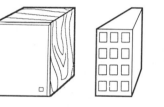

DISCOVERY ACTIVITY

1. On a 3 by 5 card, draw four lines parallel to the short ends of the card. Fold the card and tape the ends to form the open-ended, three-dimensional figure shown in the second diagram. Put it on a sheet of paper and trace around the bottom to make two pentagonal ends. Cut these out and tape them to top and bottom to make the closed figure shown in the third diagram.

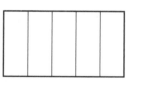
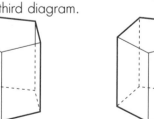

2. Take more 3 by 5 cards and make similar models in which the top and bottom are triangular, square, and so on.
3. Are the top and bottom ends parts of parallel planes?
4. Are the ends congruent?
5. What can you say about the vertical edges (the lines where you folded or taped)?

The top, bottom, and sides of the model you made in the Discovery Activity are *polygonal regions* and the three-dimensional figures themselves are *prisms*.

DEFINITION 11-1-1: **A POLYGONAL REGION is a polygon together with all the points inside the polygon.**

DEFINITION 11-1-2: **Suppose two congruent polygons are situated in parallel planes so that all the line segments joining corresponding vertices are parallel. The union of the two congruent polygonal regions and all the line segments joining corresponding points of the polygons is a PRISM.**

Any three-dimensional figure that can be formed in this way is a prism. The parallel polygonal regions in the definition are called the *bases* of the prism. Line segments joining corresponding vertices of the bases are called **lateral edges.** The sides of the bases are called **base edges**. Any two corresponding base edges and the lateral edges that connect them determine a parallelogram-shaped region called a **lateral face** of the prism. The point where two base edges intersect a lateral edge is called a **vertex** (plural: vertices).

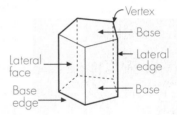

If the lateral edges of a prism are perpendicular to the bases, the prism is a **right prism.** Otherwise, it is an **oblique** prism. A prism is named according to the shape of its bases.

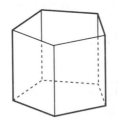 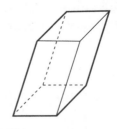 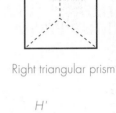

Right pentagonal prism   Oblique square prism   Right triangular prism

Example: The figure shown is an oblique triangular prism.

        Bases: △FGH and △F'G'H'

Lateral faces: □FGG'F', □GHH'G', and □HFF'H'

        Edges: $\overline{FG}$, $\overline{GH}$, $\overline{FH}$, $\overline{F'G'}$, $\overline{G'H'}$, $\overline{H'F'}$, $\overline{FF'}$, $\overline{GG'}$, and $\overline{HH'}$

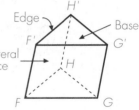

CLASS ACTIVITY

Identify each edge, base, and lateral face.

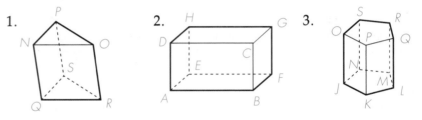

1.   2.   3.

Carefully draw each figure.

4.   An oblique rectangular prism 5.   A right hexagonal prism

CYLINDERS

Cylinders are formed very much like prisms, but their bases are curved regions.

DEFINITION 11-1-3: **A CYLINDER is three-dimensional figure consisting of two congruent curved regions in parallel planes and the line segments joining corresponding points on the curves that determine the regions. Segments joining corresponding points of the curves are parallel.**

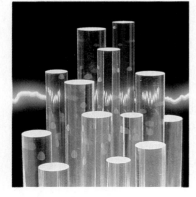

If the curved regions are circular, the cylinder is a **circular cylinder.** The segment joining the centers of the bases is called the **axis.** If the axis is perpendicular to the bases, the cylinder is a **right** circular cylinder. Otherwise, it is an **oblique** circular cylinder.

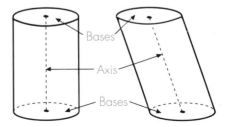

SURFACE AREA OF A RIGHT PRISM

With enough information you can find the **total surface area** of a right prism. Just find the total area of all lateral faces (the **lateral area**) and the area of the two bases.

Example: Find the total surface area of the right triangular prism. There are 3 rectangles and 2 right triangles.

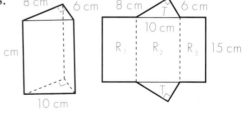

Area of bases = $2T = 2\left(\frac{1}{2}bh\right) = 2 \times \frac{1}{2} \times 8 \times 6$, or 48 cm²
Lateral area = $R_1 + R_2 + R_3$
$= (15 \times 8) + (15 \times 10) + (15 \times 6)$
$= 15(8 + 10 + 6)$ (Use the property ab + ac + ad = a(b + c + d).)
$= 15(24)$, or 360 cm²
Total surface area = 48 + 360, or 408 cm²

In the example, a shortcut was used to find the lateral area:
$(15 \times 8) + (15 \times 10) + (15 \times 6) =$
$15(8 + 10 + 6)$
Notice that 15 is the distance between the planes that contain the bases. Also, $(8 + 10 + 6)$ is the **perimeter** of the base. The distance between the planes containing the bases is called the **height** of the prism or cylinder.

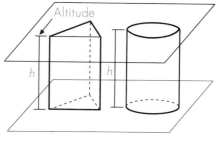

THEOREM 11-1-1   **If the height of a prism is h and each base has perimeter p and area B, the formula for the total surface area TA is TA = 2B + ph.**

SURFACE AREA OF A RIGHT CIRCULAR CYLINDER

The lateral area of a right circular cylinder is found in the same way as the lateral area of a right prism.

Lateral area = circumference of base × height of cylinder
$= 2\pi r \times h$, or $2\pi rh$
Area of bases = 2 × (area of each base)
$= 2 \times \pi r^2$, or $2\pi r^2$

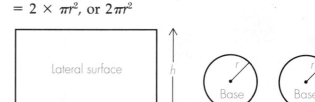

The total surface area of a right circular cylinder is found in the same way as the total surface area of a right prism:
Total area = 2 × (area of base) + lateral area

THEOREM 11-1-2   **If the height of a right circular cylinder is h and the radius of each circular base is r, then the formula for the total surface area is: Total area = $2\pi r^2 + 2\pi rh$**

CLASS ACTIVITY

Refer to the trapezoidal prism at the right. The height of the trapezoid is 1.5. Use a calculator to find the indicated region.

1. Find the area of each trapezoidal base. (Hint: Review Section 9.5.)
2. Find the perimeter of trapezoid *ABCD*.
3. Find the lateral area of the prism.
4. Find the total area of the prism.

Refer to the right circular cylinder at the right. The radius of each base is 2.2 in.

5. Find the area of each circular base.
6. Find the circumference of each base.
7. Find the lateral area of the cylinder.
8. Find the total area of the cylinder.

2.2 in.

5 in.

HOME ACTIVITY

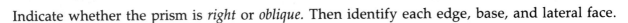

Indicate whether the prism is *right* or *oblique*. Then identify each edge, base, and lateral face.

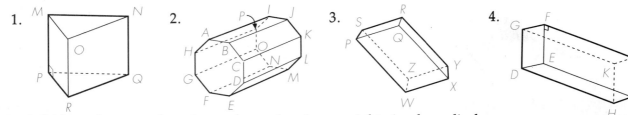

1. 2. 3. 4.

Find the lateral area and total area for each prism or right circular cylinder.

5. 6. 7. 8.

7 4 10

r = 3 cm 8 cm

5 13 12 20

e = 4 in.

9. 10. 11. 12.

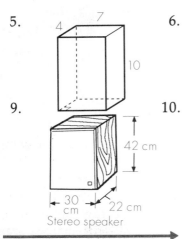

42 cm 30 cm 22 cm
Stereo speaker

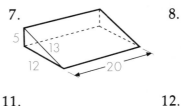

r = 12 m 40 m
Oil storage tank

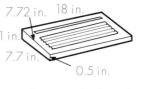

7.72 in. 18 in. 1 in. 7.7 in. 0.5 in.
Computer keyboard

r = 9 in. 12 in.
Hassock

CRITICAL THINKING

13. Which cylinder has the larger lateral area?
14. Which cylinder has the larger total area?

2 4 2
4

SECTION
11.2

Volume of Prisms and Cylinders

After completing this section, you should understand
▲ what volume is.
▲ how to find the volume of prisms and cylinders.

You are familiar with right rectangular prisms such as those shown here.

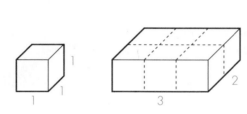

These prisms occupy space. How do you tell how much space they occupy? You can think of the larger prisms being filled with **unit cubes,** like the one on the left.

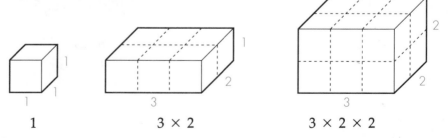

| 1 | 3 × 2 | 3 × 2 × 2 |

The prism on the left is a **cube** because all twelve of its edges are equal. It is a **unit cube** because each edge measures one unit of length. If the unit of length is the centimeter, then the space occupied by the unit cube is 1 **cubic centimeter.** In the figures just above, the prisms occupy a space of 1 cubic centimeter (1 cm³), 6 cubic centimeters (6 cm³), and 12 cubic centimeters (12 cm³), respectively. These numbers are the respective **volumes** of the three figures.

There are cubic units other then the cubic centimeter, such as the cubic meter (m^3), the cubic foot (ft^3), and so on.

DEFINITION 11-2-1: **The VOLUME of a figure is the number of cubic units that the figure contains.**

You can find how many unit cubes will fit inside a right rectangular prism if you know its *length*, *width*, and *height*.

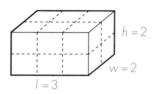
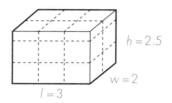

volume = number of cubes   volume = number of cubes
= 3 × 2 × 2   = 3 × 2 × 2.5
= 12 cubic units   = 15 cubic units

This suggests the following postulate.

POSTULATE 11-2-1 **The volume of a right rectangular prism is the product of its length, width, and height: $V = lwh$.**

Examples: Find the volume of each right rectangular prism.

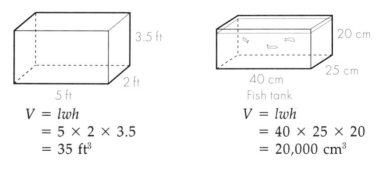

$V = lwh$ $V = lwh$
= 5 × 2 × 3.5   = 40 × 25 × 20
= 35 ft^3 = 20,000 cm^3

SOLIDS WITH EQUAL VOLUMES

At the right are shown two prisms that have the same height and the same dimensions for the base. How do the volumes compare? The next Discovery Activity will help you to answer this question.

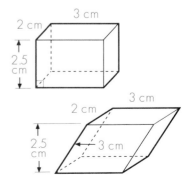

DISCOVERY ACTIVITY

This Activity may be done either in actuality or in your imagination.

1. Open a ream (500 sheets) of copier paper and place the ream flat on top of a table, so that the ream retains its original shape as a right rectangular prism.

2. Gently tap one end of the ream so that the ream slants, as shown in the second figure.

3. How does changing the shape of the ream affect the space it occupies? In other words, how does it affect its volume?

As a result of the Discovery Activity, you should see that the volume of the ream of paper will not change. This illustrates an important idea known as **Cavalieri's Principle.**

POSTULATE 11-2-2 **(Cavalieri's Principle) If two solid figures have equal heights and bases of the same area, and if every plane parallel to the bases always cuts off two cross-sections of equal area, then the two solids have equal volume.**

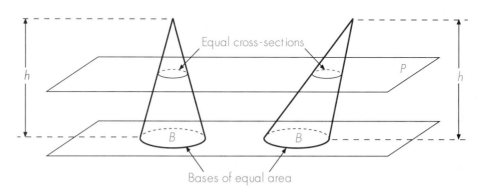

Cavalieri's Principle can be used to compare two prisms that have equal heights and bases of the same area.

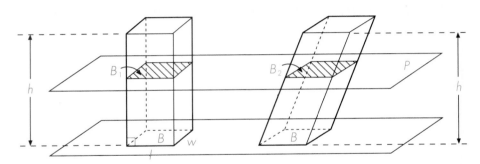

In the figure above, P is any plane that is parallel to the bases of the prisms. The plane cuts each prism in a cross-section. The left prism is a right rectangular prism with dimensions *l, w,* and *h.* The area of the base of each prism is *B.* It can be shown that the area B_1 of the left cross-section is equal to *B,* the area of the base of the prism at the left. Similarly, the area B_2 can be shown to equal *B,* the area of the base of the prism at the right. Therefore, $B_1 = B_2$. Thus, the two prisms obey Cavalieri's Principle. The same is true even when the bases are shapes other than rectangles.

THEOREM 11-2-1 **The volume of a prism is the product of the area *B* of one of its bases and the height *h* of the prism: $V = Bh$.**

Examples: Find the volume of each prism.

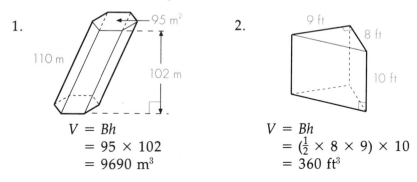

1.

$V = Bh$
$= 95 \times 102$
$= 9690 \text{ m}^3$

2.

$V = Bh$
$= (\frac{1}{2} \times 8 \times 9) \times 10$
$= 360 \text{ ft}^3$

CLASS ACTIVITY

Find the volume of each prism.

1.

$B = 63 \text{ in.}^2$

17 in.

2.

6 cm
5.2 cm 5.2 cm
11 cm
7 cm 6 cm

THE VOLUME OF A CYLINDER

Two cylinders that have the same height and bases of equal area can be studied in the same manner as the two prisms on the previous page.

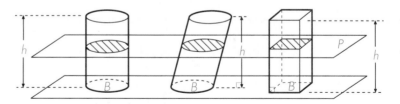

In the diagram above, the prism and the two cylinders all have bases with the same area B. They also have the same height h. It can be shown that in plane P, which is parallel to the bases of the two cylinders, all the cross-sections of the two cylinders have the same area. So, the cylinders obey Cavalieri's Principle. This leads to the following theorem.

THEOREM 11-2-2 **The volume of a circular cylinder is the product of the area B of one of its bases and the height h of the cylinder: $V = Bh$.**

Example: Find the volume of the circular cylinder.

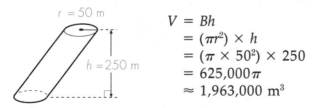

$$V = Bh$$
$$= (\pi r^2) \times h$$
$$= (\pi \times 50^2) \times 250$$
$$= 625{,}000\pi$$
$$\approx 1{,}963{,}000 \text{ m}^3$$

To the nearest thousand cubic meters, V is 1,963,000 m³.

CLASS ACTIVITY

Use a calculator to find the volume of each cylinder. Answer to the nearest whole unit.

1. $d = 6$ yd

    $h = 14$ yd

2. $r = 1.5$ ft

    $h = 10$ ft

HOME ACTIVITY

In this activity, use your calculator as needed. Find the volume of each prism. Answer to the nearest whole unit.

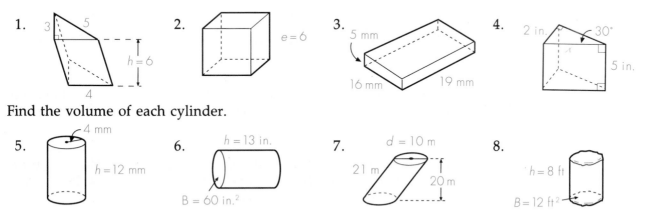

1. 3 5 h = 6 4

2. e = 6

3. 5 mm 16 mm 19 mm

4. 2 in. 30° 5 in.

Find the volume of each cylinder.

5. 4 mm h = 12 mm

6. h = 13 in. B = 60 in.²

7. d = 10 m 21 m 20 m

8. h = 8 ft B = 12 ft²

9. Find the volume of the gasoline storage tank at the right. Give your answer in gallons. (1 ft³ = 7.5 gal)

d = 40 ft h = 47 ft

10. Find the volume of the barn.

4 yd 7 yd 20 yd 16 yd

11. A basketball court has a volume of 126,000 ft³. The floor dimensions are 84 ft by 50 ft. How high is the ceiling?

12. The length of a diameter of an automobile cylinder is called the *bore*. The distance the piston moves in the cylinder is the *stroke*. The engine capacity of a car is the combined volume of all its cylinders. Find the engine capacity of a 6-cylinder engine if the bore is 3.88 in. and the stroke is 3.25 in.

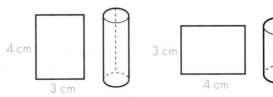

Stroke

Bore

CRITICAL THINKING

13. A cylinder can be modeled using a rectangular piece of paper in two ways, as shown below. Which cylinder has the greater volume?

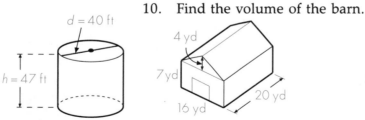

4 cm 3 cm

First cylinder

3 cm 4 cm

Second cylinder

SECTION
11.3

Area of Pyramids and Cones

After completing this section, you should understand
▲ what a pyramid is.
▲ what a cone is.
▲ how to calculate the surface area of a pyramid.
▲ how to calculate the surface area of a cone.

A closed, three-dimensional figure made up entirely of polygonal regions is called a **polyhedron.** Each of these figures is a polyhedron.

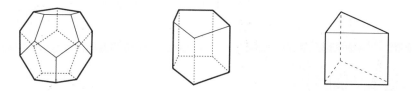

You have already studied one special kind of polyhedron, namely prisms. The figures that follow are examples of another special kind of polyhedron.

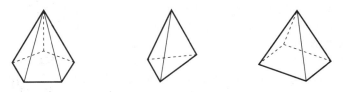

This kind of polyhedron is called a **pyramid.**

PYRAMIDS

To make a pyramid, start with a polygonal region and a point P not in the plane of the region. Then connect that point to the vertices of the region by means of line segments.

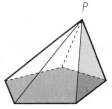

413

Each side of the polygonal region and the line segments connecting its endpoints to P determine a triangular region. All of these triangular regions together with the polygonal region you started with form a pyramid. You can also think of the pyramid in the manner described in this definition.

NOTEBOOK

DEFINITION 11-3-1: **A PYRAMID is a polyhedron formed by a polygonal region and all the line segments connecting a point P not in the plane of the region to the points of the polygon that determine the region.**

The polygonal region mentioned in the definition is the *base* of the pyramid. Point P is the *vertex* of the pyramid. Segments joining the vertex P to the vertices of the base are called *lateral edges*. Each triangular region determined by two lateral edges and an edge of the base is a *lateral face*. The perpendicular distance from the vertex P to the plane of the base is the *height* of the pyramid. The pyramid is called by the kind of polygon it has as its base.

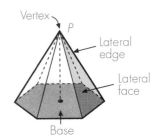

Vertex — P

Lateral edge

Lateral face

Base

DISCOVERY ACTIVITY

Consider a pyramid whose base is a regular pentagon. Suppose all the lateral faces are congruent isosceles triangles. Let e be the length of each side of the base. Let s be the height to one of the triangular lateral faces.

1. What is the area of the lateral face PAB?

2. What is the area of each of the other lateral faces?

3. In terms of the lateral faces, what does $\frac{1}{2} \times 5e \times s$ represent?

4. In terms of the base, what does 5e represent?

In the Discovery Activity, the pyramid was a special kind of pyramid called a *regular* pyramid.

A **regular pyramid** is a pyramid that has a regular polygonal region as its base and lateral edges that are all of equal length. The height s of each lateral face of a regular pyramid is called the **slant height** of the pyramid.

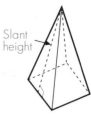

Slant height

From the Discovery Activity, you may have discovered how the slant height of a pyramid and the perimeter of the base can be used to express the **lateral area** of the pyramid.

THEOREM 11-3-1 **If p is the perimeter of the base of a regular pyramid and s is the slant height, the lateral area LA is given by the formula $LA = \frac{1}{2}ps$.**

Recall that a pyramid is identified by the shape of its base, as illustrated below.

      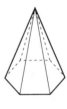

Triangular pyramid **(Tetrahedron)**   Regular quadrangular pyramid   Hexagonal pyramid   Regular hexagonal pyramid

Example: Find the lateral area of the regular square pyramid below.

Lateral area $= \frac{1}{2}ps$

$\qquad = \frac{1}{2}(4 \times 7) \times 10$

$\qquad = 140 \text{ mm}^2$

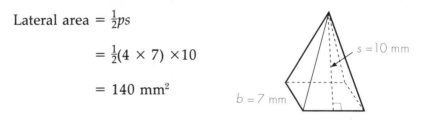

$s = 10 \text{ mm}$

$b = 7 \text{ mm}$

The **total area** of a pyramid is the sum of its lateral area and the area of the base.

THEOREM 11-3-2 **If B is the area of the base of a regular pyramid, p the perimeter of the base, and s the slant height of the pyramid, then the total area (TA) is given by the formula $TA = B + \frac{1}{2}ps$.**

Example: Find the total area of the regular tetrahedron at the right.

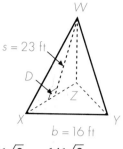

The base XYZ is an equilateral triangle. The altitude ZD of the triangular base divides the triangle into two congruent 30°-60° right triangles (see Section 7.3). So $XD = 8$ and $ZD = 8\sqrt{3}$.

Area of base of pyramid $= \frac{1}{2} \times 16 \times 8\sqrt{3} = 64\sqrt{3}$

Total area of pyramid $= B + \frac{1}{2}ps$
$= 64\sqrt{3} + \frac{1}{2} \times (3 \times 16) \times 23$
$= 64\sqrt{3} + 552$
$\approx 64 \times 1.732 + 552$
$\approx 110.8 + 552$, or about 663 ft²

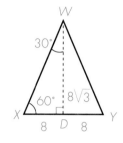

CLASS ACTIVITY

Find the total area of each regular pyramid.

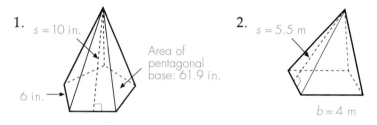

1. $s = 10$ in.
   Area of pentagonal base: 61.9 in.
   6 in.

2. $s = 5.5$ m
   $b = 4$ m

CONES

Each figure below has a vertex not in plane K and consists of a closed region and all the segments joining the vertex to points on the boundary of the region.

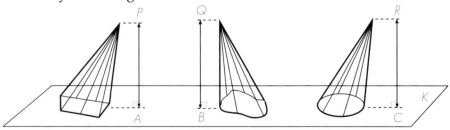

In the left figure, the region in plane K is polygonal. The figure is a pyramid. In the second figure, the boundary of the region in plane K is an irregular closed curve. Because the boundary is curved, the figure is a **cone.** Thus, a cone is similar to a pyramid except that the boundary of its base is curved. In the third figure, the region in plane K is circular. The cone is a **circular cone.**

DEFINITION 11-3-2 **A cone** is a figure formed by a closed curve region and all the line segments joining points on the boundary of the region to a point not in the plane of the region. This point is called the **vertex** of the cone. The perpendicular distance from the vertex to the plane of the base is

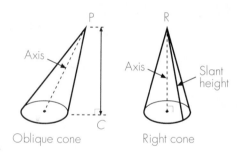

the **height** of the cone. Two kinds of circular cones are shown at the right, an *oblique cone* and a *right cone*. The segment joining the vertex to the center of the base is called the **axis** of the cone. In a right cone, the axis is perpendicular to the base. The right cone has both a height and a **slant height,** which is the length of a line segment that joins the vertex to the circle.

The pie-shaped figure at the left below is called a *sector* of a circle. It can be shown that the area of the sector is $\frac{1}{2}$ the product of the arc length *AB* and the radius *s* of the circle from which the sector was cut. When the sector is folded so that line segments *AP* and *BP* coincide, a right circular cone is formed. The lateral area of the cone is equal to the area of the sector.

Lateral area = Area of sector

$$= \frac{1}{2} \times \text{arc length} \times \text{radius}$$
$$= \frac{1}{2} \times 2\pi r \times s$$
$$= \pi rs$$

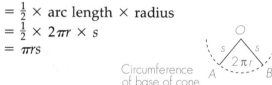

Circumference of base of cone

Notice that the circumference of the cone's base equals the arc length of the circle from which the sector was cut.

THEOREM 11-3-3: **The lateral area LA of a right circular cone with a base of radius *r* and slant height *s* is given by LA = πrs. *r* is the radius of the base and *s* is the slant.**

THEOREM 11-3-4: **The total area TA of a right circular cone is the sum of the area of the base and the lateral area. If *r* is the radius of the base and *s* is the slant height, the total area is given by the formula TA = $\pi r^2 + \pi rs$.**

Example: Find the total area.

Total area = $\pi r^2 + \pi rs$
$$= \pi \times 3^2 + \pi \times 3 \times 5$$
$$= 9\pi + 15\pi$$
$$= 24\pi$$
$$\approx 24 \times 3.14, \text{ or about } 75 \text{ mm}^2$$

r = 3 mm   *s* = 5 mm

CLASS ACTIVITY

Use a calculator to find the total area of each right circular cone.

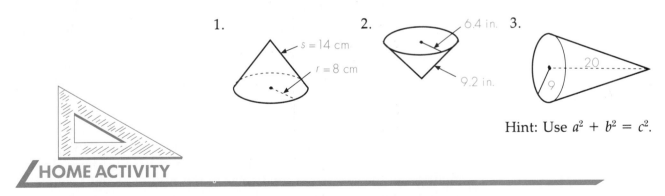

1. $s = 14$ cm
    $r = 8$ cm

2. 6.4 in.
    9.2 in.

3. 20
    9

Hint: Use $a^2 + b^2 = c^2$.

HOME ACTIVITY

Find the total area of each regular pyramid or right circular cone.

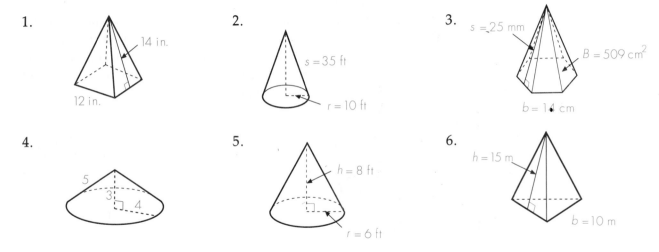

1. 14 in.
    12 in.

2. $s = 35$ ft
    $r = 10$ ft

3. $s = 25$ mm
    $B = 509$ cm^2
    $b = 14$ cm

4. 5
    3
    4

5. $h = 8$ ft
    $r = 6$ ft

6. $h = 15$ m
    $b = 10$ m

7. The top of the Washington Monument in Washington, D.C. consists of a regular square pyramid with a height of 55 ft. The length of a side of the base of the pyramid is about 34.4 ft. Find the lateral area of the pyramid.

8. The roof of a house has the shape of a right circular cone. Its height is 24 ft and the diameter of the base is 20 ft. Shingles for the roof cost $24 per bundle and 3 bundles are needed for every 100 ft^2. Find the cost of the shingles.

CRITICAL THINKING

9. A 5-12-13 right triangle is rotated about the hypotenuse to form the double cone shown at the right. Find the area of the common base of the two cones.

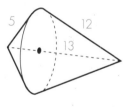

5   12
13

SECTION 11.4
Volume of Pyramids and Cones

After completing this section, you should understand
▲ how to find the volume of a pyramid.
▲ how to find the volume of a cone.

You know how to calculate the total area and the volume of a right prism and a right circular cylinder. You also know how to calculate the total area of a regular pyramid and a right circular cone. Now you will learn how to calculate the volume of both a pyramid and a cone.

MAKING A PRISM FROM THREE PYRAMIDS

In the diagram below, three pyramids of the same volume and shape are colored red, white, and blue, respectively. From their initial positions, the two outside pyramids approach the middle pyramid, as shown.

Initial position

Pyramids come together

Polyhedron is formed

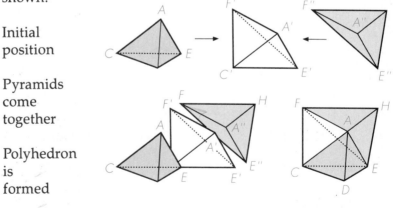

Notice how the outside pyramids fit with the middle pyramid.

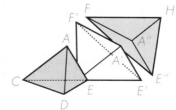

419

Face *ACE* of the red pyramid exactly fits onto face *A'C'E'* of the white pyramid.

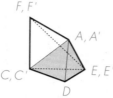

Face *A"E"F* of the blue pyramid exactly fits onto face *A'E'F'* of the white pyramid.

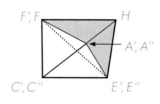

The 3 pyramids come together to form a polyhedron. Is the polyhedron a prism? The Discovery Activity below will deal with that question.

DISCOVERY ACTIVITY

Instead of assembling three pyramids into one solid figure (hoping that it is a triangular prism), do the reverse. Start with a triangular prism and partition it into pyramids as shown below.

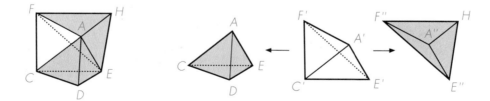

1. What kind of quadrilateral is *CEHF*, the back face of the prism?

2. Are triangles *CEF* and *EFH* congruent? How do you know?

3. Taking *A* as the vertex of pyramids *ACEF* and *AEHF*, do you think that the two pyramids have the same height? the same volume?

4. Are triangles *CDE* and *AHF* congruent? How do you know?

5. Do you think that pyramid *EAHF*, with vertex *E*, has the same height as pyramid *ACDE*, with vertex *A*? Do the two pyramids have the same volume?

6. Do all 3 pyramids have the same volume?

As a result of the Discovery Activity, you may have been able to conclude that a triangular prism can be partitioned into three pyramids of the same volume. Here is a summary of the proof that this is so.

The back face of the prism, quadrilateral *CEHF*, is a parallelogram, since the lateral faces of all prisms are parallelograms. Therefore, diagonal *EF* divides the parallelogram into two congruent triangles *CEF* and *EHF*. Two pyramids, *ACEF* and *AEHF*, then have congruent bases. They also have the same altitude (not shown) drawn from vertex *A* to the plane of their bases. It can be shown that if two pyramids have the same height and bases with the same area, then they have the same volume. Therefore, pyramids *ACEH* and *AEHF* have the same volume. A similar argument can be used to show that pyramids *EAHF* and *ACDE* have the same volume. Since all three pyramids have the same volume *V*, you can conclude that:

$3 \times$ volume of 1 pyramid = volume of triangular prism

or   volume of 1 pyramid = $\frac{1}{3}$ volume of triangular prism

NOTEBOOK

THEOREM 11-4-1 **The volume of a triangular pyramid is equal to one-third of the product of the area of the base and the height. If *B* is the area of the base and *h* is the height, then the volume is given by the formula $V = \frac{1}{3} Bh$.**

Any pyramid can be partitioned into triangular pyramids, as illustrated. Therefore, Theorem 11-4-1 can be extended to cover all pyramids.

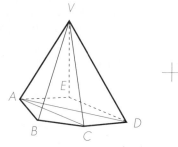

NOTEBOOK

THEOREM 11-4-2 **The volume of any pyramid is equal to one-third of the product of the area of the base and the height. The formula is $V = \frac{1}{3} Bh$.**

Examples: Find the volume of each pyramid.

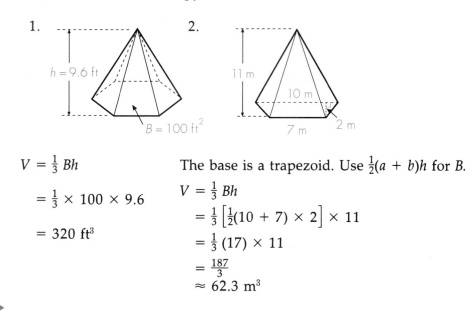

1.

$h = 9.6$ ft

$B = 100$ ft^2

2.

11 m

10 m

7 m   2 m

$V = \frac{1}{3} Bh$

$= \frac{1}{3} \times 100 \times 9.6$

$= 320$ ft^3

The base is a trapezoid. Use $\frac{1}{2}(a + b)h$ for B.

$V = \frac{1}{3} Bh$

$= \frac{1}{3} \left[\frac{1}{2}(10 + 7) \times 2 \right] \times 11$

$= \frac{1}{3} (17) \times 11$

$= \frac{187}{3}$

≈ 62.3 m^3

CLASS ACTIVITY

Find the volume of each pyramid.

1.

$h = 11$ m

7.5 m

8 m

2.

$B = 114$ in.^{2}

$h = 11$ in.

THE VOLUME OF A CONE

If you select equally spaced points around the circular base of a right cone, you can put a regular polygon inside the base of the cone. By connecting the vertices of this polygon to the vertex of the cone you can fit a regular pyramid inside the cone. The more sides the regular polygon has, the closer its area is to that of the circular base of the cone. Also, the closer the volume of the regular pyramid is to the volume of the cone. The height h of the pyramids and cone stay the same.

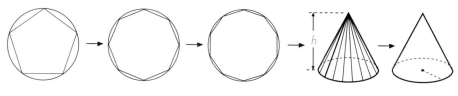

5 sides   8 sides   10 sides

Since the volume of each pyramid is $V = \frac{1}{3}$(area of polygon)h, these volumes are approaching closer and closer to $\frac{1}{3}Bh$, where B is the area of the base of the cone. This thinking leads to the following theorem.

THEOREM 11-4-3: **The volume of a right circular is equal to one-third the product of the area of the base and the height of the cone: $V = \frac{1}{3}Bh = \frac{1}{3}\pi r^2 h$.**

Example: Find the volume of the right circular cone below.

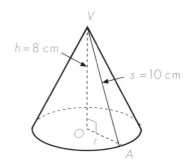

You need to find the radius of the base so you can use the formula for the volume of a cone. Use the Pythagorean Theorem.

$$r^2 + 8^2 = 10^2$$
$$r^2 + 64 = 100$$
$$r^2 = 36$$
$$r = 6$$

Next, use Theorem 11-4-3.

$$V = \frac{1}{3}\pi r^2 h$$
$$= \frac{1}{3}\pi \times 6^2 \times 8$$
$$= 96\pi \approx 301 \text{ cm}^2$$

CLASS ACTIVITY

Use a calculator to find the volume of each cone.

1.

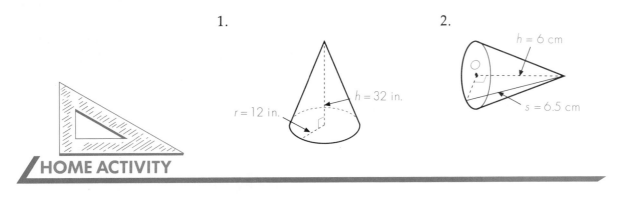

$h = 32$ in.

$r = 12$ in.

2.

$h = 6$ cm

$s = 6.5$ cm

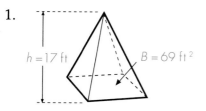

HOME ACTIVITY

Find the volume of each pyramid.

1.

$h = 17$ ft

$B = 69$ ft^2

2.

$h = 5$ m

2.3 m   6 m

3.

$h = 7$ in.

4 in.   4 in.

Find the volume of each cone.

4.

$h = 20$ m

$B = 300$ m^2

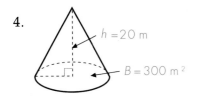

5.

$2\frac{1}{3}$ km

$h = 8\frac{1}{2}$ km

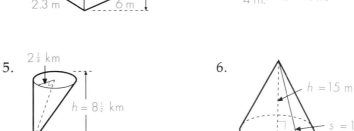

6.

$h = 15$ m

$s = 17$ m

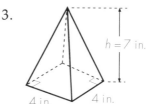

7. The Great Pyramid of Egypt has a square base that is 230 m long. Its height is 147 m. Find the volume.

8. A conical container is used to collect rain water. Its height is 4 ft and the diameter of the base measures 7 ft. How many gallons of water will it collect? (1 ft^3 = 7.5 gal)

9. Alex and Pierre pitched a tent that had the shape of a pyramid. The rectangular base measured 2.3 m by 2.6 m. The volume was 4.2 m^3. Find the height.

CRITICAL THINKING

10. A pile of grain has the shape of a right circular cone with a height of 3 ft. The base has a circumference of 12.5 ft. Find the volume of the pile.

SECTION
11.5

Applications

After completing this section, you should understand how
▲ prisms and cylinders are used in everyday life.
▲ pyramids and cones are used in everyday life.

You have studied prisms, pyramids, cylinders, and cones in the earlier sections of this chapter. In this section, you will increase your familiarity with these figures and their applications.

Examples of prisms can be found everywhere around you, in the form of buildings, commercial products, and decorative features in your surroundings. One such example is the *swimming pool*.

Example: Most swimming pools have both a shallow end and a deep end. The bottoms of the two sections are connected by a gradual slope that allows the bather to choose the water depth that he or she prefers. One such pool is shown below.

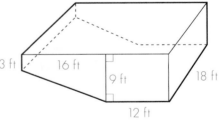

1. What kind of polygon is the shape of the two bases?

2. Find the volume of water that the pool can hold.

3. When the pool is being filled, water flows in at the rate of 67 gal/min. How long does it take to fill up the pool? There are 7.5 gallons in 1 ft³. This can be written as 7.5 gal/ft³.

To answer question 1, think of the front side as a base of a prism.

425

The pool is a prism with two quadrilaterals as its bases. Next consider how to find the volume of this prism. The base can be separated into a trapezoid and a rectangle, so let B_1 = the area of the trapezoid and B_2 = the area of the rectangle. So

$$
\begin{aligned}
V &= Bh \\
&= (B_1 + B_2)h \\
&= \left[\tfrac{1}{2}(3 + 9) \times 16 + (9 \times 12)\right] \times 18 \\
&= (96 + 108) \times 18 \\
&= 204 \times 18 \\
&= 3672 \text{ ft}^3
\end{aligned}
$$

To find how long it will take to fill the pool, rewrite 3672 ft³ as $\frac{3672 \text{ ft}^3}{1}$ and 7.5 gal/ft³ as $\frac{7.5 \text{ gal}}{1 \text{ ft}^3}$. Then multiply.

$$
\frac{3672 \text{ ft}^3}{1} \times \frac{7.5 \text{ gal}}{1 \text{ ft}^3} \times = \frac{27{,}540 \text{ gal}}{1 \times 1}, \text{ or } 27{,}540 \text{ gal}
$$

The rate of flow of water can be written in the following ways.

$$
67 \text{ gal/min} \qquad \frac{67 \text{ gal}}{1 \text{ min}} \qquad \frac{1 \text{ min}}{67 \text{ gal}}
$$

Use the third of these ways together with the number of gallons to find the number of minutes required to fill the pool.

$$
\frac{27{,}540 \text{ gal}}{1} \times \frac{1 \text{ min}}{67 \text{ gal}} = \frac{411 \text{ min}}{1}, \text{ or } 411 \text{ min}
$$

It takes about 411 min to fill the pool. The number of hours to fill the pool is 411 ÷ 60, or about 6.9 h.

In the example, the expression $\frac{7.5 \text{ gal}}{1 \text{ ft}^3}$ is called a *conversion factor*, because it can be used to convert cubic feet to gallons (and vice versa). Notice that conversion factors and similar expressions (such as water-flow rates) can be written and multiplied as if they were fractions. This technique helps you organize your calculations so that your final answer will be automatically expressed in the correct kind of units.

CLASS ACTIVITY

A swimming pool has the shape of a rectangular prism 25 ft long, 12 ft wide, and 7 ft deep. It costs 72 cents to run 1000 gal of water into the pool.
1. Find the volume of the pool.
2. Find the cost of filling the pool.

SHAPES OTHER THAN PRISMS

Sometimes large storage structures are cylindrical in shape. Examples are farm silos and gasoline storage tanks.

Example: The county plans to paint the outside of a 65 ft-high cylindrical water tank that has a diameter of 6 ft.

One can of paint covers 425 ft² and costs $15.56. The county officials estimate that the labor cost is 5 times the cost of the paint. How much is the total estimated cost of painting the tower?

Since the diameter is 6 ft, the radius of the tank is 3 ft.

$$\text{Lateral area of tower} = 2\pi rh$$
$$= 2 \times \pi \times 3 \times 65$$
$$= 390\pi$$
$$\approx 390 \times 3.14$$
$$\approx 1225 \text{ ft}^3$$

$$\text{Number of cans of paint} = \frac{1225 \text{ ft}^3}{1} \times \frac{1 \text{ can of paint}}{425 \text{ ft}^3}$$

$$= \frac{1225 \times 1 \text{ cans of paint}}{1 \times 425}$$

$$= 2.88 \text{ cans of paint}$$

The actual number of cans of paint needed is the next higher whole number. There will be 3 cans of paint needed.

$$\text{Cost of paint} = 3 \times 15.56, \text{ or } \$46.68$$
$$\text{Cost of labor} = 5 \times \text{cost of paint}$$
$$= 5 \times 46.68, \text{ or } \$233.40$$
$$\text{Total cost} = \text{cost of paint} + \text{cost of labor}$$
$$= 46.68 + 233.40$$
$$= 280.08, \text{ or about } \$280$$

Example: The Great Pyramid of Egypt is about 480 ft high and has a square base with a side that is about 755 ft long. Suppose that a modern building (with the shape of a square prism) has the same height as the Great Pyramid and the same dimensions. Assume that each floor of a building is 10 ft high and that a city block is a square that measures 300 ft on each side.

a. How many floors would the building have?
b. How many city blocks would the building occupy?

a. Number of floors = $480 \text{ ft} \times \frac{1 \text{ floor}}{10 \text{ ft}}$

$$= \frac{480 \times 1 \text{ floor}}{10}, \text{ or } 48 \text{ floors}$$

b. Number of blocks occupied by Great Pyramid =

$$\frac{1 \text{ city block}}{300 \times 300 \text{ ft}^2} \times \frac{755 \times 755 \text{ ft}^2}{\text{Base of Pyramid}} =$$

$$\frac{755 \times 755 \times 1}{300 \times 300} \times \frac{\text{city blocks}}{\text{Base of Pyramid}} = 6.33 \text{ blocks/Pyramid base}$$

or slightly more than 6 city blocks in area.

Example: A lumber company plans to plant new trees on the side of a mountain that has the shape of a right circular cone. The mountain is 3000 ft high and has a base with a circumference of 18 mi. (1 mi = 5280 ft) Each tree is to have 100 ft² of space. To the nearest hundred thousand, how many trees can be planted?
First, find the radius of the base.

$$C = 2\pi r$$
$$18 = 2\pi r$$
$$\frac{18}{2\pi} = r$$
$$2.86 \approx r$$
$$r \approx 2.86 \text{ mi}$$
$$\approx 2.86 \text{ mi} \times \frac{5280 \text{ ft}}{1 \text{ mi}} \approx 15,100 \text{ ft}$$

Next, find the area of the side of the mountain. The slant height is found by using the Pythagorean Theorem.

$$r^2 + h^2 = s^2$$
$$15,100^2 + 3000^2 = s^2$$
$$237,010,000 = s^2$$

Take the square root of each side of the equation.
$$15,395 \approx s, \text{ or } s \approx 15,395 \text{ ft}$$

Lateral area of right circular cone $= \pi rs$
$$\approx \pi \times 15,100 \times 15,395$$
$$\approx 730,308,000 \text{ ft}^2$$

Number of trees $= 730,308,000 \text{ ft}^2 \times \frac{1 \text{ tree}}{100 \text{ ft}^2} \approx 7,303,080,$ or about 7,300,000 trees.

CLASS ACTIVITY

Use a calculator for Exercises 1 and 2.

1. Find the volume of a large, cylindrical can of flour that is 30 cm high and that has a base with a diameter of 24 cm.

2. A conical scoop is used to empty the can of flour of Exercise 1. The height of the cone is 12 cm and the diameter of the base is 12 cm. How many scoops are required?

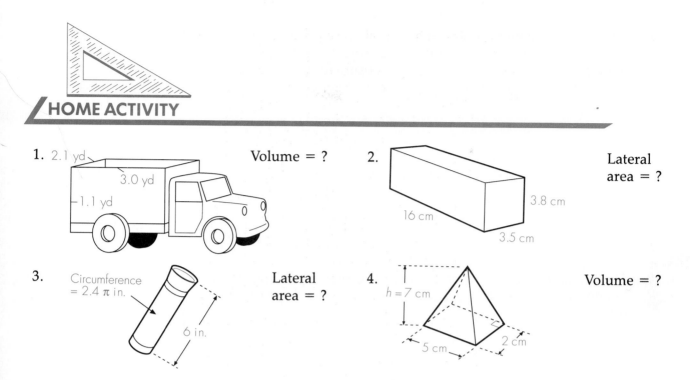

HOME ACTIVITY

1. Volume = ?

2. Lateral area = ?

3. Lateral area = ?

4. Volume = ?

5. A pencil factory is ordering paint for a batch of pencils that it is shipping. To prepare the order, the factory needs to know the lateral area of each of its pencils. Assume that a pencil is a regular hexagonal prism with each side of the base 0.125 in. long. Each lateral edge is 7 in. long. Find the lateral area of a pencil.

6. The inside radius of the graduated cylinder at the right is 1.5 cm. How much will the level of liquid rise if 20 cm³ of additional liquid are poured in?

7. A large Christmas tree is to be erected in a local community. The height is to be 30 ft and the diameter of the base is to be 15 ft. There is to be an average of one ornament for each square foot of the tree's surface. Assuming that the tree is a right circular cone, find the approximate number of ornaments required.

8. In banks, coins are separated by type and then wrapped into cylindrical stacks. Assume that a quarter is 0.0675 in. thick and has a diameter of 0.95 in. What is the area of a paper coin wrapper for a roll of 10 dollars worth of quarters? Ignore the extra paper that will be needed to overlap the first layer and the ends of the roll.

9. The botanical laboratory shown has the shape of one-half of a right circular cylinder. Except for several semicircular frames, the entire structure is translucent in order to provide the plants with the largest possible amount of exposure to light. Find the total area of the structure that allows the passage of light. Ignore the thickness of the frames.

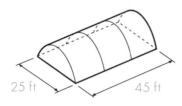

25 ft   45 ft

10. The triangular prism shows the original design of the roof of a house. To save materials, the builder changed the design. How many square feet of area are saved by constructing the second roof? (Include the ends.)

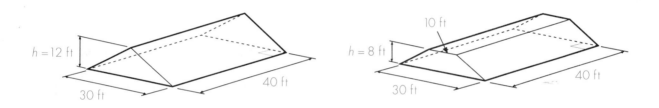

$h = 12$ ft   40 ft   30 ft   10 ft $h = 8$ ft   40 ft   30 ft

CRITICAL THINKING

11. Two squares have been removed from opposite ends of a chess board, as shown. Roberto is attempting to cover the remaining squares with a set of dominoes. Each domino can cover two adjacent squares. Can Roberto cover all the squares? Justify your answer.

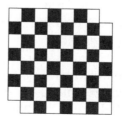

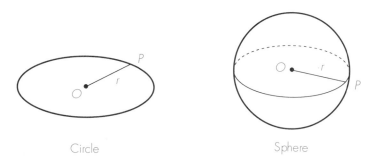

Area and Volume of a Sphere

After completing this section, you should understand
▲ how to find the volume of a sphere.
▲ how to find the area of a sphere.

In a *plane*, a circle is the set of points equidistant from a point called the center. In *space*, the set of such points is a **sphere.**

Circle   Sphere

DEFINITION 11-6-1   **A sphere** is the set of all points equidistant from a point called the **center.** The distance from any point of the sphere to the center is the **radius** of the sphere.

At the right, a sphere is inscribed in a cylinder. The radius of the base of the cylinder is *r*, the same as the radius of the sphere. The height of the cylinder is the same as the diameter of the sphere, 2*r*. Over 2200 years ago, Archimedes, the

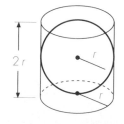

greatest mathematician of his time, discovered a relationship between the above two figures. He regarded the relationship as so important that he requested that the figures be etched on his tomb. This is the relationship he discovered: $\dfrac{\text{Volume of inscribed sphere}}{\text{Volume of cylinder}} = \dfrac{2}{3}$

431

REPORT 11-6-1 Prepare a report on Archimedes explaining and illustrating the following:

1. the time in history when he lived

2. his mathematical accomplishments

Sources:
Bell, Eric T. *Men of Mathematics*. New York: Simon and Schuster, 1937.
Eves, Howard. *An Introduction to the History of Mathematics*. New York: Holt, Rinehart, and Winston, 1976.
Kline, Morris. *Mathematical Thought from Ancient to Modern Times*. New York: Oxford University Press, 1972.

You can use the relationship that Archimedes found to derive a formula for the volume of a sphere.

$$\frac{\text{Volume of inscribed sphere}}{\text{Volume of cylinder}} = \frac{2}{3}$$

$$\text{Volume of sphere} = \frac{2}{3} \text{ Volume of cylinder}$$

$$= \frac{2}{3} \times \pi r^2 h$$

$$= \frac{2}{3} \times \pi r^2 \times 2r$$

$$= \frac{2}{3} \times 2 \times \pi r^2 \times r, \text{ or } \frac{4}{3} \pi r^3$$

THEOREM 11-6-1 **The formula for the volume of a sphere is**
Volume of sphere $= \frac{4}{3} \pi r^3$,
where r is the radius of the sphere.

Example: Find the volume of the sphere of radius 5.

$$V = \frac{4}{3} \pi r^3$$

$$= \frac{4}{3} \times \pi \times 5^3$$

$$= \frac{4}{3} \times \pi \times 125$$

$$= \frac{500}{3} \pi$$

$$\approx 167 \times 3.14, \text{ or about } 523 \text{ mm}^3$$

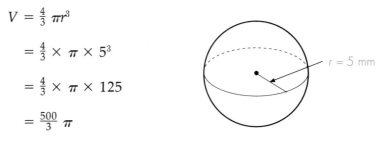

$r = 5$ mm

AREA OF A SPHERE

The following Discovery Activity will help you find the formula for the area of a sphere.

DISCOVERY ACTIVITY

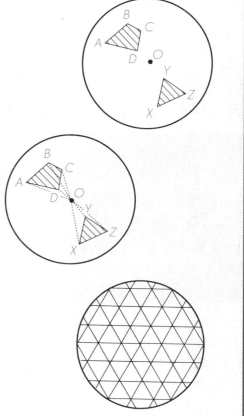

In the figure at the right, points *A*, *B*, *C*, and *D* are four points of a sphere. Similarly, *X*, *Y*, and *Z* are three points of the sphere. The points in each set are connected by line segments as shown to create quadrilateral *ABCD* and △ *XYZ*. Then seven other line segments are formed by connecting each vertex of the two polygons to point *O*, the center of the sphere.

1. What kind of solid figure is polyhedron *OXYZ*? polyhedron *OABCD*? How do you know?

2. Use your answer to question 1 to tell the formula for the volume of each of the two figures.

Suppose that instead of using just two polygons, you had 1000 very small polygons and thus 1000 pyramids packing the inside of the sphere.

3. Would the surface of the entire sphere have more area than the total area of the bases of the pyramids? less area? the same area?

4. Would the sphere have more volume than the total volume of the pyramids? less volume? the same volume?

5. What would be the answers to Questions 3 and 4 if the number of small pyramids were increased to 10,000? to 1,000,000?

In the Discovery Activity, you may have discovered something that happens as the number of pyramids packing the sphere increases and their size decreases. The total area of their bases approximates the area of the sphere more closely. Also, their total volume approximates the volume of the sphere more closely.

You can use these ideas to find a formula for the volume of a sphere.

The volume of a small pyramid with height h and base of B_1 is:

$$\tfrac{1}{3} B_1 h$$

Suppose that the total number of pyramids is n.
If you add up the volumes of all the pyramids, then you get the following result.

$$\tfrac{1}{3} B_1 h + \tfrac{1}{3} B_2 h + \tfrac{1}{3} B_3 h + \ldots + \tfrac{1}{3} B_n h$$

The three dots (. . .) mean "and so on."
If the number of pyramids is very large, then you can make three reasonable assumptions:

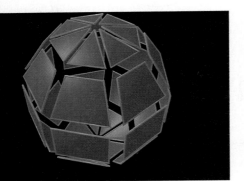

1. The volume of the sphere is very close to the volume of all the pyramids.
2. The height of each pyramid is very close to the radius of the sphere.
3. The area of the sphere is very close to the sum of all the base areas.

Volume of sphere $\approx \tfrac{1}{3} B_1 h + \tfrac{1}{3} B_2 h + \ldots + \tfrac{1}{3} B_n h$ (Assumpt. 1)

$$\tfrac{4}{3} \pi r^3 \approx \tfrac{1}{3} h (B_1 + B_2 + B_3 + \ldots + B_n) \text{ (Algebra Prop.)}$$

$$\tfrac{4}{3} \pi r^3 \approx \tfrac{1}{3} r (B_1 + B_2 + B_3 + \ldots + B_n) \quad \text{(Assumpt. 2)}$$

$$\tfrac{4}{3} \pi r^3 \approx \tfrac{1}{3} r \times \text{(total of base areas)} \quad \text{(Assumpt. 3)}$$

Next, multiply each side by 3 and divide each side by r.

$$4 \pi r^2 \approx \text{total of base areas}$$

This result suggests the following theorem.

THEOREM 11-6-2   **The formula for the surface area A of a sphere of radius r is $A = 4\pi r^2$.**

Example: Find the area of the sphere at the right.

$$A = 4\pi r^2$$

$$= 4 \times \pi \times 5^2$$

$$= 100\,\pi$$

$$\approx 100 \times 3.14, \text{ or about } 314 \text{ mm}^2$$

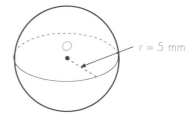

$r = 5$ mm

CLASS ACTIVITY

Use a calculator to find the volume and area of each sphere.

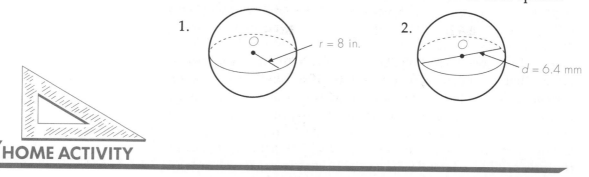

1. $r = 8$ in.

2. $d = 6.4$ mm

HOME ACTIVITY

Find the volume and area of each sphere.

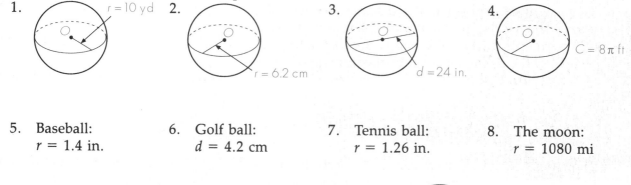

1. $r = 10$ yd

2. $r = 6.2$ cm

3. $d = 24$ in.

4. $C = 8\pi$ ft

5. Baseball:
   $r = 1.4$ in.

6. Golf ball:
   $d = 4.2$ cm

7. Tennis ball:
   $r = 1.26$ in.

8. The moon:
   $r = 1080$ mi

9. The radius of the earth is about 7900 mi. About 70% of its surface is water. If the world's population is 5,400,000,000, what is the average number of people for each square mile of land?

Great circles

10. If a plane passes through a sphere at its center, then the cross-section that is formed is called a **great circle** (see diagram at the right). What is the relationship between the area of a great circle of a sphere and the surface area of the sphere?

CRITICAL THINKING

11. A great circle of a sphere divides a circle into two **hemispheres** of equal volume. The cylinder, cone, and hemisphere at the right all have radius r; r is also the height of the cylinder and cone. Show that the volume of the hemisphere is the average of the volume of the cylinder and cone.

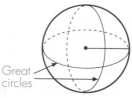

11.1 A **prism** is a closed figure in space. Its bases are two congruent polygonal regions in parallel planes, and the segments joining corresponding vertices are parallel segments called lateral edges. Two corresponding sides of the bases and the lateral edges that connect them determine a parallelogram-shaped region called a lateral face. Each point where a base edge intersects a lateral edge is a vertex of the prism. If the lateral edges are perpendicular to the bases, the prism is a right prism. Otherwise it is oblique. In the figure, $ABCDE$ is the top base of the prism. $\overline{EJ}$ is a lateral edge. $\square AEJF$ is a lateral face. Points F and D are two vertices. The prism shown is a right prism.

The height of a prism is the distance between the planes of the bases. If a right prism has bases of area B and perimeter p, then the lateral area LA and total area TA of the prism are given by the formula $LA = ph$ and $TA = 2B + ph$.

Area = B square units

The bases of a **cylinder** are congruent curved regions that lie in parallel planes. The segments joining corresponding points on the boundries of these regions are all parallel. If the axis of a circular cylinder is perpendicular to the bases, the cylinder is a right cylinder. Otherwise it is oblique. The height of a cylinder is the distance between the planes of the bases. In the figure, the left cylinder has regions 1 and 2 as bases. The second cylinder is a right cylinder. Both cylinders have a height of 9.

11.2 The volume of a space figure is the number of cubic units it contains. A right rectangular prism of length l, width w, and height h has volume $V = lwh$. The volume of a prism whose bases have area B and whose height is h is $V = Bh$. The volume of a cylinder whose bases have area B and whose height is h is $V = Bh$.

$$V = 120 \text{ cm}^3 \qquad V = 78 \text{ m}^3 \qquad V = 52\pi \approx 163 \text{ m}^3$$

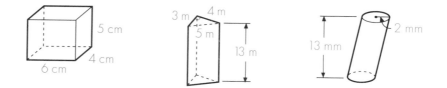

11.3 A pyramid has a polygonal region for its base. All the points of the polygon are joined to a point P (the vertex) not in the plane of the base. The segments joining the vertices of the base to P are lateral edges of the pyramid. The lateral faces of a pyramid are triangular regions. The height of a pyramid is the distance from the vertex of the pyramid to the plane of the base. If the base of a pyramid is a regular polygonal region and all the lateral edges are of equal length, the pyramid is a regular pyramid. The slant height of a regular pyramid is the height of a lateral face.

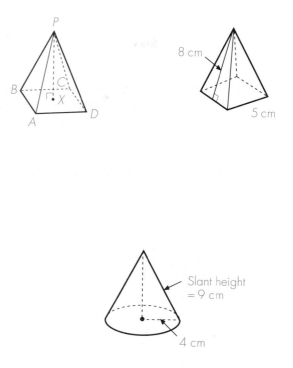

Cones are formed much the same way as pyramids, except the base is a region with a curved boundary. The line segment joining the vertex to the center of a circular cone is the axis of the cone. If the axis is perpendicular to base, the cone is a right cone. The height of a cone is the distance from the vertex to the plane of the base. For a right circular cone, the length of a segment from the vertex to a point on the circle is the slant height.

For a right circular cone whose base has radius r and whose slant height is s, $LA = \pi rs$ and $TA = \pi r^2 + \pi rs$. For the right circular cone in the diagram, $LA = \pi \times 4 \times 9 = 36\pi$ (about 113 cm^2) and $TA = \pi \times 4^2 + \pi \times 4 \times 9 = 52\pi$ (about 163 cm^2).

11.4 If a pyramid has a base of area B and a height of h, its volume is $V = \frac{1}{3}Bh$. If a cone has a height of h and a base of radius r, its volume is $V = \frac{1}{3}\pi r^2 h$. In the diagram, the pyramid is a square pyramid. So $V = \frac{1}{3} \times 25 \times 10 = 83.3$ cm^3. For the circular cone, $V = \frac{1}{3} \times \pi \times 3^2 \times 10 = 30\pi$, or about 94 cm^3.

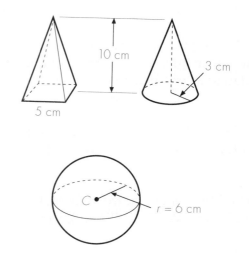

11.6 A sphere is the set of all points that are a given distance r from a certain point C. Point C is the center of the sphere and r is the radius of the sphere. The volume V of the sphere is given by $V = \frac{4}{3}\pi r^3$ and the area of the sphere is $A = 4\pi r^2$. For the sphere in the diagram, $V = \frac{4}{3} \times \pi \times 6^3 = 288\pi$.
$$A = 4 \times \pi \times 6^2 = 144\pi.$$

The simplest way to draw three-dimensional figures with the LOGO turtle is to use the x-y plane and the SETXY command.

Example: Write a procedure using the command SETXY to tell the turtle to draw a right rectangular prism whose length is 40 units, width is 20 units, and height is 15 units. Then write a procedure to compute the volume of the prism.

The RECPRISM procedure below tells the turtle to draw the right rectangular prism.

```
TO RECPRISM
  SETXY 40 0
  SETXY 50 10
  SETXY 50 20
  SETXY 10 20
  SETXY 0 10
  SETXY 40 10
  SETXY 50 20
  PU SETXY 40 10 PD
  SETXY 40 0
  PU HOME PD
  SETXY 0 10
END
```

Since this is a three-dimensional figure, the width in the figure drawn by the turtle will take perspective into account. The output is shown below.

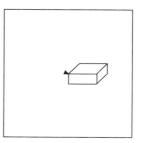

The procedure to find the volume of a right rectangular prism is given below.

```
TO VOLRECPRI :LENGTH :WIDTH :HEIGHT
  OUTPUT :LENGTH * :WIDTH * :HEIGHT
END
```

To find the volume of the right rectangular prism whose length, width, and height are given above, simply type VOLRECPRI 40 20 15. The output will be 12000.

Modify the RECPRISM procedure to tell the turtle to draw each right rectangular prism described in Exercises 1 through 5. Then use the VOLRECPRI procedure to compute the volume of each figure.

1. length = 25, width = 15, height = 20
2. length = 30, width = 18, height = 25
3. length = 48, width = 13, height = 36
4. length = 86, width = 9, height = 55
5. length = 70, width = 29, height = 61

Write procedures to compute the volume of each figure described in Exercises 6 through 8. Give each procedure the indicated name.

6. a right triangular prism; VOLTRIPRI
7. a triangular pyramid; VOLTRIPYR
8. a quadrangular pyramid; VOLQUADPYR

Use the VOLTRIPRI procedure from Exercise 6 to compute the volume of each right triangular prism.

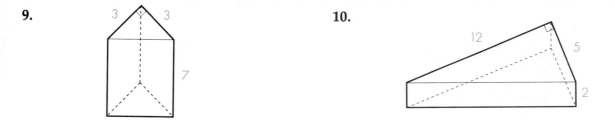

9.   10.

Use the VOLTRIPYR procedure from Exercise 7 to compute the volume of each triangular pyramid.

11.   12.

Use the VOLQUADPYR procedure from Exercise 8 to compute the volume of each regular square pyramid.

13.   14.

Find the lateral area and total area for each prism or right circular cylinder.

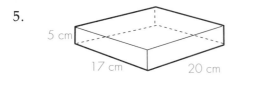

1. 2 m, 6.5 m, 4 m, base

2. r = 2 in., 6 in.

3. e = 3 ft

4. 24, 26, 30

Find the volume of each prism or cylinder.

5.

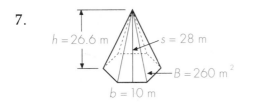

5 cm, 17 cm, 20 cm

6.

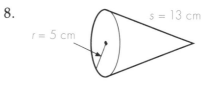

r = 10 yd, h = 25 yd

Find the total area and volume of each regular pyramid or right circular cone.

7.

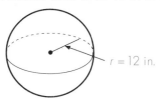

h = 26.6 m, s = 28 m, B = 260 m², b = 10 m

8.

s = 13 cm, r = 5 cm

Find the volume and area of each sphere.

9.

r = 12 in.

10.

C = 10 π ft

A company is considering buying a water tank. One type is cylindrical, the other spherical. The radius of the sphere is 10 ft. The radius of a base of the cylinder is 8 ft and its height is 30 ft.

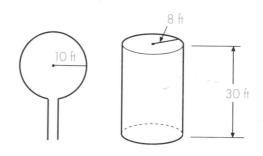

8 ft, 10 ft, 30 ft

11. Which tank holds more? How many cubic feet more?

12. How many more gallons does the larger tank hold? (1 ft³ = 7.5 gal)

Find the lateral area and total area for each prism or right circular cylinder.

1.

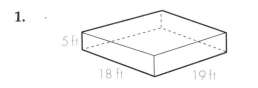

5 ft   18 ft   19 ft

2.

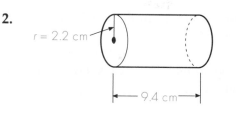

r = 2.2 cm   9.4 cm

Find the volume of each prism or cylinder.

3.

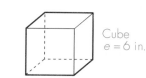

Cube
e = 6 in.

4.

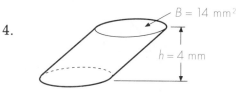

B = 14 mm²   h = 4 mm

Find the total area and volume of each regular pyramid or right circular cone.

5.

6 cm   7 cm   h = 6.3 cm

6.

s = 10 yd   h = 6 yd

Find the volume and area of each sphere.

7.

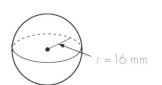

r = 16 mm

8.

d = 10 mm

9. The Vertical Assembly Building at the John F. Kennedy Space Center in Florida has the shape of a rectangular prism. The height is 525 ft, and the dimensions of the floor area are 716 ft by 518 ft. Find the lateral area of the building.

10. A cylindrical swimming pool has a diameter of 50 ft and a height of 8 ft. If the pool is being filled at the rate of 60 gal/min, how long will it take to fill the pool? (7.5 gal = 1 ft³)

Study these examples.

In the coordinate plane, find the set of points that satisfy the given condition. Then describe the location of the points.

1. The x-coordinate is always equal to the y-coordinate.

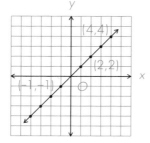

2. The y-coordinate is always equal to $2\frac{1}{2}$.

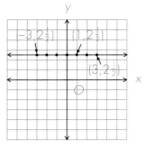

These are all the points on the line that bisects the angle formed by the positive x- and y-axes and the vertical angle.

These are all the points on the horizontal line $2\frac{1}{2}$ units above the x-axis.

Example: Find and describe the set of all points that satisfy *both* of these two following conditions.

a. The sum of the coordinates is 7.

b. The x-coordinate is 6.

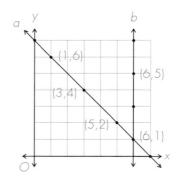

The set consists of a single point at (6, 1).

Follow Examples 1 and 2 in solving these problems. In the coordinate plane, find the set of points that satisfy the given condition. Then describe the location of the points.

1. The x-coordinate equals the opposite of the y-coordinate.

2. The y-coordinate equals twice the x-coordinate.

3. The y-coordinate is the absolute value of the x-coordinate.

Follow Example 3 in solving these problems. Find and describe the set of all points that satisfy both of the given conditions.

4. Condition a: The y-coordinate is 2 greater than the x-coordinate. Condition b: The sum of the x- and y-coordinates is 5.

5. Condition a: The x-coordinate is 4. Condition b: The absolute value of the y-coordinate is 2.

6. Condition a: The absolute value of the x-coordinate is 3. Condition b: The absolute value of the y-coordinate is 3.

442

12

LOCUS AND TRANSFORMATIONS

A method of solution is perfect if we can foresee from the start, and even prove, that following that method we shall attain our aim.

Gottfried W. Leibnitz

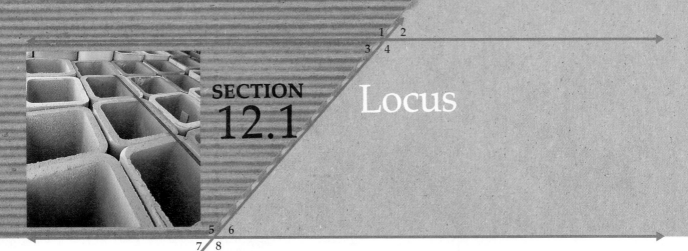

After completing this section, you should understand
▲ the definition of geometric locus.
▲ how to determine a locus that satisfies given conditions.

The word **locus** is Latin, meaning place. In this chapter you will investigate geometric places, or **loci** (pronounced lō sī), that are determined by sets of points that satisfy given conditions. Loci can be two-dimensional or three-dimensional, on one plane or in space.

This square is a two-dimensional plane figure. Its sides are the locus of a set of points that satisfy the conditions for this particular square.

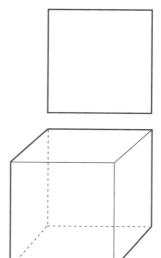

This drawing of a cube is a representation of a three-dimensional or space figure, such as a block. Its edges and faces are the locus of a set of points that satisfy the conditions for this particular cube.

DEFINITION 12-1-1 **A geometric LOCUS is the set of all points that satisfy the given conditions.**

Example: What is the set of points on a football field that is 50 yards from each goal line?

The 50-yard line is the set of points that matches this condition. The 50-yard line is the locus.

To help you find a locus, use the following steps.

1. Read and reread the problem until you understand the conditions your locus must satisfy.
2. Locate several points that satisfy the conditions.
3. Do the points form a pattern? If not, locate several more.
4. What is the pattern?
5. Draw a solution set.
6. Write a statement describing the locus.
7. Try to prove your answer.
   a. Prove that every point in your proposed locus satisfies the conditions,

      and

      prove that every point that satisfies the conditions is a point in your locus;

      or
   b. Prove that every point that does not satisfy your proposed locus does not satisfy the conditions.

DISCOVERY ACTIVITY

1. Use a ruler to draw a large angle. Label the angle ABC, with B as the vertex.
2. Locate five points that are equally distant from $\overrightarrow{BA}$ and $\overrightarrow{BC}$.
3. Imagine the number of points increasing to many points, and describe the pattern or set of these points.
4. What can you conclude about these points? Write a statement to describe the locus.

You may have concluded that the locus of points equidistant from the rays of an angle is the angle bisector. The next step is to prove that your locus satisfies the given conditions. You can write a two-part proof to show this.

Part 1: Any point D on the angle bisector is equidistant from $\overrightarrow{BA}$ and $\overrightarrow{BC}$.

Given: $\angle ABC$ with $\overrightarrow{BD}$ the angle bisector

Justify: D is equidistant from $\overrightarrow{BA}$ and $\overrightarrow{BC}$.

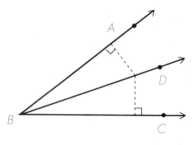

You know this is true from Theorem 6-6-3: If a point is on the angle bisector, then the point is equidistant from the sides of the angle.

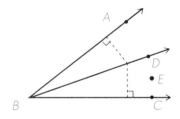

Part 2: Show that any point not equidistant from $\overrightarrow{BA}$ and $\overrightarrow{BC}$ is not on the angle bisector *or* any point not on the angle bisector is not equidistant from $\overrightarrow{BA}$ and $\overrightarrow{BC}$.

Given: ∠*ABC* with point *E* not equidistant from $\overrightarrow{BA}$ and $\overrightarrow{BC}$

Justify: Point *E* is not on the angle bisector.

You can use indirect reasoning to prove this. There are only two cases for point *E*: either *E* is on the angle bisector or *E* is not on the angle bisector. Assume that *E* is on the angle bisector. This means *E* is equidistant from $\overrightarrow{BA}$ and $\overrightarrow{BC}$, by Theorem 6-6-3. This contradicts the given information—that *E* is *not* equidistant from $\overrightarrow{BA}$ and $\overrightarrow{BC}$. Since there are only two cases for *E*, and since the case where *E* is on the angle bisector leads to a contradiction of the given, the other case must be true: point *E* is not on the angle bisector. The proof of the two parts justifies that the locus is the angle bisector.

THEOREM 12-1-1   **The locus of points equidistant from the sides of an angle is the angle bisector.**

CLASS ACTIVITY

1. Find the locus of points in a plane equidistant from the endpoints of a line segment. Draw the locus for a segment.

2. Describe the locus in Exercise 1 by completing this statement: The locus of points equidistant from the endpoints of a line segment is _____ .

   Write a two-part proof to show that the locus you found in Exercise 1 satisfies the conditions.

3. Given: $\overline{AB}$ with perpendicular bisector $\overleftrightarrow{CD}$
   Justify: *D* is equidistant from *A* and *B*.

4. Refer to the figure for Exercise 3.
   Given: $\overline{AB}$ with perpendicular bisector $\overline{CD}$ and point *E* not equidistant from *A* and *B*
   Justify: Point *E* is not on the perpendicular bisector.

THEOREM 12-1-2   **The locus of points equidistant from the endpoints of a line segment is the perpendicular bisector of the line segment.**

MORE LOCUS THEOREMS

Finding the locus of points that satisfies certain conditions gives you an interesting way of looking at some familiar geometric figures. Each of the following procedures will result in a locus theorem that states in a new way information that you may already know.

1. Find the locus of points in a plane that are a set distance from a line.

    a. Locate four points above a line *AB* and four below the line that are the set distance (use 1 inch) from the line.
    b. Draw in the color locus for all possible points.
    c. Describe the locus.

THEOREM 12-1-3   **The locus of points in a plane that are a given distance from a line on the plane is two parallel lines.**

2. What is the locus of points in a plane that are equidistant from a point on the plane?

    a. Mark a point at the center of a sheet of paper. Label it 0.
    b. Locate a dozen points 3 cm from point 0.
    c. Draw the locus for all possible points in color.
    d. Describe the locus.

THEOREM 12-1-4   **The locus of points a given distance from a point on a plane is a circle.**

3. Follow the directions to find this locus.

    a. Draw a 3-inch line segment and label it *AB*.
    b. Through point *A*, draw three lines, with one line perpendicular to $\overline{AB}$.
    c. From point *B*, draw lines that are perpendicular to the three lines through *A*. Label the points of intersection, C_1, C_2, C_3.
    d. Repeat Steps b and c, drawing additional lines through *A*.
    e. Describe the locus.

THEOREM 12-1-5   **The locus of the vertex (*C*) of all right triangles with a given hypotenuse ($\overline{AB}$) is a circle with diameter $\overline{AB}$ minus the endpoints *A* and *B* of the hypotenuse.**

REPORT 12-1-1   Find out what a cycloid is and describe its physical features. How is it used in amusement parks?

Suggested source:
Johnson, R.E., Kiokemeister, F.L. *Calculus with Analytic Geometry.* Boston: Allyn and Bacon, Inc., 1969.

HOME ACTIVITY

Follow the steps listed on page 445 to help you determine the following loci. Prove your conclusions only when you are asked to do so.

Find and describe the locus for Exercises 1–4. Then write a LOGO procedure to tell the turtle to draw the figure and the locus.

1. Draw a 3-inch line segment, $\overline{AB}$. Find all points equidistant from A and B.

2. Draw $\angle ABC$. Find all points equidistant from $\overrightarrow{BA}$ and $\overrightarrow{BC}$.

3. Draw line AB. Find all points 3 centimeters from $\overleftrightarrow{AB}$.

4. Find all points 0.75 inch from a point A.

5. Find the locus of points that are equidistant from two parallel lines.

6. What is the locus of all points in a plane that are equidistant from two intersecting lines?

7. What is the locus of all points in a plane that are equidistant from a line segment?

On graph paper, draw the locus.

8. $y = 3x + 1$

9. $y = |x|$

10. $y < x + 2$

11. $x^2 + y^2 = 25$

CRITICAL THINKING

12. The conditions for Exercises 1, 3, 4, 5, and 7 can also be satisfied in three dimensions by space figures. Draw the three-dimensional picture and describe the locus.

SECTION
12.2

Transformations

After completing this section, you should understand
▲ how to find the translation image of a figure.
▲ how to find the reflection of a figure and locate the line of reflection.
▲ how to find the rotation image of a figure.

The meaning of the word **transformation** implies change—of size, shape, or position. In this section, you will be investigating transformations of geometric figures. These transformations may change a figure's position or size, but they will not change the shape.

NOTEBOOK

DEFINITION 12-2-1 **A TRANSFORMATION is a way of creating a mapping or image of a geometric figure in a plane, preserving shape but not necessarily size.**

We will investigate three types of transformations: translations, reflections, and rotations.

TRANSLATIONS

A **translation** can be thought of as a sliding of a figure to a new position.

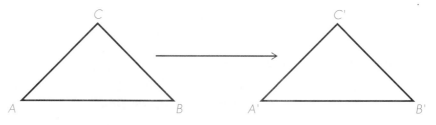

NOTEBOOK

DEFINITION 12-2-2 **A TRANSLATION is a transformation that slides the figure from one position to another, preserving congruency.**

449

From the definition of transformation, you know that the shape of the translated figure will be the same as the shape of the original figure.

Example: $A'B'C'D'$ is the translation image of rectangle $ABCD$.

Measure the distance between A and A', B and B', C and C', and D and D'. What do you observe?

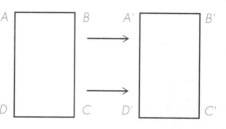

DISCOVERY ACTIVITY

1. On one half of a sheet of tracing paper, draw a concave quadrilateral. Label the vertices $ABCD$.
2. Fold the paper in half and trace the figure on the other half. Label this figure $A'B'C'D'$.
3. Measure the distance between A and the fold line. Then measure the distance between points A and A'. What do you notice?
4. Do the same for BB', CC', and DD'. What do you notice about the relationship of the fold line to each of these line segments?
5. Find the midpoint of one of the sides of $ABCD$. Label it E. Then locate E' on $A'B'C'D'$. Fold the paper along the fold line. Where does point E fall?
6. Draw $\overline{AA'}$, $\overline{BB'}$, $\overline{CC'}$, $\overline{DD'}$, and $\overline{EE'}$. What can you observe about these line segments?

REFLECTIONS

You may have found that when a figure is reflected across a line, the line of reflection is the perpendicular bisector of line segments joining corresponding points on the figure and its reflection.

DEFINITION 12-2-3 **A REFLECTION is a transformation such that a line (mirror) is the perpendicular bisector of the segments joining corresponding points.**

You can see why the reflection of a figure is often called a mirror image. This type of transformation can also illustrate why the phrase "preserving shape but not necessarily size" is part of the definition. Some mirrors magnify a reflected image, and this is still a transformation.

Example: $\overline{A'B'}$ is a reflection of $\overline{AB}$ across line m. What do you think might be true of $\overline{AA'}$ and $\overline{BB'}$?

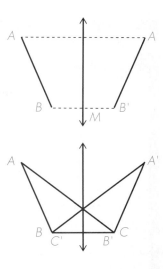

Example: $\triangle ABC$ is reflected across line m. Its image is $\triangle A'B'C'$.

For $\triangle ABC$ and $\triangle A'B'C'$, do you think $\overline{AB} \cong \overline{A'B'}$, $\overline{BC} \cong \overline{B'C'}$, and $\overline{AC} \cong \overline{A'C'}$? Why?

What do you think might be true for $m\angle ABC$ and $m\angle A'B'C'$? How could you show this?

CLASS ACTIVITY

Tell whether each transformation is a translation or a reflection.

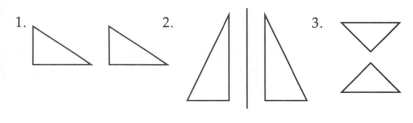

1. 2. 3.

4. Name the reflection of points P, Q, and S across line m.

5. Copy the figure and show its reflection across line m. Then write a LOGO procedure to tell the turtle to draw the figure and its reflection across line m.

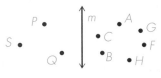

6. $\triangle X'Y'Z'$ is a reflection of $\triangle XYZ$. Copy the figures. Then find and draw the line of reflection.

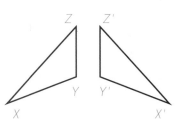

DISCOVERY ACTIVITY

1. Draw a right triangle on a sheet of graph paper. Label it *ABC*.

2. Trace △*ABC* on a sheet of tracing paper and label the tracing △*A'B'C'*. With a pin or the point of a compass, hold point *A'* on top of *A* and rotate △*A'B'C'* 90° clockwise. Mark points *A'B'C'* on the graph paper.

3. Rotate △*A'B'C'* through another 90° and mark the points *A"B"C"*. Do this one more time.

4. What do you notice about △*ABC* and △*A'B'C'* after three rotations?

ROTATIONS

You may have found that a rotation through 360° brings a figure back to its original position.

DEFINITION 12-2-4   **A ROTATION is a transformation that maps each point in a figure *A* onto the corresponding point in figure *A'* by revolving figure *A* about a point.**

Example: △*A'B'C'* is the image of △*ABC* under a rotation of 45° about point *P*.

The center of rotation can be anywhere in a plane.

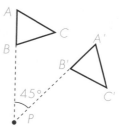

CLASS ACTIVITY

Identify the transformation as a translation, a reflection, or a rotation.

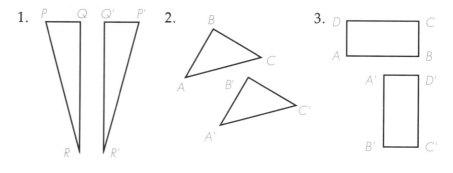

Identify each transformation.

4.

5.

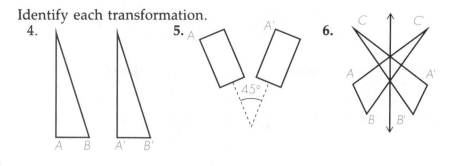

6.

REFLECTIONS AND A GAME

The drawing shows a billiard table with a white ball and a red ball. A player wants to hit the white ball and cause it to strike the red ball. But the white ball must first hit side AB of the table before it hits the red ball. To do this, the player uses reflections.

The player visually estimates the point R'. R' is the reflection of R across $\overline{AB}$. The player then aims at R'. The white ball hits the table at T and rebounds to strike the red ball. The path of the white ball after it hits the table at T is the reflection of the straight line TR' across $\overline{AB}$.

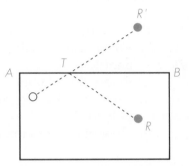

HOME ACTIVITY

Identify the transformation as a translation, a reflection, or a rotation.

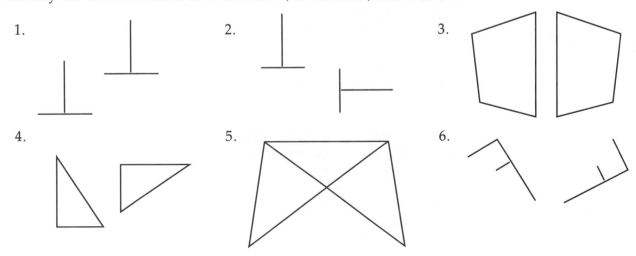

1.

2.

3.

4.

5.

6.

7. List the capital letters of the alphabet that have the same image when reflected across a line. Tell whether the line of reflection is vertical or horizontal.

8. Copy the figure and draw its reflection across line *m*. Then write a LOGO procedure to tell the turtle to draw the figure and its reflection across line *m*.

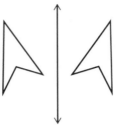

9. Use a reflection to determine where ball *W* should hit side *CD* in order to hit ball *R*. (Trace the diagram.)

10. Make a construction to find the point where ball *W* should strike side $\overline{AC}$, then side $\overline{AB}$, and then rebound to hit ball *R*. Hint: Reflect both balls and then draw the straight line.

11. a. Construct an equilateral triangle with sides of 1 inch. Label it *ABC*.
    b. Rotate the triangle 60° counterclockwise about vertex *A*. Mark the points of the vertices *B* and *C*.
    c. Rotate the triangle another 60° and mark the vertices.
    d. Continue the 60° rotation until a regular polygon is formed. What is the name of the polygon?
    e. How many rotations were there?

12. Repeat the steps in Exercise 11 for a 1-inch square and a rotation angle of 90°.
    Plot the following equations on graph paper. Name the type of transformation.

13. $y = x$ and $y = x + 2$

14. $y = x$ and $y = -x$

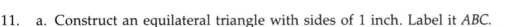

CRITICAL THINKING

Trace △*XYZ* and parallel lines *m* and *m'*. Reflect △*XYZ* across line *m*, then across line *m'*.

15. What can you conclude about the way the final reflection is related to the original figure?

16. What other type of transformation does this show?

Trace the flag and intersecting lines *m* and *m'*. Reflect the flag across line *m*, then across line *m'*.

17. What can you conclude about the way the final reflection is related to the original figure?

18. What other type of transformation does this show?

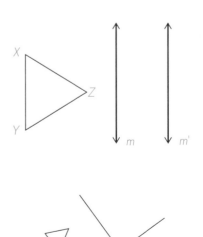

SECTION 12.3 Symmetry

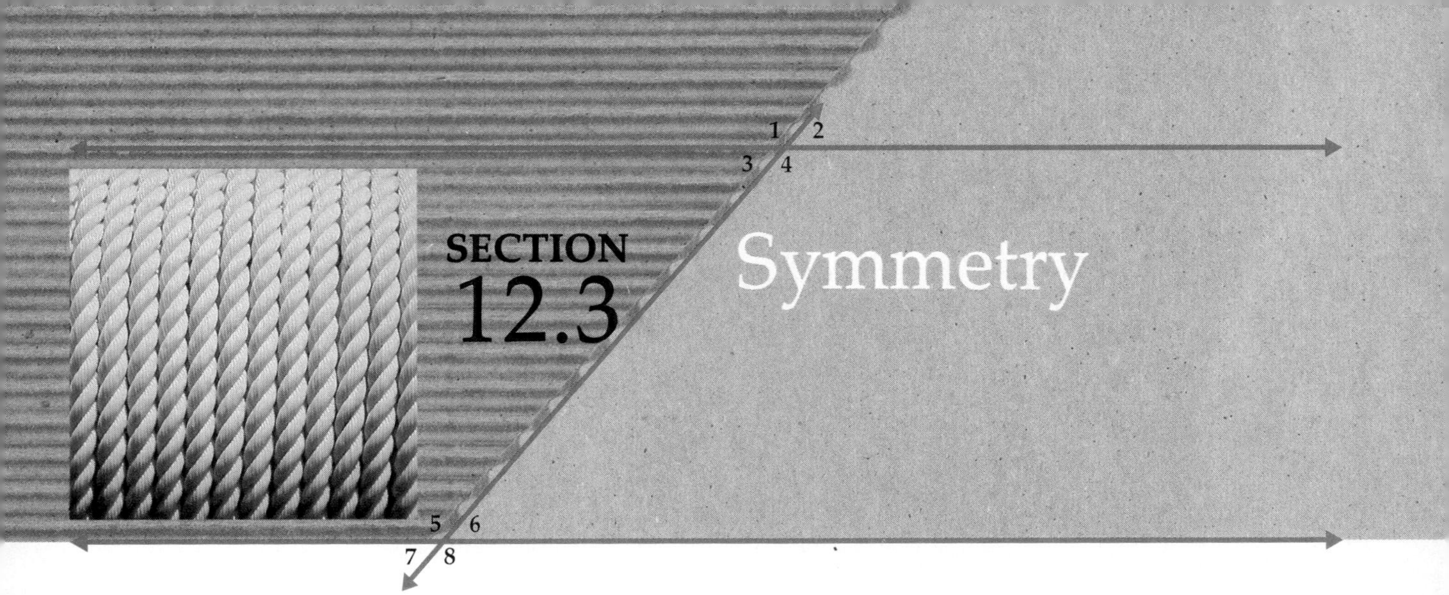

After completing this section, you should understand
▲ how to determine whether a figure has line symmetry.
▲ how to find lines of symmetry.
▲ how to determine whether a figure has point or rotation symmetry.

A special kind of transformation occurs when a figure is mapped onto itself. In the picture above, the figure is reflected across a line; the part on the left side is reflected on the right side.

◀ DISCOVERY ACTIVITY

1. Fold a sheet of paper in half and label the halves A and B.

2. Beginning at the fold, draw a design or picture on half A. End the drawing at some point on the fold.

3. Now fold the sheet, hold it to the light, and trace the design on the back of half B. Open the sheet and trace the design on the front side of B.

4. What do you notice about the two halves of the design?

5. Is this transformation a translation, a reflection, or a rotation? How do you know?

You may have found that you reproduced, or made an exact copy of, your original design from side A on side B.

NOTEBOOK

DEFINITION 12-3-1 **SYMMETRY is a transformation that maps a figure onto itself.**

455

The word symmetry comes from the Greek words meaning same measure.

In Section 12.2 you investigated three kinds of transformations.

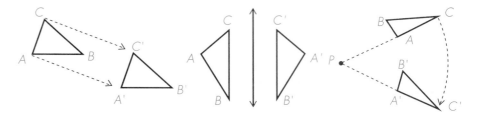

Symmetry is a reflection transformation. There are two types of symmetries—line symmetry and point, or rotation, symmetry.

DEFINITION 12-3-2 **LINE SYMMETRY is a reflection transformation that maps a figure onto itself.**

These designs are symmetric with respect to a line. The set of points on one side of the line is a reflection of the set of points on the other side.

If you look at your surroundings, you can see that line symmetry occurs in nature, art, architecture, and living things. Symmetry is involved in the design of most of the objects you see and use daily—furniture, clothing, utensils, tools, buildings, methods of transportation, and so on.

Some figures have only one line of symmetry. Others may have more than one or many lines of symmetry.

Example: Dotted lines mark the lines of symmetry in this regular octagon.

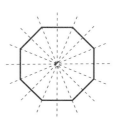

CLASS ACTIVITY

Tell which figures have line symmetry. Then find all lines of symmetry.

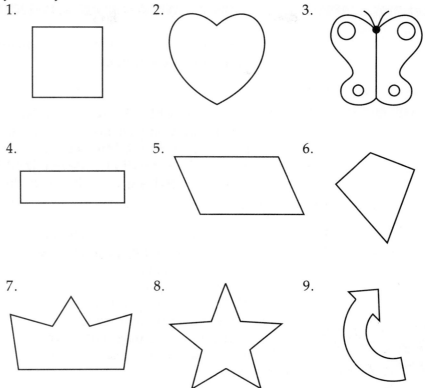

1.

2.

3.

4.

5.

6.

7.

8.

9.

POINT SYMMETRY

Some figures can be rotated to map the figure onto itself.

DISCOVERY ACTIVITY

1. Draw an equilateral triangle *ABC*. Find and draw the lines of symmetry. Mark the point of intersection *O*.
2. What is the measure of the smallest angle between any two lines of symmetry at point *O*? Why?
3. Trace △*ABC* and its lines of symmetry on a sheet of tracing paper. Label the tracing *A'B'C'*.
4. Place △*A'B'C'* over △*ABC* and use a pin or compass point to hold the figures at point *O*.
5. Through how many degrees must you rotate △*A'B'C'* to map the triangle onto itself?
6. What happens when you rotate △*A'B'C'* through only 60°? Is the figure mapped onto itself for a 60° angle of rotation?

7. What can you conclude about the size of the angle of rotation that will provide point symmetry for △ABC?

You may have found that an equilateral triangle will be mapped onto itself after a rotation of 120°.

DEFINITION 12-3-3 **POINT SYMMETRY or ROTATION SYMMETRY is a rotation transformation that maps a figure onto itself. The point is determined by the intersection of two or more line symmetries. The angle of rotation is determined by the angle formed by the line symmetries or a multiple of the angle.**

Example: This square has four line symmetries. The smallest angle of rotation is 90°. Why?
What if the square is only rotated 45°? Will it have point symmetry? Copy the figure and try it.
Will the square have point symmetry if it is rotated 180°? Why?

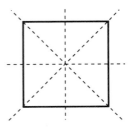

CLASS ACTIVITY

Each figure below has point symmetry. Write a LOGO procedure to tell the turtle to draw each figure.

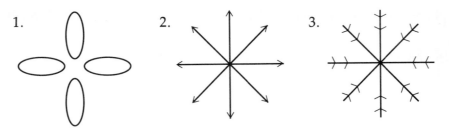

1.   2.   3.

4. Which figures have point symmetry?

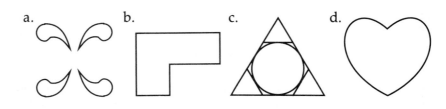

a.   b.   c.   d.

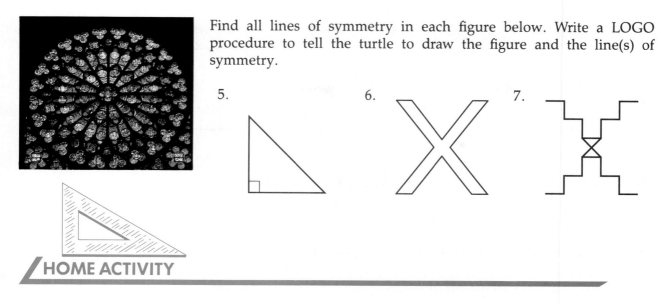

Find all lines of symmetry in each figure below. Write a LOGO procedure to tell the turtle to draw the figure and the line(s) of symmetry.

5.   6.   7.

HOME ACTIVITY

1. Copy the figures below. Classify each shape and label it.

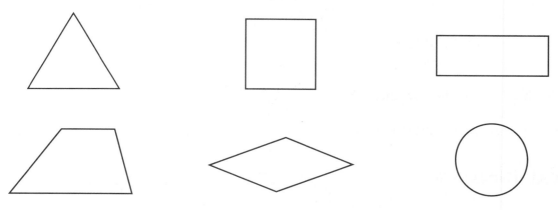

2. Identify the figures in Exercise 1 that have line symmetry. Draw the line(s) of symmetry.

3. Identify the figures in Exercise 1 that have point symmetry. Indicate the point of rotation. Find the smallest angle of rotation.

On graph paper, plot the equations. Identify each as a translation, reflection, or rotation.

4. $y = 2x + 2$ and $y = 2x + 1$

5. $y = 4x$, $y = x$, and $y = \frac{1}{4}x$

6. $y = \frac{1}{2}x$, and $y = \frac{1}{2}x + 2$

7. $y = x^2$ and $x = y^2$

8. Segment with endpoints $(-4, 2)$ and $(-2, 4)$ and segment with endpoints $(2, 4)$ and $(4, 2)$.

Below are the shapes of some common traffic signs.

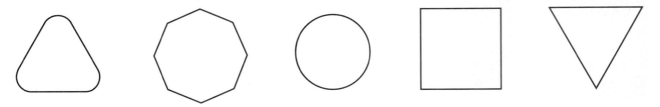

9. Which signs have line symmetry? Draw the lines of symmetry.

10. Which signs have point symmetry?

11. What is the angle of rotation for each figure with point symmetry?

The following objects are found in nature.

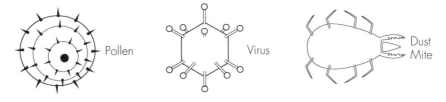

12. Which ones have line symmetry?

13. Which ones have point symmetry?

CRITICAL THINKING

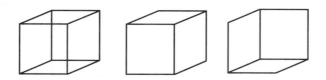

These three drawings represent a solid cube. The first drawing shows all the edges, which makes the cube harder to recognize. The second and third drawings show the cube with the hidden edges removed.

The figure to the right represents three intersecting blocks.

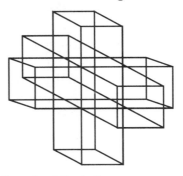

14. Which lines should be removed to make the blocks easier to visualize? (HINT: Trace the edges you think should be seen.)

15. How many ways can you visualize the blocks?

SECTION
12.4
Applications

After completing this section, you should understand
▲ how to solve problems by using definitions and theorems about loci and transformations.

The geometry of loci and transformations is used by architects, engineers, artists, designers, mapmakers, and computer programmers. You use it when you decorate a room or rearrange furniture. Planning a team strategy in baseball or soccer, hitting a tennis ball so that your opponent cannot return it—both involve the use of loci and transformations.

PARABOLAS

The following activity illustrates a very practical application using the concept of locus.

DISCOVERY ACTIVITY

1. Use a ruler to draw a line about 4 inches from the bottom of a sheet of unlined paper and parallel to the lower edge. Label the line AB.

2. Mark point F about $\frac{3}{4}$ inch above the line, halfway between A and B.

3. Draw or construct a perpendicular from F to $\overline{AB}$. Where it intersects the line, label the point C.

4. Bisect $\overline{FC}$ and label the midpoint P_1.

5. Point P_1 is equidistant from point F and from $\overline{AB}$.

6. Draw the set of points 1 inch from $\overline{AB}$.

7. Draw the set of points 1 inch from F.
8. Label the intersections of the points in steps 6 and 7 as P_2.
9. What can you say about the relationship between points P_2 and $\overline{AB}$?
10. You now have three points that are just as far from F as they are from $\overline{AB}$. Locate additional points. Follow steps 6 and 7 for these conditions:
    a. points $\frac{1}{2}$ inch from F and $\overline{AB}$; intersections labeled P_3
    b. points $\frac{3}{4}$ inch from F and $\overline{AB}$; intersections labeled P_4
    c. points $1\frac{1}{2}$ inches from F and $\overline{AB}$; intersections labeled P_5
    d. points 2 inches from F and $\overline{AB}$; intersections labeled P_6
11. Draw a curve through the points from left to right with a colored marker. Your curve should look like this.

You may have drawn an upward curve that seems to continue on indefinitely. This plane curve is called a **parabola.** Named by the ancient Greeks, the parabola is similar to the curved path followed by a ball thrown into the air.

DEFINITION 12-4-1 **A PARABOLA is the set of points on a plane equidistant from a line and a point not on the line.**

Now draw the same curve rotated 90° to the right about $\overline{FC}$. Imagine that the curve is spinning about the line $\overline{FC}$. Draw the three-dimensional picture of what you would see.

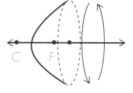

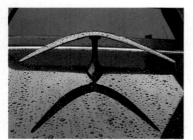

The automobile industry uses a parabolic curve in the design and manufacture of over 40 million headlights each year.

If the light bulb is placed at the focus point F, the light rays will be reflected out through the lens parallel to each other. The best time to observe this property is when you are driving in a car on a foggy night.

A satellite dish also has a parabolic shape. In this case, the rays, or television signals, are received instead of reflected.

CLASS ACTIVITY

1. Draw two 7-centimeter line segments, $\overline{AB}$ and $\overline{BC}$, that have a common endpoint, B, and form an angle of less than 180°.
2. Starting at point A (A is 1), number the centimeters 1 to 7.
3. Starting at point C (C is 7), number the centimeters 7 to 1.
4. Now connect the points, 1 to 1, 2 to 2, . . . 7 to 7.
5. What have you drawn?

DISCOVERY ACTIVITY

Suppose that a water-treatment plant is to be built near a river and shared by two cities. A pipeline will carry water from the plant to each city. To keep costs down, the cities want to install pipeline over the shortest possible distance. At what point along the river should the water-treatment plant (P) be located?

1. Use a centimeter ruler to make a scale drawing of this map. (Scale: 1 cm = 2 km)

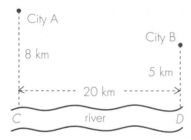
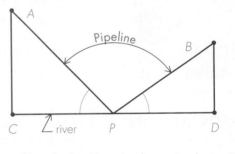

2. Use the scale drawing and centimeter ruler to estimate the total length of the pipeline from the plant to both cities—$AP + PB$—when P is located at point C. Record the length in the table where $CP = 0$ km. Use a protractor to measure $\angle 1$ (the angle at which the pipeline approaches the river) and $\angle 2$ (the angle at which the pipeline leaves the river).

CP	Total Length AP + PB	m∠1	m∠2
0 km	8 km + PB		
4 km			
8 km			
12 km			
16 km			
20 km			

3. Now move the possible location of P along the river toward point D, so that $CP = 4$ km. Repeat step 2 and record the length and angle measures.
4. Repeat step 2 for the other locations of plant P that are listed in the table.

5. From the table, decide where the water treatment plant should be located so that the pipeline length ($AP + PB$) is as short as possible.
6. What did you observe about m∠1 and m∠2 where the length ($AP + PB$) is the shortest?

You may have discovered that when length ($AP + PB$) is shortest, the measures of ∠1 and ∠2 are equal. Instead of experimentally finding the shortest path from City A to the river to City B, you can use a reflection transformation to find the distance.

CLASS ACTIVITY

1. Use your scale drawing from the Discovery Activity. Reflect City B across the river R.
2. Draw a line to connect A to B'. Label the intersection of $\overline{AB'}$ with line R, Q. Let P be any other point on line R.

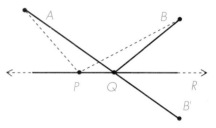

3. Measure and compare ($AP + PB$) and ($AP + PB'$).
4. Measure and compare ($AQ + QB$) and ($AQ + QB'$).
5. Compare ($AP + PB$) and ($AQ + QB$). Which length is less?
6. How does ($AQ + QB$) compare to your result for the shortest pipeline in the Discovery Activity?
7. Explain why reflecting City B across the river and finding the straight-line distance between City A and City B' gives the shortest distance between the two cities.

MORE TRANSFORMATIONS

Transformations are often used in the design of flags. Identify the types of transformation in each flag.

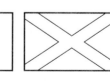

Bangladesh   Yugoslavia   France   Jamaica

Because most flags are rectangular, any rotation transformation in their design will be 180°.

Suppose the flags were square instead of rectangular. What transformations do they use in their designs? What is the angle of rotation?

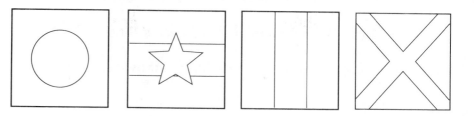

CLASS ACTIVITY

How could you use a map to find the locus of the sets of points?

1. All cities 4 miles from a given highway

2. All towns or cities 30 miles from your town

3. All houses or buildings 3 miles from your school building

4. Find a picture of your state flag. Tell whether its design contains any transformations and identify them.

HOME ACTIVITY

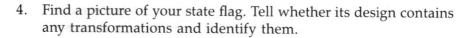

1. What is the locus of points described by the tip of the minute hand of a clock in one hour?

2. What is the locus of points described by the entire length of a minute hand of a clock in one hour?

3. What is the locus of points described by the entire length of a minute hand of a clock in 15 minutes?

4. How would you arrange the chairs in a room so that you would be equidistant from each person sitting in a chair? Where would your chair be?

Construct a circle with a radius of 1 inch.

5. What is the locus of points 1.5 inches from the circle?

6. What is the locus of points 1 inch from the circle?

7. What is the locus of points 0.5 inch from the circle?

8. What is the locus of points 2 centimeters from a point on a plane? in space?

9. What kind of transformation preserves parallelism?

10. What kind of transformation is like a mirror image?

11. What kind of transformation is illustrated by a spinning wheel?

Trace the flags.

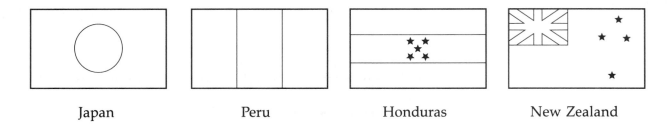

| Japan | Peru | Honduras | New Zealand |

12. Which flags have line symmetry? Draw the lines of symmetry.

13. Which flags have point symmetry? What is the angle of rotation?

Write a LOGO procedure that uses the procedure XYPLANE from page 118 and the command SETXY to tell the turtle to graph each figure. Name the transformation. (→ means mapped onto.)

14. $(-3, 5) \rightarrow (3, 5)$

15. $y = x \rightarrow y = -x$

16. $A(-3, 5), B(-1, 7) \rightarrow A'(-3 + 3, 5 + 3), B'(-1 + 3, 7 + 3)$

17. $A(-4, 0), B(-2, 6), C(0, 0) \rightarrow A'(4, 0), B'(2, 6), C'(0, 0) \rightarrow A''4, 0), B''(2, -6), C''(0, 0)$

18. $y = 2x - 1 \rightarrow y = 2x + 1$

19. $y = x^2 \rightarrow x = y^2$

CRITICAL THINKING

20. Explain what it means to say that a cube has plane symmetry. Use a model to help you.

21. Is a cube symmetric about more than one plane? How many?

22. Does a sphere have any plane symmetries? Explain.

23. Does a cylinder have any plane symmetries? Draw a picture to explain your answer.

Logic for Decision Making

After completing this section, you should understand

▲ how a conditional statement is related to its converse, inverse, and contrapositive.

▲ how to determine the truth value of a statement.

In Section 12.1 you found that, in determining a locus, you had to prove that every point in the locus satisfied the given conditions and that every point that satisfied the conditions was a point in the locus. In other words, you proved the truth of a statement and its converse.

Statement: If A, then B.
If a point is on an angle bisector, then it is equidistant from the sides of the angle.
Converse: If B, then A.
If a point is equidistant from the sides of an angle, then it is on the angle bisector.

You also had to prove one of two other possibilities—the inverse or the contrapositive of the statement.

Inverse: If not A, then not B.
If a point is not on an angle bisector, then it is not equidistant from the sides of the angle.
Contrapositive: If not B, then not A.
If a point is not equidistant from the sides of an angle, then it is not on the angle bisector.

Recall that a statement can be true or false, but it cannot be both true and false at the same time.

Example: "Watch out!" is neither true nor false, so it is not considered to be a statement.

Logic demonstrates that new statements can be constructed from a given statement and that the truth or falsity (truth value) of a new statement can be determined from the truth or falsity of the original statement.

DISCOVERY ACTIVITY

1. Tell whether the statement is true or false.
   a. If a figure is a square, then it is a rectangle.
   b. If a line segment is a chord of a circle, then it is a diameter.
   c. If a triangle is equilateral, then all three sides have the same length.
2. Write the converse of each statement. Is it true or false?
3. Write the inverse of each statement. Is it true or false?
4. Write the contrapositive of each statement. Is it true or false?
5. Copy and complete the table by writing true or false for each statement.

	Statement	Converse	Inverse	Contrapositive
a.	true	false		
b.				
c.				

6. Study the table of truth values for each statement and its converse, inverse, and contrapositive. What do you observe about the relationships between the statement and its contrapositive? the converse and the inverse?

You may have found that the truth values of a statement and its contrapositive are the same, and that the truth values of the converse and the inverse are the same. You can show this in a diagram.

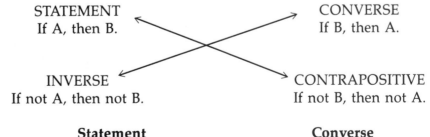

STATEMENT
If A, then B.

CONVERSE
If B, then A.

INVERSE
If not A, then not B.

CONTRAPOSITIVE
If not B, then not A.

Example:

Statement
If an animal has four legs, then it is a dog.

Converse
If an animal is a dog, then it has four legs.

Inverse
If an animal does not have four legs, then it is not a dog.

Contrapositive
If an animal is not a dog, then it does not have four legs.

1. Which statement has the same truth value as the original statement? Are these two statements true or false?

2. Which statement has the same truth value as the converse? Are these two statements true or false?

LOGIC AND PROOFS

In Section 12.1, you used logical statements and indirect reasoning to prove that the locus of points equidistant from the endpoints of a line segment is the perpendicular bisector of the segment. You can use sets of logical statements to prove this theorem: the locus of points in a plane equidistant from a point on the plane is a circle.

First, write the theorem in if-then, or conditional, form. Then write the converse, inverse, and contrapositive.

STATEMENT: If a set of points in a plane is equidistant from a given point, then the points are a circle.

CONVERSE: If a set of points is a circle, then the points are equidistant from a given point.

INVERSE: If a set of points is not equidistant from a given point, then the points are not a circle.

CONTRAPOSITIVE: If a set of points is not a circle, then the points are not equidistant from a given point.

Which statements must you prove true to show that the theorem is true?

CLASS ACTIVITY

For each statement, write the converse, inverse, and contrapositive. Classify each statement as true or false. Then use a computer and a geometric drawing tool to draw figures to justify each answer.

1. If a triangle is equilateral, then it is right.

2. If two angles are vertical angles, then their measures are equal.

3. If a triangle is equilateral, then it is isosceles.

4. If a figure is a polygon, then it is a triangle.

5. If a figure has four equal sides, then it is a square.

HOME ACTIVITY

For each statement, write the converse, inverse, and contrapositive.

1. If the sun is not shining, then it is night.

2. If a triangle is equilateral, then it is equiangular.

3. If two angles are right angles, then their measures are equal.

4. If a triangle is equilateral, then it is acute.

5. If I stay at home, then I don't pitch at the game.

6. If a figure is a quadrilateral, then its diagonals are equal.

7. If two lines are parallel, then they are coplanar.

8. If two triangles have the same size and the same shape, then they are congruent.

9. If a quadrilateral has two pairs of parallel sides, then it is a rectangle.

10. If it is snowing, then the temperature is 32° or less.

11. For exercises 1–10, classify each statement and its converse, inverse, and contrapositive as true or false. Identify the pairs of statements in each set that have the same truth value.

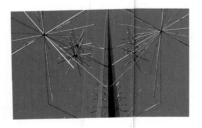

CRITICAL THINKING

12. It's Sunday, and I'm trying to plan my errands so that I can make just one trip downtown on one afternoon this week. I have to visit the dentist, get bread at the bakery, and pick up fresh tomatoes at the farmers' market. The dentist's office is open only on Tuesday, Wednesday, and Friday. The bakery is closed on Wednesday and Saturday afternoons. The farmers' market is not open on Tuesday or Thursday. What afternoon should I go downtown?

12.1 The locus of all points equidistant from a point on a plane is a circle.

$OB = OC$ so B and C are on the circle.
B and C are on the circle, so $\overline{OB} = \overline{OC}$.

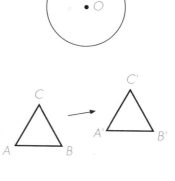

12.2 Identify the transformation pictured.

$\triangle A'B'C'$ is the translation of $\triangle ABC$.

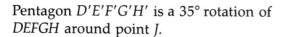

Pentagon $D'E'F'G'H'$ is a 35° rotation of $DEFGH$ around point J.

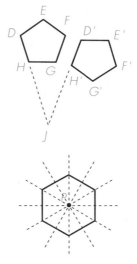

12.3 The regular hexagon demonstrates what two types of symmetry?

Line symmetry along six lines (reflection across the lines) and point, or rotation, symmetry (rotations of 60° around point P).

12.4 Describe the locus of points equidistant from point F and the x-axis.

A parabola

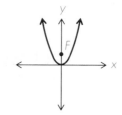

12.5 How are the truth values of these statements related?

Statement: If a figure is a square, then it is a rhombus.
Converse: If a figure is a rhombus, then it is a square.
Inverse: If a figure is not a square, then it is not a rhombus.
Contrapositive: If a figure is not a rhombus, then it is not a square.

Statement and contrapositive have the same truth value (true); inverse and converse have the same truth value (false).

Tesselations are patterns made up of figures that are translated over and over again to fill a plane so that there are no gaps or overlaps. Tesselations can be formed using one figure or a combination of different figures. An example of a tesselation made up of equilateral triangles is shown below.

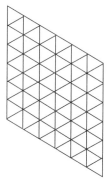

The LOGO turtle can be used to draw tesselations. First write a procedure to tell the turtle to draw the smallest part of the tesselation that is translated. Then use the REPEAT command to translate the pattern and draw the tesselation.

Example: Write a procedure to tell the turtle to draw the tesselation shown above.

The small part of the tesselation that is translated is shown below.

The procedure FIGURE given below tells the turtle to draw this figure.

```
TO FIGURE
   REPEAT 2[FD 10 RT 120]
   FD 10
   REPEAT 2[LT 120 FD 10]
END
```

The procedure TRANSLATE given below puts the turtle in position to translate the figure.

```
TO TRANSLATE
  PU LT 120 FD 10 RT 120 BK 10 PD
END
```

The procedure COLUMN given below tells the turtle to draw a column of the figures.

```
TO COLUMN
    REPEAT 5[FIGURE TRANSLATE]
END
```

The procedure TESS given below tells the turtle to draw the tesselation.

```
TO TESS
    REPEAT 5[COLUMN PU FD 30 PD]
END
```

Write a LOGO procedure to tell the turtle to draw each tesselation.

1.

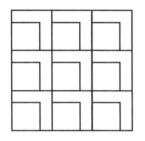

2.

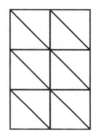

3.

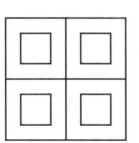

4.

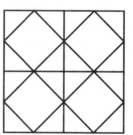

5.

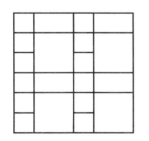

6.

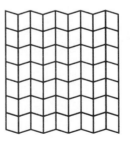

7. Draw your own tesselation. Write a LOGO procedure to tell the turtle to draw your tesselation.

1. What is the locus of all points within a circle that are equidistant from the endpoints of a given chord?

2. Find the locus of the centers of all the circles that are tangent to two parallel lines.

3. Draw a circle with a radius of 2.5 centimeters. Draw and describe the locus of all points 1.5 centimeters from the circle.

Tell whether the transformation is a translation, reflection, or rotation.

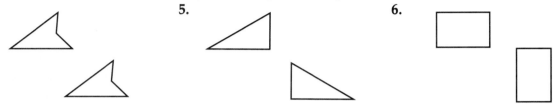

4.   **5.**   **6.**

Does the figure have line symmetry? Trace the figure and draw the line(s) of symmetry.

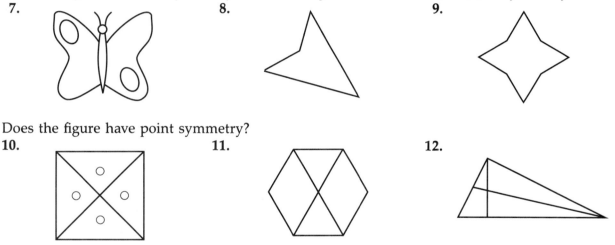

7.   **8.**   **9.**

Does the figure have point symmetry?

10.   **11.**   **12.**

13. The drawing shows a billiard table with balls W and R. Where should a player aim if she wants ball W to hit side $\overline{AB}$, then strike ball R? Draw the picture.

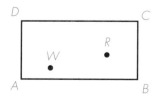

14. Write the converse, inverse, and contrapositive for this statement, and tell whether each is true or false: If the sum of the measures of two angles is 180°, then the angles are supplementary.

1. Describe the locus of the midpoints of all the radii of a given circle.
2. What is the locus of the centers of all the circles tangent to two intersecting lines?

Draw a line segment $\overline{AB}$ 3.5 centimeters long.
3. Find and describe the locus of all points equidistant from the endpoints of $\overline{AB}$.
4. Find and describe the locus of all points that are 2 centimeters from $\overline{AB}$.

Classify the transformation(s) that will make the figure on the left map onto the figure on the right.

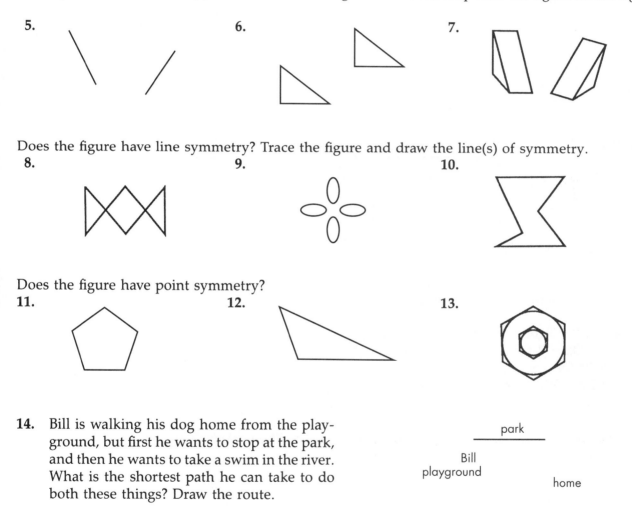

5.

6.

7.

Does the figure have line symmetry? Trace the figure and draw the line(s) of symmetry.
8.   9.   10.

Does the figure have point symmetry?
11.   12.   13.

14. Bill is walking his dog home from the playground, but first he wants to stop at the park, and then he wants to take a swim in the river. What is the shortest path he can take to do both these things? Draw the route.

park

Bill
playground

home

river

15. Write the converse, inverse, and contrapositive for this statement: If an animal lives in water, then it is a fish. Tell whether each statement is true or false, and identify the pairs of statements with the same truth value.

ALGEBRA SKILLS

The set of all solutions of an equation in two variables is called the *solution set* of the equation. Since a solution set is a set of ordered pairs, you can graph it on the x-y-axes. To graph an equation, first find its solution set. Make a table of values to find several ordered pairs that satisfy the equation. Then plot the ordered pairs on the x-y-axes.

The graph of a linear equation in two variables is a straight line.

Examples: $y = x + 2$ $x = 3$

This equation is equivalent to $x + 0y = 3$, so for any value of y, $x = 3$.

x	y
-2	0
-1	1
0	2
1	3
2	4
3	5

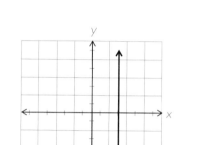

Graph the equation for the given values.

1. $y = 3x - 1$ for $x = -3, -2, -1, 0, 1, 2, 3$
2. $y = -3x - 1$ for $x = -3, -2, -1, 0, 1, 2, 3$
3. $y = |x|$ for $-3 < x < 3$
4. $-3(2x - 3y) = 18$ for $-4 < x < 4$
5. $y = x^2$ for $-3 < x < 3$

Find five elements of the solution set for each equation. Then graph the equation.

6. $x + y = 0$ 7. $y = x + 5$

8. $x + y = -2$ 9. $2x + y = 0$ 10. $y = -4x$

11. $2(x + 5) + 7(x - y) = 4x - 8y$
12. $20 = 3(-x - y) - 5(-3y + 4x)$

13. What is the equation for the set of points for which the x value is twice the y value? Draw the graph.

13

THE NEW GEOMETRY

Each problem that I solved became a rule which served afterwards to solve other problems.

René Descartes

SECTION 13.1

What Is a Point and What Is a Line?

After completing this section, you should understand
▲ how the term *point* is defined in coordinate geometry.
▲ how equations are used to define *lines*.

The seventeenth-century mathematician René Descartes is credited with introducing coordinates into the study of geometry. All at once it became possible to translate questions about shapes into questions about numbers and equations. In other words, questions about geometry became questions in algebra. This was the great inspiration behind Descartes's "new" geometry.

You already know that coordinate geometry uses ordered pairs of real numbers to talk about points.

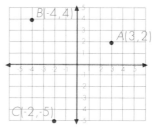

Example: In the figure at the right, you can speak of point *A* or of point (3,2). Point *B* is (−4,4), and point *C* is (−2,−5).

Of course it is always helpful to have diagrams to refer to. But in the "new" geometry—coordinate geometry—it is possible to define what a point is by using numbers only.

DEFINITION 13-1-1   **A POINT is an ordered pair (*x,y*) of real numbers.**

You learned in Chapter 12 that a geometric locus is a set consisting of all the points that satisfy a certain condition. In coordinate geometry, the condition is often an equation. The locus is the **graph** of the equation.

DISCOVERY ACTIVITY

See what you can discover about the graph of $y = 2x + 1$. You can find points on the graph by picking values for x and finding the values for y. For example if $x = 3$, then $y = 2 \cdot 3 + 1$, or 7. (3,7) is a point on the graph, since $y = 2x + 1$ will be true when you replace x with 3 and y with 7.

1. Copy and complete this table to find six points on the graph of the equation $y = 2x + 1$.
2. Draw and label x- and y-axes on a piece of graph paper. Mark the points for the ordered pairs listed in the last column of the table.

x	y	Point
3	7	(3,7)
$2\frac{1}{2}$		
1		
0		
-2		
-4		

3. What do you notice about the points you marked?

You may have discovered that the points you marked in the coordinate plane all lie on the same line l. If you use more values for x to find more points on the graph of $y = 2x + 1$, you will find that they are on the same line. The graph of $y = 2x + 1$ is the line l.

CLASS ACTIVITY

For each equation, make a table like the one in the Discovery Activity. Use the same values for x. Graph the ordered pairs in the coordinate plane. If the points for the equation seem to be on the same line, draw the line.

1. $y = 2x - 3$
2. $y = 0x + 4$
3. $y = 2x + 1$

For each equation, make a table like the one in the Discovery Activity. Use the x values given. Mark the points in the coordinate plane. If the points seem to be on the same line, draw the line.

4. $y = -3x + 2; x = 2\frac{1}{2}, 1, 0, -\frac{1}{2}, -1, -2$

5. $y = -\frac{1}{2}x - 4; x = 6, 1, 0, -1, -3, -4$

Each equation in the Class Activity has the form $y = mx + b$, where m and b are real numbers. Also, the graph of each equation is a line. If an equation has the form $y = mx + b$, where m and b are real numbers, then the graph is a straight line.

Is the reverse also true? If you have a line in the coordinate plane, is there an equation $y = mx + b$ whose graph is the given line? To find out, pick a line in the coordinate plane, say the line determined by $(-4,-3)$ and $(3,5)$. If these points are on the graph of $y = mx + b$, then $y = mx + b$ must be true when you put the coordinates in place of x and y.

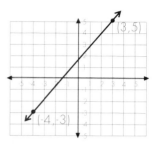

$$-3 = m(-4) + b$$

$$5 = m(3) + b$$

Subtract the second equation from the first.

$$-8 = -7m$$
$$m = \tfrac{-8}{-7} = \tfrac{8}{7}$$

Put $\tfrac{8}{7}$ in place of m in $-3 = m(-4) + b$.

$$-3 = \tfrac{8}{7}(-4) + b$$
$$-3 = \tfrac{-32}{7} + b$$
$$b = -3 + \tfrac{32}{7} = \tfrac{11}{7}$$

So the line through $(-4,-3)$ and $(3,5)$ is the graph of $y = \tfrac{8}{7}x + \tfrac{11}{7}$.

You can use this method for finding an equation for every *nonvertical* line in the coordinate plane. To see why the method fails for vertical lines, try using it to find an equation of the form $y = mx + b$ for the line passing through $(2,3)$ and $(2,-5)$.

$$3 = m(2) + b$$
$$-5 = m(2) + b$$

There are no real numbers m and b that will let $m(2) + b$ have two different values (3 and -5) at the same time. Therefore, there is no equation of the form $y = mx + b$ for this line.

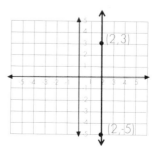

An equation for the vertical line through (2,3) and (2,−5) is $x = 2$. Every vertical line has an equation of the form $x = a$, where a is a real number. All nonvertical lines have an equation of the form $y = mx + b$. You can use these ideas to state a "new geometry" definition of the term *line*.

NOTEBOOK

DEFINITION 13-1-2 **A LINE is the set of all ordered pairs (points) in the coordinate plane that satisfy the equation of the form $y = mx + b$ or of the form $x = a$, where m, b, and a are real numbers.**

CLASS ACTIVITY

For each line, write an equation of the form $y = mx + b$ or of the form $x = a$.

1.

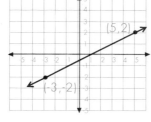

2.

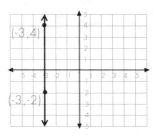

3.

4.

Use the XYPLANE procedure from page 118 to tell the turtle to draw the line determined by each pair of points. Then write an equation for the line.

5. (1,1) and (−1,−5)   6. (−2,6) and (7,−3)

7. (7,0) and (7,4)   8. (0,−3) and (4,−11)

9. (−5,−2) and (−5,3)   10. $\left(\frac{1}{2},−6\right)$ and $\left(\frac{1}{2},5\right)$

HOME ACTIVITY

Graph each equation in the coordinate plane. (First make a table of ordered pairs.)

1. $y = 2x + 6$ 2. $y = -x - 3$ 3. $y = 3x - 6$
4. $x = -4$ 5. $x = 5$ 6. $y = -3x + 5$

Tell whether the graph of the equation will be a vertical or a nonvertical line. You do not have to draw the graph.

7. $y = 8x - 2$ 8. $y = -\frac{3}{2}x + 6$ 9. $x = -9$

For each line, write an equation of the form $y = mx + b$ or of the form $x = a$.

10.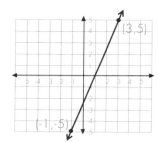

11.

Use the XYPLANE procedure from page 118 to tell the turtle to draw the line determined by each pair of points. Then write the equation.
12. (0,0) and (5,5) 13. (−1,2) and (3,−4)
14. (5,0) and (5,8) 15. (6,5) and (7,4)

CRITICAL THINKING

16. Explain what is wrong with the following "proof" that $2 = 0$. Assume that the numbers x and y are equal.

Steps	Reasons
1. $x = y$	1. Given
2. $x^2 = y^2$	2. Squares of equal are equal.
3. $x^2 - y^2 = 0$	3. Subtract y^2 from both sides.
4. $(x + y)(x - y) = 0$	4. Factoring
5. $x + y = 0$	5. Divide both sides by $x - y$.
6. $x + x = 0$	6. Substitution
7. $2x = 0$	7. Simplification
8. $2 = 0$	8. Divide both sides by x.

SECTION 13.2

The Easy Way to Graph a Line

After completing this section, you should understand
▲ the slope of a line.
▲ how to find the slope of a line given two points on the line.
▲ the *y*-intercept of a line.

You have seen that the graph of any equation of the form $y = mx + b$, where *m* and *b* are real numbers, is a nonvertical line. The numbers *m* and *b* give useful information about the graph. The number *m* in the equation $y = mx + b$ has a special name.

NOTEBOOK

DEFINITION 13-2-1: **The number *m* in $y = mx + b$ is the SLOPE of the line graph.**

If (x_1, y_1) and (x_2, y_2) are any two points on the graph $y = mx + b$, then these two equations are true:

$$y_2 = mx_2 + b$$
$$y_1 = mx_1 + b$$

Subtract the second equation from the first and you get

$$y_2 - y_1 = mx_2 - mx_1$$
$$\text{or } y_2 - y_1 = m(x_2 - x_1).$$

Divide both sides by $(x_2 - x_1)$ and you find that

$$m = \frac{y_2 - y_1}{x_2 - x_1}.$$

This reasoning gives you the following theorem.

483

THEOREM 13-2-1: If (x_1, y_1) and (x_2, y_2) are two points on the graph of $y = mx + b$, then the slope m of the graph is equal to

$$\frac{y_2 - y_1}{x_2 - x_1}.$$

Think of getting from (1,2) to (5,4) in the coordinate plane. You can start at (1,2), *run* four units parallel to the *x*-axis, then *rise* two units parallel to the *y*-axis. You arrive at (5,4). The ratio of *rise* to *run* equals the slope of the line passing through the two points.

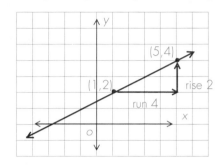

If you use the word **run** to mean the change in *x* and **rise** to mean the change in *y*, then you can say that the slope of any nonvertical line is the ratio of rise to run. This ratio is constant because it does not depend on which two points of the line you select.

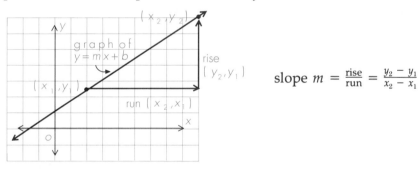

$$\text{slope } m = \frac{\text{rise}}{\text{run}} = \frac{y_2 - y_1}{x_2 - x_1}$$

The greater the rise is in comparison to the run, the *steeper* the line is. The steeper the line is, the greater the absolute value of its slope. The Discovery Activity will help you find out more about the slope of a line.

DISCOVERY ACTIVITY

Each diagram shows a nonvertical line and two points on the line. Study the diagrams and answer the questions.

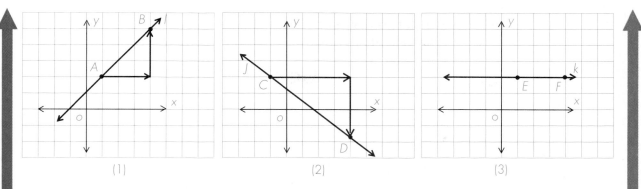

(1)   (2)   (3)

1. In diagram 1, the run is positive. Is the *rise* positive, negative, or zero? Is the *slope* positive, negative, or zero? Does line *l* go up or down as you move from left to right?

2. In diagram 2, the run is positive. Is the *rise* positive, negative, or zero? Is the *slope* positive, negative, or zero? Does line *j* go up or down as you move from left to right?

3. In diagram 3, the run is again positive. Is the *rise* positive, negative, or zero? Is the *slope* positive, negative, or zero? Does line *k* go up or down as you move from left to right?

You may have discovered that as you move from left to right (parallel to the *x*-axis), lines with positive slope go up, lines with negative slope go down, and lines with zero slope stay parallel to the *x*-axis.

None of the diagrams in the Discovery Activity were vertical. What happens if the line is vertical?

For a vertical line such as the line *m*, the run is zero and the rise is positive. The ratio rise/run cannot be found, because in algebra division by zero is not allowed: **a vertical line has no slope.**

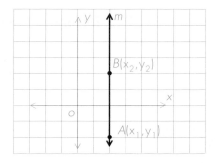

CLASS ACTIVITY

If the graph of the equation is a nonvertical line, give the slope. If the graph is a vertical line, answer "no slope."

1. $y = 7x - 5$ 2. $y = -3x + 4$

3. $x = 6$ 4. $y = -2$

Use the information in the graphs to find the slope of each nonvertical line. Tell whether the line goes up, down, or stays level or parallel to the x-axis, as you move from left to right in the coordinate plane.

5.

6.

7.

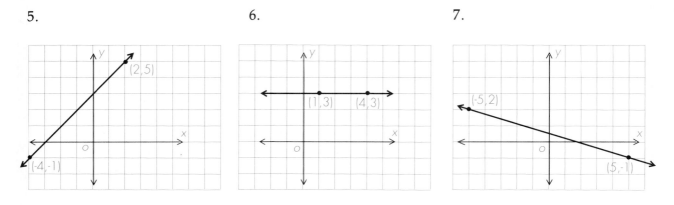

Find the slope of the line passing through the given points.

8. (3,6) and (8,6)

9. (5,7) and (9,−12)

10. (−1,−2) and (3,−10)

11. $\left(\frac{1}{2},\frac{1}{2}\right)$ and (3,−4)

THE Y-INTERCEPT OF A LINE

Every nonvertical line in the coordinate plane intersects the y-axis. The y-coordinate of the point of intersection has a special name.

DEFINITION 13-2-2: **The Y-INTERCEPT of a nonvertical line in the coordinate plane is the y-coordinate of the point where the line intersects the y-axis.**

Example: In the diagram, line l intersects the y-axis at (0,−2). Its y-intercept is −2. Line l intersects the y-axis at $\left(0,3\frac{1}{2}\right)$. Its y-intercept is $3\frac{1}{2}$.

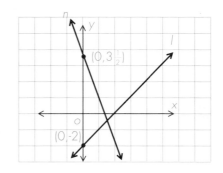

If you know an equation for a nonvertical line, it is easy to find its y-intercept. The x-coordinate of the point where the graph crosses the y-axis is $x = 0$. Put 0 in place of x in the equation and solve for y.

Example: Find the y-intercept of the graph of $y = -5x + 3$.

When you put 0 in place of x, you get

$$y = -5 \cdot 0 + 3 = 3.$$

The y-intercept is 3. The graph crosses the y-axis at (0,3).

If you put 0 in place of x in $y = mx + b$, you get

$$y = m \cdot 0 + b = b.$$

THEOREM 13-2-2:   **For any real numbers m and b, the y-intercept of the line graph of $y = mx + b$ is b, which means that the line crosses the y-axis at (0,b).**

Any equation that can be put into the form $y = mx + b$ has a graph that is a nonvertical line of slope m and y-intercept b.

Example: Graph $3x + 2y = y - 1$.

You can collect the terms with y on the left side, and the terms with x on the right side. You get

$$y = -3x - 1 \text{ or } y = -3x + (-1).$$

The graph will be the line that has slope -3 and y-intercept -1.

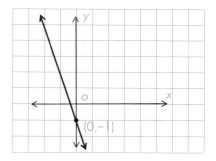

CLASS ACTIVITY

Give the slope and y-intercept of each line.

1. $y = -\frac{1}{2}x + 7$ 2. $y = 5x - 6$ 3. $x + y = 4$

4. $3x + 2y = 0$ 5. $y = -10$ 6. $y = 2x + \left(\frac{-12}{13}\right)$

Marion has a quick way for graphing any equation of the form $y = mx + b$. She marks the point where the line will cross the y-axis. She goes right one unit and then up or down as many units as the slope indicates. She draws the line through the point she arrives at and the point where she started. Use her method to graph each of the following equations.

7. $y = -2x + 3$ 8. $y = \frac{5}{2}x + 4$ 9. $y = -4x + (-2)$

HOME ACTIVITY

Use the information in the diagrams to find the slope of each nonvertical line. Tell whether the line goes up or down or stays level as you move from left to right in the coordinate plane.

1.　　　　　　　　　　　2.　　　　　　　　　　　3.

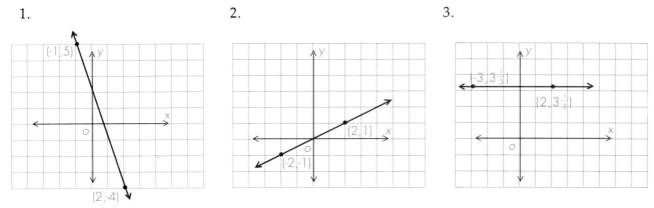

Give the slope and y-intercept of each line. If the line is vertical, write "no slope." Then use the XYPLANE procedure from page 118 in a Logo procedure to tell the turtle to draw the graph of each equation.

4. $y = 4x + 15$　　5. $y = -3x + \left(-\frac{1}{2}\right)$　6. $y = -x + 6$

7. $y = 7x - 18$　　8. $x = -9$　　9. $y = 3\frac{2}{5}$

10. $2x + y = \frac{4}{3}$　　11. $6x - 2y = 16$　　12. $x + 3 = 0$

Draw the graph of each equation.

13. $y = \frac{2}{3}x + 2$　　　　14. $y = 1\frac{1}{2}$　　　　15. $5x + 10y = 20$

Write an equation for the line that has the given slope and y-intercept.

16. Slope: 10
    y-intercept: 3

17. Slope: -6
    y-intercept: $2\frac{1}{2}$

18. Slope: 0
    y-intercept: -7

CRITICAL THINKING

19. In the town of Wilson, two-thirds of the adult men are married to three-fifths of the adult women. What fraction of the adults in Wilson are married? Explain how you arrived at your answer.

SECTION 13.3

Slopes of Perpendicular Lines

After completing this section, you should understand .
▲ how the slopes of parallel lines are related.
▲ how the slopes of perpendicular lines are related.

In Section 13.2, you learned that the slope of a line gives information about how steeply a line rises and falls as you move from left to right in the coordinate plane. This suggests that the slopes of two lines might provide information about whether the lines are parallel.

SLOPES OF PARALLEL LINES

Two vertical lines, such as $x = -1$ and $x = 3$, are parallel. Their slopes cannot be compared, however, since they do not have slopes.

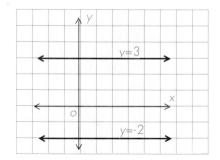

What about two horizontal lines, such as $y = -2$ and $y = 3$? The lines are parallel. Their equations can be written as follows:

$$y = 0x + (-2)$$
$$y = 0x + 3$$

This tells you that the slopes are both equal to 0.

The Discovery Activity will help you see what the situation is for parallel lines that are neither vertical nor horizontal.

489

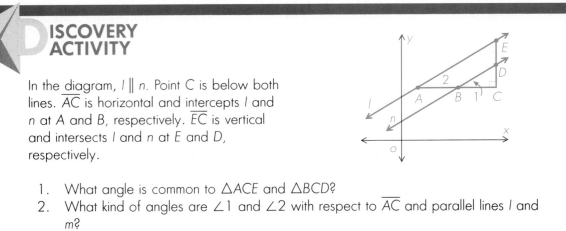

DISCOVERY ACTIVITY

In the diagram, $l \parallel n$. Point C is below both lines. $\overline{AC}$ is horizontal and intercepts l and n at A and B, respectively. $\overline{EC}$ is vertical and intersects l and n at E and D, respectively.

1. What angle is common to $\triangle ACE$ and $\triangle BCD$?
2. What kind of angles are $\angle 1$ and $\angle 2$ with respect to $\overline{AC}$ and parallel lines l and m?
3. $\triangle AEC$ is similar to $\triangle BDC$. Why?
4. Slope of $l = CE/AC$ and slope of $n = CD/BC$. How does the ratio CE/AC compare with the ratio CD/BC? How do you know?
5. How does the slope of l compare with the slope of n?

You may have discovered that if two lines are parallel and are neither horizontal nor vertical, then their slopes are equal. You know that any two horizontal lines have slopes equal to 0. This leads to the following theorem.

THEOREM 13-3-1: **If two nonvertical lines in the coordinate plane are parallel, then their slopes are equal.**

The converse of this theorem is also true: if the slopes of two lines are equal, then the lines are nonvertical and parallel.

Example: The lines $y = 2x + 3$ and $y = 2x - \frac{1}{2}$ are parallel, because both have a slope of 2.

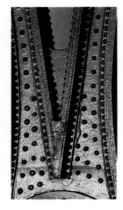

CLASS ACTIVITY

Tell whether the given lines are parallel.

1. $y = 7x + 8$ 2. $y = -3x - 1$ 3. $y = \frac{1}{2}x + 6$
   $y = 8x + 7$ $y = -3x + 15$ $2y - x = -8$

4. The line determined by $(0,0)$ and $(2,-1)$, and the line determined by $(-4,4)$ and $(4,0)$
5. The line $y = 3x + (-5)$ and the line determined by $(1,8)$ and $(2,5)$
6. The line $x = 3$ and the line $x = -11$

SLOPES OF PERPENDICULAR LINES

In the coordinate plane, sketch two perpendicular lines, neither of which is vertical. You will quickly see that one slopes downward from left to right and the other upward. This gives you some idea of how their slopes are related. One is positive and the other is negative.

DISCOVERY ACTIVITY

Consider these three pairs of lines.

$$\begin{cases} y = 3x - 1 \\ y = -\frac{1}{3}x + \frac{7}{3} \end{cases} \qquad \begin{cases} y = \frac{2}{5}x + 3 \\ y = -\frac{5}{2}x + 3 \end{cases} \qquad \begin{cases} y = \frac{4}{7}x \\ y = -\frac{7}{4}x + 2 \end{cases}$$

None of these lines are vertical.

1. Graph each pair of lines on a coordinate plane.

2. Measure the angle formed by each pair of lines. What kind of angle is it?

3. Which pairs of lines are perpendicular?

4. Multiply the slopes of the lines in each pair. What do you get each time?

5. What do you think may be true of the slopes of two nonvertical perpendicular lines?

When you drew the graphs for the Discovery Activity, you may have found that all the pairs of lines form 90° angles. So the lines in each pair are perpendicular. When you multiplied the slopes of the lines, you probably discovered that, in each case, the product is -1.

THEOREM 13-3-2: **If l_1 and l_2 are perpendicular lines with slopes of m_1 and m_2 respectively, then $m_1 m_2 = -1$.**

The converse of this theorem is also true: if lines l_1 and l_2 have slopes of m_1 and m_2 and if $m_1 m_2 = -1$, then l_1 is perpendicular to l_2.

Example: Show that the line l through $(-2,3)$ and $(1,5)$ is perpendicular to the line n through $(-2,3)$ and $(2,-3)$.

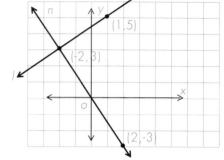

Slope of $n = \dfrac{3 - (-3)}{-2 - 2} = \dfrac{6}{-4}$, or $\dfrac{-3}{2}$

Slope of $l = \dfrac{5 - 3}{1 - (-2)} = \dfrac{2}{3}$

$\dfrac{-3}{2} \cdot \dfrac{2}{3} = -1$, so $l \perp n$.

Example: Show that the lines $y = 2x + 4$ and $y = -\frac{3}{2}x + 1$ are *not* perpendicular.

The product of their slopes is $2 \cdot \left(\dfrac{-3}{2}\right) = -3$. Since the product of the slopes is not -1, the lines cannot be perpendicular.

CLASS ACTIVITY

Tell whether the given pair of lines are perpendicular.

1. $y = 7x + 5$
   $y = \frac{1}{7}x - 2$

2. $y = \frac{3}{4}x + 6$
   $y = -\frac{4}{3}x + 1$

3. $2x + 6y = 3$
   $-3x + y = 8$

4. The vertices of a quadrilateral are $A(-2,1)$, $B(2,5)$, $C(6,1)$, and $D(2,-3)$. Tell what kind of quadrilateral $ABCD$ is and justify your answer.

5. The vertices of a quadrilateral are $W(-1,0)$, $X(2,4)$, $Y(7,4)$, and $Z(4,0)$. Show that $WXYZ$ is not a square.

6. Show that the diagonals of the quadrilateral $WXYZ$ of Exercise 5 are perpendicular.

7. The vertices of $\triangle PQR$ are $P(-2,0)$, $Q(0,2)$, and $R(3,-3)$. Show that $\triangle PQR$ is isosceles and that the line $y = -x$ is perpendicular to and bisects the base of $\triangle PQR$.

HOME ACTIVITY

Tell whether the given lines are parallel.

1. $y = 7x - 1$
   $y = 7x + 10$

2. $y = -4x + 6$
   $y = \frac{1}{4}x + 2$

3. $y = 10$
   $y = \frac{-1}{10}$

4. $y = \frac{6}{5}x + 12$
   $y = \frac{6}{5}x - 18$

5. $7x + 3y = 10$
   $y = \frac{7}{3}x - 2$

6. $4x + 5y + 6 = 0$
   $4x + 5y - 8 = 0$

7. The line determined by $(-4,-2)$ and $(-3,-5)$, and the line determined by $(1,4)$ and $(3,-2)$

8. The line $y = 5x + 9$ and the line determined by $(2,-1)$ and $(3,3)$

Refer to the figure $ABCD$.

9. What is the slope of $\overline{AB}$? of $\overline{DC}$?

10. What is the slope of $\overline{AD}$? of $\overline{BC}$?

11. What kind of quadrilateral is $ABCD$? How do you know?

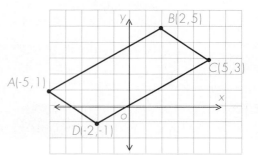

Tell whether the given lines are perpendicular.

12. $y = \frac{2}{3}x + 1$
    $y = -\frac{3}{2}x + 8$

13. $y = \frac{8}{9}x - 4$
    $y = \frac{9}{8}x + 4$

14. $y = 6$
    $y = \frac{1}{2}x - \frac{1}{6}$

15. The line determined by $(-5,4)$ and $(4,-5)$, the line determined by $(10,10)$ and $(-10,-10)$

Refer to $\triangle ABC$.

16. Find the slope of each side of $\triangle ABC$.
17. Is $\triangle ABC$ a right triangle? How do you know?
18. If you drew a line passing through A and perpendicular to BC, what would its slope be?
19. The vertices of $\triangle KLM$ are $K(-1,3)$, $L(3,6)$, and $M(2,-1)$. Show that $\triangle KLM$ is an isosceles right triangle.

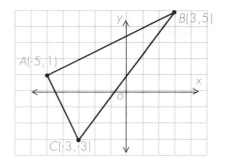

Use the XYPLANE procedure from page 118 in a Logo procedure to tell the turtle to draw the line determined by each pair of points. Tell whether each pair of lines appear to be parallel, perpendicular, or neither. Justify your answer.

20. (−6,7) and (−4,4)
    (−7,4) and (−4,6)

21. (1,1) and (4,4)
    (3,−1) and (6,2)

22. (−6,−1) and (−1,4)
    (−1,−2) and (0,3)

23. (−3,0) and (3,1)
    (−3,−3) and (3,−2)

24. Use what you know about slopes to show that figure *ABCD* is a trapezoid. Is it an *isosceles* trapezoid? How do you know?

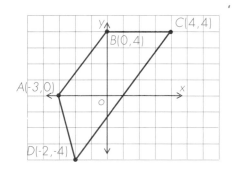

25. Use what you know about slopes to show that quadrilateral *PQRS* is a rectangle.

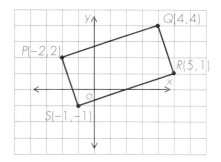

CRITICAL THINKING

How many times is the digit seven printed in page numbers if a book has the following number of pages? Assume that the pages are numbered consecutively, starting with 1.

26. 10 pages

27. 30 pages

28. 70 pages

29. 80 pages

30. 100 pages

31. 200 pages

SECTION 13.4

Circles

After completing this section, you should understand
▲ how to write an equation for a circle whose center is the origin.
▲ how to graph equations of the form $x^2 + y^2 = r^2$.

You have seen something of how coordinate geometry can deal with lines. Now you will see how it can deal with curves.

CIRCLES

You will recall that a circle is the set of all points in a plane that are equidistant from a given point (the center of the circle).

DISCOVERY ACTIVITY

On a piece of graph paper, draw x- and y-axes. Mark and label the five points shown in the plane at the right. Take a compass and draw the circle that has its center at the origin and that passes through (5,0).

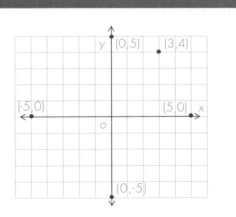

1. What is the length of a radius of the circle?
2. Does the circle pass through (3,4)?
3. Mark and label seven other points that have integer coordinates and that lie on the circle.
4. For each point you labeled in your diagram, square the coordinates and add the two squares. What number do you get each time?
5. If (x,y) is any point on the circle, what value do you think $x^2 + y^2$ will have?

In Chapter 7 you learned that the distance between any two points $A(x_1, y_1)$ and $B(x_2, y_2)$ is given by the distance formula:

$$d = \sqrt{(x_2 - x_1)^2 + (y_2 - y_1)^2}$$

Apply this formula to the situation in the Discovery Activity.

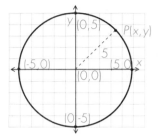

Let $P(x,y)$ be a point on the circle with center $(0,0)$ and radius 5. Because it is on the circle, the distance from (x,y) to $(0,0)$ must be 5. Using the distance formula to express the distance from (x,y) to $(0,0)$, you get the following:

$$\sqrt{(x - 0)^2 + (y - 0)^2} = 5$$

$$\sqrt{x^2 + y^2} = 5$$

Square both sides of the last equation and you get

$$x^2 + y^2 = 25$$

Clearly you could use any positive real number r for the radius. This reasoning leads to the following theorem:

NOTEBOOK

THEOREM 13-4-1: **The points $P(x,y)$ whose coordinates satisfy the equation $x^2 + y^2 = r^2$, where r is a positive real number, form a circle of radius r whose center is $(0,0)$.**

Example: Describe the set of points in the coordinate plane whose coordinates satisfy $x^2 + y^2 = 64$.

Since $64 = 8^2$, the points satisfy $x^2 + y^2 = 8^2$. The set of points is the circle with center $(0,0)$ and radius 8.

Example: Write an equation for the circle with center $(0,0)$ and radius 13.

Use 13 for r in $x^2 + y^2 = r^2$. You get $x^2 + y^2 = 13^2$, or $x^2 + y^2 = 169$.

CLASS ACTIVITY

Refer to circles 1 and 2.

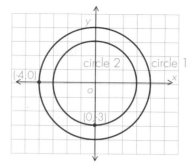

1. Write an equation for circle 1.

2. Write an equation for circle 2.

3. Which of the two circles does $(2, 2\sqrt{3})$ lie on?

4. Does $(-1, 2\sqrt{2})$ lie on either of the circles? If so, which one?

Write an equation for each circle.

5. The circle with center $(0,0)$ and radius 7

6. The circle with center $(0,0)$ and radius $\frac{9}{2}$

7. The circle with center $(0,0)$ and radius 100

8. The circle with center $(0,0)$ and radius 13.

9. Consider the circle $x^2 + y^2 = 36$. Write the ordered pairs for the points on this circle that have integer x-coordinates. Give the y-coordinates as decimals to the nearest hundredth. Use a calculator.

10. Tell which of these three circles lies inside the other two.

$$\text{Circle 1: } x^2 + y^2 = 9$$
$$\text{Circle 2: } x^2 + y^2 = (\sqrt{6})^2$$
$$\text{Circle 3: } x^2 + y^2 = 4^2$$

TANGENTS TO CIRCLES

Recall that a line is tangent to a circle if it intersects the circle in just one point. You saw in Theorem 10-4-4 that a tangent to a circle is perpendicular to the radius drawn to the point where the tangent intersects the circle. The converse of Theorem 10-4-4 is also true. In other words, if you draw a line perpendicular to a radius $\overline{OP}$ at the point P (where O is the center of the circle), then the line is tangent to the circle.

Consider the circle $x^2 + y^2 = 25$ and the line $y = -\frac{3}{4}x + \frac{25}{4}$.

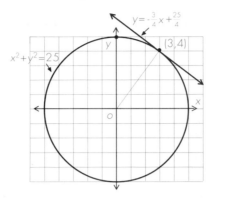

Suppose you draw the radius from the center of the circle to $(3,4)$, which is a point on the circle. The point $(3,4)$ is also on the line $y = -\frac{3}{4}x + \frac{25}{4}$. To show that this is true, just put 3 in place of x and 4 in place of y.

$$4 = -\frac{3}{4} \cdot 3 + \frac{25}{4}$$
$$4 = -\frac{9}{4} + \frac{25}{4}$$
$$4 = \frac{16}{4}$$
$$4 = 4$$

The slope of the radius is $\frac{4-0}{3-0}$, or $\frac{4}{3}$. The slope of the given line is $-\frac{3}{4}$. The product of these two slopes is $-\frac{12}{12}$, or -1. Therefore $y = -\frac{3}{4}x + \frac{25}{4}$ is perpendicular to the radius at the point $(3,4)$. So the given line is indeed tangent to the circle.

CLASS ACTIVITY

Write an equation for the given circle.

1. The circle with center $(0,0)$ and radius 6

2. The circle with center $(0,0)$ and radius $\sqrt{15}$

3. The circle with center $(0,0)$ and radius $\frac{3}{4}$

4. The circle with center $(0,0)$ and radius 18

5. The circle with center $(0,0)$ and passing through the point $(5,1)$

6. The circle with center $(0,0)$ and passing through the point $(5\sqrt{2}, 5\sqrt{2})$.

HOME ACTIVITY

A circle has center (0,0) and radius 5.

1. Write an equation for the circle.

2. Is $(-4,3)$ a point on the circle?

3. What is the slope of the line determined by $(0,0)$ and $(-4,3)$?

4. Which of these lines are perpendicular to the line through $(0,0)$ and $(-4,3)$?

   a. $y = \frac{4}{3}x + \frac{20}{3}$ b. $y = \frac{4}{3}x + \frac{25}{3}$ c. $y = \frac{4}{3}x + 30$

5. Which of the lines in Exercise 4 is tangent to the given circle? How do you know that it is a tangent?

6. Write an equation for the circle tangent to the line determined by $(0,6)$ and $(6,0)$ and with its center at $(0,0)$.

Use a calculator to help you determine whether the given point is on the circle whose equation is $x^2 + y^2 = 746$.

7. $(18,19)$ 8. $(21,14)$ 9. $(11,25)$

10. $(-12,24)$ 11. $(27,\sqrt{17})$ 12. $(\sqrt{15},30)$

Graph the equation.

13. $x^2 + y^2 = 49$ 14. $x^2 + y^2 = 4$ 15. $x^2 + y^2 = 20$

16. $x^2 + y^2 = \left(\frac{11}{2}\right)^2$ 17. $x^2 + y^2 = \frac{100}{9}$ 18. $x^2 + y^2 = 64$

Tell whether the given point is on the circle shown in the graph.

19. $(2,2)$ 20. $(-3,\sqrt{7})$

21. $(-4,0)$ 22. $(1,\sqrt{15})$

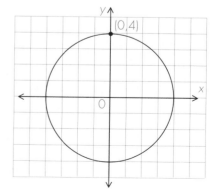

Consider the circle with center (0,0) and radius 13.

23. Write an equation for the circle.

24. Is (5,12) a point on the circle?

25. What is the slope of the line determined by (0,0) and (5,12)?

26. Does the line $y = -\frac{5}{12}x + \frac{169}{12}$ pass through the point (5,12)?

27. Is the line $y = -\frac{5}{12}x + \frac{169}{12}$ tangent to the given circle?

CRITICAL THINKING

This puzzle is known as the Tower of Hanoi problem. There are three rings of different sizes on the left peg of a board that has three pegs. The smallest ring is on the top, the largest ring is on the bottom.

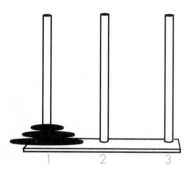

The goal is to transfer rings one at a time from one peg to another until all the rings are on the peg on the right. A larger ring can never be placed on top of a smaller ring.

28. If there were only one ring on peg 1, how many moves would be required?

29. If there were two rings on peg 1, how many moves would be required?

30. For the three rings shown in the diagram, how many moves are required?

31. Describe a pattern that tells how many moves are required for n rings.

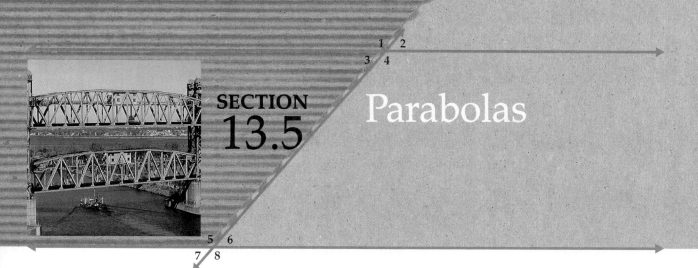

SECTION 13.5   Parabolas

After completing this section, you should understand
▲   how to find and graph an equation for a parabola.

In Section 12.4 you learned that in a plane, a parabola is a set consisting of all points equidistant from a given line and a point not on the line. You can use this description to find equations of parabolas.

FINDING EQUATIONS OF PARABOLAS

Point $(0,1)$ is on the y-axis. It lies one unit above the origin. Line $y = -1$ is a horizontal line one unit below the x-axis. This point and line determine a parabola.

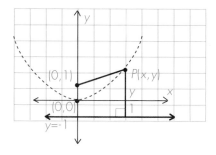

If $P(x,y)$ is a point on this parabola, then it must be equidistant from $(0,1)$ and the line $y = -1$. You can express the two distances as follows:

distance from (x,y) to $y = -1$ is $y + 1$;
distance from (x,y) to $(0,1)$ is $\sqrt{(x - 0)^2 + (y - 1)^2}$.

These distances are equal, so

$$y + 1 = \sqrt{(x - 0)^2 + (y - 1)^2}.$$
$$(y + 1)^2 = (x - 0)^2 + (y - 1)^2$$
$$y^2 + 2y + 1 = x^2 + y^2 - 2y + 1$$
$$4y = x^2$$
$$y = \tfrac{1}{4}x^2$$

Therefore an equation for the parabola is $y = \tfrac{1}{4}x^2$.

501

The choice of the point (0,1) and the line $y = -1$ made the algebra easier. Also it guaranteed that the low point of the parabola would pass through the origin.

CLASS ACTIVITY

Find an equation for the parabola determined by the given point and line.

1. The point (0,2) and the line $y = -2$

2. The point (0,3) and the line $y = -3$

3. The point (0,4) and the line $y = -4$

4. The point $(0,\frac{1}{4})$ and the line $y = -\frac{1}{4}$

5. The point $(0,\frac{1}{8})$ and the line $y = -\frac{1}{8}$

The method used so far can be applied to find an equation for the parabola determined by any point $(0,a)$, where $a > 0$, and the line $y = -a$.

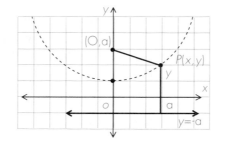

$$y + a = \sqrt{(x - 0)^2 + (y - a)^2}$$
$$(y + a)^2 = (x - 0)^2 + (y - a)^2$$
$$y^2 + 2ay + a^2 = x^2 + y^2 - 2ay + a^2$$
$$4ay = x^2$$
$$y = \frac{x^2}{4a}, \text{ or } y = \frac{1}{4a}x^2$$

THEOREM 13-5-1: **An equation for the parabola determined by $(0,a)$ and the line $y = -a$, where $a > 0$, is $y = \frac{1}{4a}x^2$.**

GRAPHING PARABOLAS

You have seen how to find equations for parabolas. Now you will get some experience in drawing their graphs.

DISCOVERY ACTIVITY

1. Make a table of values of x and y for the equation $y = x^2$. Use these values for x:
    $-3, -2, -1, \frac{-1}{2}, 0, \frac{1}{2}, 1, 2, 3$.

2. Graph the nine points you get from the table.

3. Connect the nine points with a smooth curve to show the parabola.

4. What could you do to get an even better, smoother graph?

5. Repeat steps 1 through 3 to graph the equation $y = -x^2$.

6. How are the graphs of $y = x^2$ and $y = -x^2$ the same? How are they different?

In the Discovery Activity, you may have found that the graph of $y = x^2$ is a parabola that opens upward. The graph of $y = -x^2$ is a parabola that opens downward.

Notice that each of the parabolas that you drew in the Discovery Activity is symmetric with respect to the y-axis. This is true because the square of any number and its opposite are equal.

Example: The points $(2,4)$ and $(-2,4)$ are both on the graph of $y = x^2$, since $2^2 = (-2)^2 = 4$. Both points are four units above the x-axis. $(2,4)$ is two units to the right of the y-axis, and $(-2,4)$ is two units to the left.

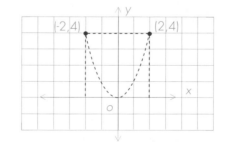

CLASS ACTIVITY

1. On the same $x-y$ plane, graph $y = \frac{1}{3}x^2$, $y = x^2$, $y = 2x^2$, and $y = 3x^2$.

2. Describe how the number by which x^2 is multiplied in the equations of Exercise 1 affects the shape of the graph.

3. Do the graphs for Exercise 1 open up or down?

4. What is the lowest point on each graph in Exercise 1?

5. Describe what you think the graphs of $y = -\frac{1}{3}x^2$, $y = -x^2$, $y = -2x^2$, and $y = -3x^2$ would look like.

Use a calculator to help you determine whether the given point is on the graph of $y = -0.75x^2$.

6. $(2, -2)$ 7. $(2, -3)$ 8. $(-3, -6.75)$

9. $(-0.5, 0.1875)$ 10. $(\sqrt{2}, -1.5)$ 11. $(-0.8, 0.5)$

12. Graph $y = -0.75x^2$. Graph the points from Exercises 6–11 to verify your answers to these exercises.

Circles and parabolas are examples of curves known as **conic sections.** Other important conic sections are the ellipse and the hyperbola. An ellipse is the locus of all points in a plane the sum of whose distances from two fixed points A and B is a given number d. A hyperbola is the locus of all points in a plane the difference of whose distances from A and B is a given number d. These curves and their properties were studied in detail by the Greek mathematicians of ancient times. The Greeks called these curves *conic* sections because they are the curves you get by slicing through a cone with a plane.

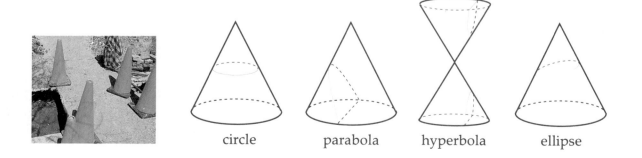

circle   parabola   hyperbola   ellipse

These curves, as it turns out, are important for understanding many things that occur in daily life and in nature. For example, when a football player kicks a goal, the football travels in a parabolic path. The orbits of the planets in the solar system are shaped like ellipses.

REPORT 13-5-1: Prepare a report on the studies the ancient Greeks made of conic sections. Include the following information:

a. the names and lifetimes of some of the mathematicians who made especially important discoveries about conic sections
b. the names of some of the more important written works about conics by these mathematicians
c. some of the special discoveries the Greek mathematicians made about conic sections

Sources:
The New Encyclopaedia Brittanica, 15th ed., s.v. "geometry."
The New Encyclopaedia Brittanica, 15th ed., s.v. "Apollonius of Perga."
Boyer, Carl B. *A History of Mathematics.* Princeton: Princeton University Press, 1985.

HOME ACTIVITY

Find an equation for the parabola determined by the given point and line.

1. The point $(0,5)$ and the line $y = -5$

2. The point $(0,\frac{1}{2})$ and the line $y = \frac{-1}{2}$

3. The point $(0,\frac{3}{4})$ and the line $y = \frac{-3}{4}$

4. The point $(0,-1)$ and the line $y = 1$

Graph each parabola. You may find it helpful to first make a table of values of x and y.

5. $y = \frac{1}{2}x^2$
6. $y = \frac{-1}{2}x^2$
7. $y = -2x^2$
8. $y = 4x^2$

Tell whether the parabola will open up or down.

9. $y = -9x^2$ 10. $y = \frac{2}{3}x^2$ 11. $y = 7x^2$ 12. $y = -4x^2$

13. $y = -6x^2$ 14. $y = 100x^2$ 15. $y = \frac{-3}{2}x^2$ 16. $y = \frac{x^2}{5}$

Consider these parabolas: $y = 25x^2$, $y = \frac{1}{10}x^2$, $y = \frac{-1}{100}x^2$, and $y = -25x^2$.

17. Which of the parabolas have exactly the same shape?

18. Which parabola is flatter than the others?

19. What point do all the parabolas pass through?

20. What line is a line of symmetry for all the parabolas?

CRITICAL THINKING

21. In the same coordinate plane, draw the graphs of $y = x^2$, $y = x^2 + 3$, and $y = (x - 2)^2$. Describe the effect that adding 3 to x^2 has on the graph of $y = x^2$. Describe the effect that replacing x with $(x - 2)$ has on the graph of $y = x^2$.

SECTION 13.6

Applications

After completing this section, you should understand
▲ how to prove theorems using coordinate geometry.

You have now seen enough of coordinate geometry to be able to use it to prove theorems.

THE MIDPOINT THEOREM

The midpoint theorem tells you that you can find the midpoint of any segment by finding the average of the x-coordinates and the average of the y-coordinates.

THEOREM 13-6-1: **(The midpoint theorem) If $P(a,b)$ and $Q(c,d)$ are any two points in the coordinate plane, then the midpoint of segment PQ is $M\left(\frac{a+c}{2}, \frac{b+d}{2}\right)$.**

CLASS ACTIVITY

To prove the midpoint theorem, you need to prove two things:
$MP = MQ$ and $MP + MQ = PQ$.

1. Use the distance formula to write algebraic expressions for MP, MQ, and PQ.

2. Use algebra to show that $MP = MQ$.

3. Use algebra to show that $MP + MQ = PQ$.

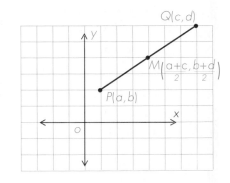

PROVING THEOREMS

The midpoint theorem is very helpful in proofs that use coordinate geometry. The first theorem you will prove is Theorem 6-6-1. You may find it interesting to compare the coordinate geometry proof with the proof in Section 6.6.

DISCOVERY ACTIVITY

In this activity, you will use coordinate geometry to prove that the segment connecting the midpoints of two sides of a triangle is parallel to the third side and is half as long as the third side.

Refer to the figure. You are given any triangle ABC. It has been placed so that A is at the origin and C is on the x-axis. This makes three of the coordinates 0 and helps keep the algebra simple.

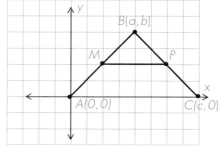

Supply the missing reasons.

Steps	Reasons
1. $\triangle ABC$ has vertices $A(0,0)$, $B(a,b)$, and $C(c,0)$.	1. Given
2. The midpoint M of $\overline{AB}$ has coordinates $\left(\frac{a}{2}, \frac{b}{2}\right)$.	2. Midpoint theorem
3. The midpoint P of $\overline{BC}$ has coordinates $\left(\frac{a+c}{2}, \frac{b}{2}\right)$.	3.
4. Slope of $\overline{AC} = 0$	4.
5. Slope of $\overline{MP} = 0$	5.
6. $\overline{AC} \parallel \overline{MP}$	6. Segments that have the same slope are parallel.
7. $AC = c - 0 = c$	7. Distance on x-axis $= \lvert x_2 - x_1 \rvert$
8. $MP = \frac{a+c}{2} - \frac{a}{2} = \frac{c}{2}$	8.
9. $MP = \frac{1}{2}AC$	9. Substitution

In the Discovery Activity you may have noticed that by placing the triangle a certain way, you can keep the algebra as simple as possible.

CLASS ACTIVITY

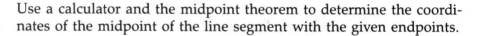

Use a calculator and the midpoint theorem to determine the coordinates of the midpoint of the line segment with the given endpoints.

1. (15.3,17.8) and (20.2,40.7)
2. (−26.42,19.8) and (37.5,−12.23)
3. (56.71,38.9) and (−64.83,−27.64)
4. (−83.51,−14.6) and (−49.382,76.15)

5. Suppose you had to use coordinate geometry to prove that the diagonals of a square are perpendicular and have the same length. Which of these placements of the square will make the proof easier?

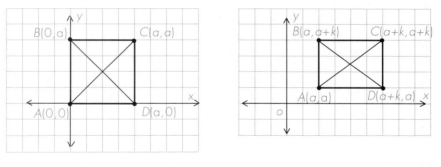

6. Refer to Exercise 5. What formula would you make use of to prove that $AC = BD$? What theorem would you use to prove that $\overline{AC} \perp \overline{BD}$?

7. Refer to figure $DEFG$ to prove that the diagonals of an isosceles trapezoid have the same length.

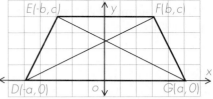

8. Refer to figure $WXYZ$ to prove that the diagonals of a parallelogram bisect each other.

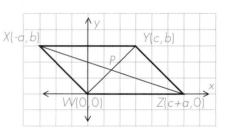

HOME ACTIVITY

Use the midpoint theorem to find the coordinates of the midpoint of the line segment having the given endpoints.

1. (1,0) and (5,4)

2. (2,6) and (8,20)

3. (−3,1) and (5,7)

4. (2,9) and (3,1)

5. (−6,−1) and (9,0)

6. (−a,b) and (a, −b)

7. Refer to figure *ABCD*. Prove that the diagonals of quadrilateral *ABCD* are perpendicular.

8. Refer to △*ABC* to show that for point *C* on the semicircle shown, $\overline{AC} \perp \overline{BC}$.

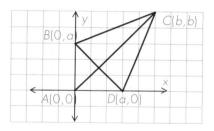

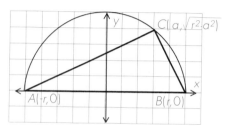

9. Refer to figure *ABCD*. Show that if you join the midpoints of consecutive sides of any quadrilateral, the resulting figure is a parallelogram. (The midpoints are *P*, *Q*, *R*, and *S*.)

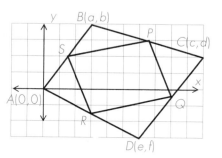

CRITICAL THINKING

10. Suppose you want to use coordinate geometry to prove that the midpoint of the hypotenuse of a right triangle is equidistant from the vertices. Explain why the diagram below would *not* be satisfactory for your proof.

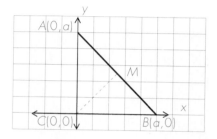

13.1 Is $y = \frac{3}{2}x + \frac{5}{2}$ an equation for line l?

Substitute the given values for x and y into the equation.
$7 = \frac{3}{2}(3) + \frac{5}{2} = 7$ $4 = \frac{3}{2}(1) + \frac{5}{2} = 4$
Because both substitutions are true,
$y = \frac{3}{2}x + \frac{5}{2}$ is a valid equation for line l.

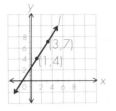

13.2 Give an equation for line n.
One equation will be in the form $y = mx + b$.
$m = \frac{y_2 - y_1}{x_2 - x_1} = \frac{0 - 4}{4 - 0} = -1$ Slope
$b = 4$ y-intercept
An equation for line n is $y = -x + 4$.

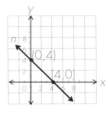

13.3 Show that lines r and t are parallel and that line s is perpendicular to them.

All three lines are in the form $y = mx + b$. Lines r and t have slopes of $\frac{1}{2}$, so they are parallel. Line s has a slope of -2. By the converse of Theorem 13-3-2, the line s is perpendicular to r and t.

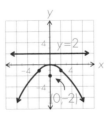

13.4 Find an equation for circle O.

Circle O has its center at $(0,0)$, so an equation for it will be in the form $x^2 + y^2 = r^2$. $x^2 + y^2 = 5^2$, or
$x^2 + y^2 = 25$ is an equation for the circle.

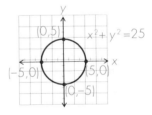

13.5 Find an equation for the parabola determined by point $(0,-2)$ and line $y = 2$.
The line and point are in the form $(0,a)$ and $y = -a$.
$y = \frac{1}{4a}x^2$ By Theorem 13-5-1
$y = \frac{1}{4(-2)}x^2 = \frac{-x^2}{8}$
An equation for the parabola is $y = \frac{-x^2}{8}$.

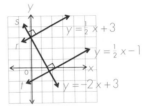

13.6 Find the midpoint M of segment XZ.

$M = \left(\frac{(-2) + 4}{2}, \frac{5 + (-4)}{2} \right)$ By the midpoint theorem

$M = \left(1, \frac{1}{2} \right)$

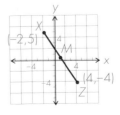

CALCULATOR ACTIVITY

You can use a graphing calculator to plot points and graph equations. Before you begin to draw a graph, you must first use the RANGE key to set the parameters of the graph. After this key is pressed, a list of values will appear on the screen, showing the number of units on each axis and an approximate scale factor for each axis. To change these values, you can use the arrow keys to move from line to line or from right to left. Then use the EXE key to lock the values into the memory of the calculator.

To plot a point, you use the SHIFT, PLOT, and EXE keys. A segment can be drawn to connect two points by pressing the SHIFT, LINE, and EXE keys.

Example: Use a graphing calculator to draw the line determined by the points (2,5) and (−4,3).

First, press the RANGE key. Set the parameters for the x-axis at −5 and 5 and the parameters for the y-axis at −6 and 6. Set the scale factor for both axes at 1. The resulting display is shown below.

XMIN: −5
XMAX: 5
SCL: 1
YMIN: −6
YMAX: 6
SCL: 1

Next, plot the points. As each point is plotted, a flashing light will appear on the screen at the location of the point.

SHIFT PLOT 2 SHIFT , 5 EXE

SHIFT PLOT (−) 4 SHIFT , 3 EXE

Then draw the line.

SHIFT LINE EXE

The graph will appear on the screen as shown below.

You can use the $\boxed{\text{GRAPH}}$ key on a graphing calculator to graph a line when you are given the equation of the line.

Example: Use a graphing calculator to draw the graph of $y = 2x - 1$.

First, set the parameters for the graph.

$\boxed{\text{RANGE}}$ XMIN: -2 XMAX: 3 SCL: 1 YMIN: -3 YMAX: 4 SCL: 1

Then press the $\boxed{\text{GRAPH}}$ key. The display on the screen is Y = . Enter the rest of the equation. You may need to use the $\boxed{\text{ALPHA}}$ key to enter the variable x. If you make a mistake entering the equation, use the arrow keys to correct your mistake. After the equation has been correctly entered, press the $\boxed{\text{EXE}}$ key. The graph will appear on the screen as shown below.

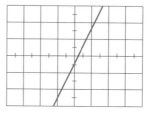

Use a graphing calculator to draw the line determined by each pair of points.

1. (5,1) and (2,4)

2. (3,−1) and (−2,2)

3. (−4,6) and (7,−2)

4. (−3,−2) and (−8,6)

Use a graphing calculator to draw the graph of each equation.

5. $y = -2$

6. $y = \frac{2}{3}x$

7. $y = x + 1$

8. $y = 3x + 3$

9. $y = 4x - 2$

10. $y = \frac{x}{2} + 4$

Graph each equation in the coordinate plane. Tell whether the line is vertical or nonvertical. If it is nonvertical, give its slope and y-intercept.

1. $x = 5\frac{1}{2}$

2. $y = 3x - 5$

3. $y = \frac{-1}{3}x + 2$

Write an equation for each figure.

4. The line with slope 7 and y-intercept -3

5. The vertical line through $\left(\frac{7}{3}, 0\right)$

6. The line that passes through $(-6, -1)$ and $(-4, 0)$

7. The circle with center $(0,0)$ and radius 9

8. The circle with center $(0,0)$ and radius $\frac{4}{3}$

Tell whether the lines go up, down, or are level as you move from left to right in the coordinate plane.

9. $y = -\frac{2}{3}x + 10$

10. $y = 11$

11. $y = \frac{1}{8}x - 6$

Tell whether the lines are parallel, perpendicular, or neither.

12. $y = 3x - 2$

$y = \frac{1}{3}x + \frac{1}{2}$

13. $y = \frac{7}{4}x + 12$

$y = -\frac{4}{7}x - 12$

14. $y = 5x - 13$

$y = 5x + 2$

15. What is the slope of any line perpendicular to $y = 9x - 1$?

16. Write an equation for the line that passes through $(5,6)$ and is perpendicular to the line $y = -4x + 3$.

Graph each equation. You may use a calculator for calculations.

17. $y = \frac{3}{2}x^2$

18. $x^2 + y^2 = 18$

19. $y = -2x^2$

Tell whether a graph of the equation is a circle or a parabola. If it is a circle, give its center and radius. If it is a parabola, tell whether it opens up or down.

20. $y = 12x^2$

21. $x^2 + y^2 = 45$

22. $x^2 + y^2 = 169$

23. $y = -\frac{7}{2}x^2$

24. Use coordinate geometry to prove that, if you connect the midpoints of a square in order as you go around the square, then the resulting quadrilateral is also a square.

Write an equation for the line that passes through the given pair of points.

1. $(-1,2)$ and $(1,4)$ **2.** $(7,5)$ and $(4,11)$ **3.** $(3,9)$ and $(-4,9)$

Write an equation for each line.

4. The line with slope -2 and y-intercept 8

5. The line with slope 3 that passes through $(0,5)$

6. The line with slope 0 that passes through $(5,-4)$

Tell whether the lines are parallel, perpendicular, or neither.

7. $y = -2x + 13$
$y = \frac{1}{2}x - 11$

8. $y = 6x + 4$
$12x + 2y = 5$

9. $y = 14x + 76$
$y = 14x - \frac{1}{3}$

The slope of line l is $\frac{3}{4}$, $l \parallel n$, and $m \perp l$.

10. Write an equation for line m.

11. Write an equation for line n.

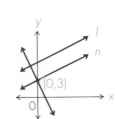

Give the center and radius of each circle.

12. $x^2 + y^2 = 100$ **13.** $x^2 + y^2 = 3$ **14.** $x^2 + y^2 = \frac{1}{64}$

Tell whether the parabola opens up or down.

15. $y = -9x^2$ **16.** $y = 10x^2$ **17.** $y = x^2$ **18.** $y = -\frac{2}{3}x^2$

19. Write an equation for the parabola that is the locus of all points equidistant from $(0,3)$ and the line $y = -3$.

Graph each equation on the coordinate plane. You may use a calculator or the table of square roots on page 562 if you wish.

20. $y = -3x + \frac{1}{2}$ **21.** $y = \frac{3}{2}x^2$ **22.** $x^2 + y^2 = 49$

23. *ABCD* is a parallelogram. Use coordinate geometry to prove that the segment $\overline{MN}$ connecting the midpoints of $\overline{DC}$ and $\overline{AB}$ is parallel to $\overline{AD}$ and $\overline{BC}$.

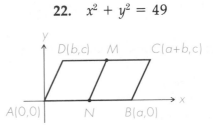

515

In coordinate geometry and algebra you sometimes need to solve equations such as $|x - 3| = 7$. To solve equations involving absolute value, you need to remember that, for any positive number, there are two numbers that have that number as their absolute value.

Example: Solve $|x - 3| = 7$
$x - 3 = 7$ or $x - 3 = -7$
$x = 10$ or $x = -4$

The equation has two solutions, 10 and -4.

Find all the solutions of each equation.

1. $|x - 5| = 9$

2. $|x + 6| = 1$

3. $|2x - 1| = 3$

4. $|5x - 10| = 0$

5. $|x + 8| = 8$

6. $|-x + 4| = 10$

7. $|\frac{1}{2}x - 6| = 7$

8. $|x + 1| = \frac{3}{4}$

9. $|3x - 8| = 13$

You may need to use some of the properties of absolute value for some equations. In the following example, you use the property that, if a and b are any real numbers, b not equal to zero, then $|a|/|b| = |a/b|$.

Example: Solve $|x - 2| = |x - 6|$.
Divide both sides by $|x - 6|$ and go on from there.
$$|x - 2| = |x - 6|$$

$$\left|\frac{x - 2}{x - 6}\right| = 1$$

$$\left|\frac{x - 2}{x - 6}\right| = 1$$

$$\frac{x - 2}{x - 6} = 1 \qquad \text{or} \qquad \frac{x - 2}{x - 6} = -1$$

$$x - 2 = x - 6 \qquad\qquad x - 2 = -x + 6$$
$$\uparrow \qquad\qquad\qquad 2x = 8$$
This equation has $\qquad\qquad x = 4$
no solution.

The equation $|x - 2| = |x - 6|$ has 4 as its only solution.

Find all the solutions of each equation.

10. $|x - 3| = |2x + 4|$

11. $|x + 5| = |x - 8|$

12. $|3x + 2| = |2x - 1|$

13. $|-2x + 3| = |x + 4|$

14

TRIGONOMETRY

To measure the unmeasurable

Anonymous

The Sine and Cosine Ratios

After completing this section, you should understand
▲ how to find sine ratios.
▲ how to find cosine ratios.

You are familiar with similar triangles such as those shown below.

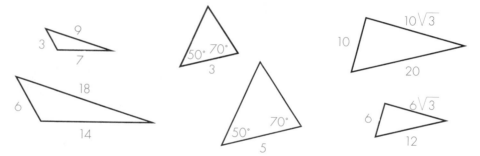

In Section 7.3 you saw how you can use certain similar right triangles, such as those with angles of 30°, 60°, and 45°, to measure distances indirectly. In this chapter you will see how to use *any* right triangle to measure distances indirectly.

SIMILAR RIGHT TRIANGLES

You know by Definition 4-3-1 that two triangles are similar if two angles of one are equal to two angles of the other. If two triangles are known to be right triangles, they automatically have one pair of equal angles (the right angles). Therefore, a pair of right triangles is similar if an acute angle of one is equal to an acute angle of the other.

Example: Among the right triangles shown below, find all the pairs of similar triangles.

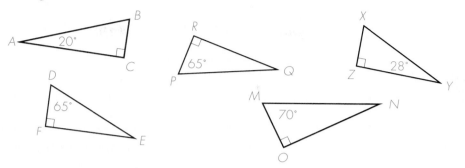

$\triangle PQR \sim \triangle DEF$, since $m\angle P = m\angle D = 65°$. In $\triangle NMO$, $m\angle N = 20°$, since $\angle M$ and $\angle N$ are complementary angles. Therefore, $\triangle ABC \sim \triangle NMO$.

In the Discovery Activity below, you will learn more about similar right triangles.

**ISCOVERY ACTIVITY**

1. Copy the table below. Then use a centimeter ruler to measure the line segments indicated in the table and shown in the figure. Measure to the nearest tenth of a centimeter. Record the measurements in the table.

$BC =$ _____	$DE =$ _____	$FG =$ _____
$AB =$ _____	$AD =$ _____	$AF =$ _____
$\dfrac{BC}{AB} =$ _____	$\dfrac{DE}{AD} =$ _____	$\dfrac{FG}{AF} =$ _____

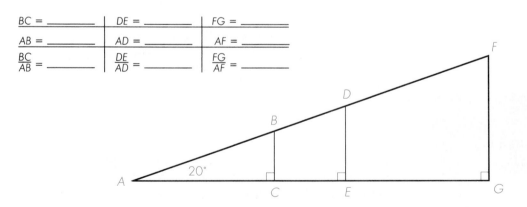

2. Use a calculator and write the ratios in the third row of the table in decimal form, to the nearest hundredth. What do you notice?

3. What can you say about the ratio of the length of the leg *opposite* the 20° angle to the length of the *hypotenuse*?

In the Discovery Activity, you may have noticed that in a right triangle with an acute angle that measures 20°, the ratio of the length of the leg opposite that angle to the length of the hypotenuse is about 0.34, regardless of the size of the right triangle. An even more accurate value is 0.3420. This number is called the sine ratio of 20°, or simply the **sine** of 20°. It is written "sin 20°." Any acute angle, not just 20°, has a constant value for its sine. For example, the sine of 25° is about 0.4226. This allows you to have the following definition.

DEFINITION 14-1-1　**The SINE of an acute angle A of a right triangle is the ratio of the length of the leg opposite the angle to the length of the hypotenuse. In symbols, this is written as**

$$\sin A = \frac{\text{length of leg opposite angle } A}{\text{length of hypotenuse}}$$

The equation for the sine ratio can be abbreviated as $\sin A = \frac{\text{opp}}{\text{hyp}}$. The sine ratio is the first of three **trigonometric ratios** that you will study in this chapter.

Examples:　1.　Find sin A.　　　　2.　Find sin A.

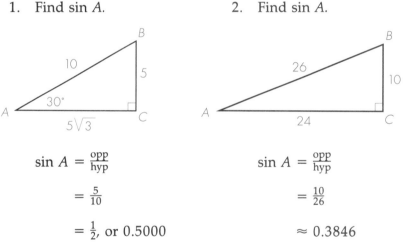

$$\sin A = \frac{\text{opp}}{\text{hyp}} \qquad\qquad \sin A = \frac{\text{opp}}{\text{hyp}}$$

$$= \frac{5}{10} \qquad\qquad\qquad = \frac{10}{26}$$

$$= \frac{1}{2}, \text{ or } 0.5000 \qquad\qquad \approx 0.3846$$

In the second example above, the "equals approximately" sign ($\approx$) is used in the last step because 0.3846 is a rounded value. From now on, for simplicity, the equality symbol ($=$) will often be used.

The second example shows that if you know the lengths of the sides of a right triangle, you can find the sines of the acute angles whether you know their measures or not.

You do not have to measure right triangles every time you need to know the sine of an angle. If you know the measure of the angle, you can find its sine by using either a *table* or a *scientific calculator*.

Example: Find the sine of 32° using a calculator or the table of trigonometric ratios on page 563.

To use a calculator, press the keys 3 2 SIN The display will show 0.52991926.

To use the table on page 563, look at the portion of it shown at the right. Read down the "degrees" column until you get to 32°, then find sin 32° in the "sin" column: sin 32° = 0.5299.

Degrees	Sin	Cos	Tan
30°	0.5000	0.8660	0.5774
31°	0.5150	0.8572	0.6009
32°	0.5299	0.8480	0.6249
33°	0.5446	0.8387	0.6494

3	2	SIN

CLASS ACTIVITY

1. Find all pairs of similar triangles from the set of triangles below.

Calculate or use the table on page 563 to find sin A. You may use a calculator.

2.
3.
4.

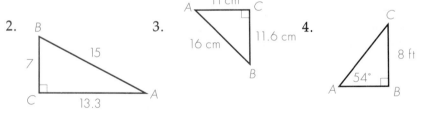

THE COSINE RATIO

If A is an acute angle of a right triangle, then the three sides of the triangle are related to A as shown at the right. Another trigonometric ratio, the *cosine* ratio, is formed by using the *adjacent* leg and the hypotenuse.

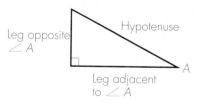

DEFINITION 14-1-2 **The COSINE of an acute angle of a right triangle is the ratio of the length of the leg adjacent to the angle to the length of the hypotenuse. In symbols, this is written as**

$$\cos A = \frac{\text{length of leg adjacent to angle } A}{\text{length of hypotenuse}}$$

The cosine formula can be abbreviated as $\cos A = \frac{\text{adj}}{\text{hyp}}$.

Example: Find cos A and cos B.

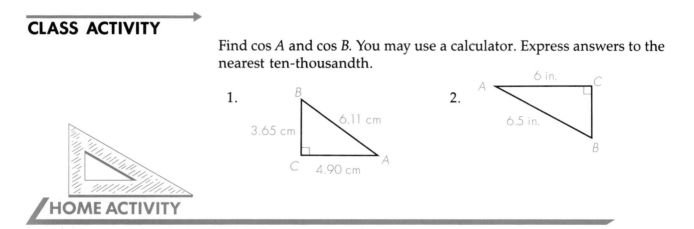

$$\cos A = \frac{\text{adj}}{\text{hyp}}$$

$$= \frac{4}{5}$$

$$= 0.8000$$

Use $a^2 + b^2 = c^2$ to find the leg adjacent to $\angle B$.

$$a^2 + 4^2 = 5^2$$

$$a^2 = 5^2 - 4^2, \text{ or } 9$$

$$a = 3$$

$$\cos B = \frac{a}{c} = \frac{3}{5} = 0.6000$$

CLASS ACTIVITY

Find cos A and cos B. You may use a calculator. Express answers to the nearest ten-thousandth.

1.

2.

HOME ACTIVITY

In these exercises, you may use a calculator or the table on page 563. Find sin A, sin B, cos A, and cos B. Express answers to the nearest ten-thousandth.

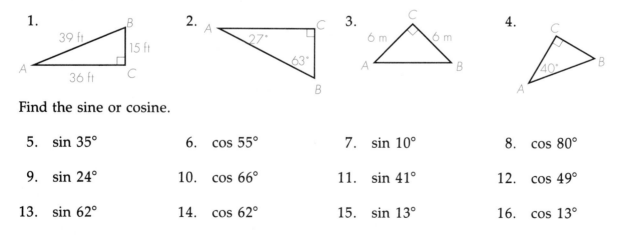

1.

2.

3.

4.

Find the sine or cosine.

5. sin 35°   6. cos 55°   7. sin 10°   8. cos 80°

9. sin 24°   10. cos 66°   11. sin 41°   12. cos 49°

13. sin 62°   14. cos 62°   15. sin 13°   16. cos 13°

CRITICAL THINKING

17. In quadrilateral ABCD, cos ∠1 = cos ∠2. Find the perimeter of the quadrilateral.

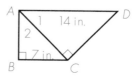

SECTION 14.2

The Tangent Ratio

After completing this section, you should understand
▲ how to use the tangent ratio.
▲ how to use trigonometry to solve triangle problems.

You have seen that the following two ratios can be defined for an acute angle A of a right triangle.

$$\sin A = \frac{opp}{hyp}$$

$$\cos A = \frac{adj}{hyp}$$

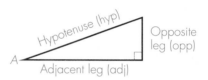

You will now learn about a third trigonometric ratio.

DISCOVERY ACTIVITY

1. Copy the table below. Then use a centimeter ruler to find the lengths of the legs of each triangle. Measure to the nearest tenth of a centimeter. Record the measurements in the table.

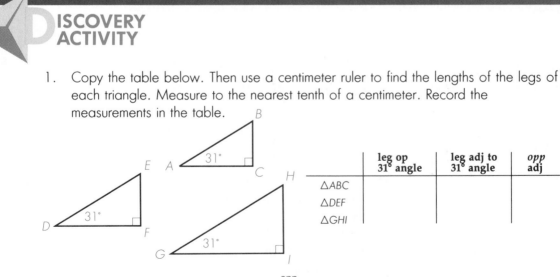

	leg op 31° angle	leg adj to 31° angle	$\frac{opp}{adj}$
$\triangle ABC$			
$\triangle DEF$			
$\triangle GHI$			

2. For each triangle, express the ratio $\frac{opp}{adj}$ as a decimal. Record the results in the table.
3. What can you say about the ratio $\frac{opp}{adj}$?

In the Discovery Activity, you may have discovered that the ratio $\frac{opp}{adj}$ has the same value (about 0.6) for all three similar triangles even though they are not of the same size. This common ratio is called the tangent ratio, or **tangent** of 31°. It is written "tan 31°."

DEFINITION 14-2-1   **The TANGENT of an acute angle A of a right triangle is the ratio of the length of the leg opposite the angle to the length of the leg adjacent to the angle. In symbols, this is written as**

$$\tan A = \frac{\text{length of leg opposite angle } A}{\text{length of leg adjacent to angle } A}$$

The equation for the tangent ratio can be abbreviated as $\tan A = \frac{opp}{adj}$. Values for the tangent ratio are found in the table on page 563.

Examples: 1. Find tan B.

2. Find tan A. Use the table on page 563.

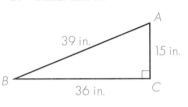

$\tan B = \frac{opp}{adj}$

$= \frac{15}{36}$

$= 0.4167$

$m\angle A + 35° = 90°$

$m\angle A = 90° - 35°$, or $55°$

From the table:

$\tan 55° = 1.428$

The table of trigonometric ratios on page 563 lets you find the sine, cosine, or tangent of an acute angle. You can also use the table to find the measure of an angle if you know its sine, cosine, or tangent.

Examples: 1. If tan A = 0.3443, what is $m\angle A$?

Use the portion of the table shown at the right. Read down the "tan" column to the number 0.3443. Then read left to 19° in the "degrees" column. You find that $m\angle A = 19°$.

Degrees	Sin	Cos	Tan
18°	0.3090	0.9511	0.3249
19°	0.3256	0.9455	0.3443
20°	0.3420	0.9397	0.3640
21°	0.3584	0.9336	0.3839
22°	0.3746	0.9272	0.4040
23°	0.3907	0.9205	0.4245

2. Given that $\tan B = 0.4167$, find $m\angle B$.

When you look for 0.4167 in the table, it is not there, so choose the table value that is closest to 0.4167. The number 0.4167 is between 0.4040 and 0.4245.

$$
\begin{array}{l}
0.4040 \\
\qquad \rightarrow \text{difference is} \\
0.4167 \qquad\quad 0.0127 \\
\qquad \rightarrow \text{difference is} \\
0.4245 \qquad\quad 0.0078
\end{array}
$$

The smaller difference is that between 0.4167 and 0.4245. Since 0.4167 is closer to 0.4245 than to 0.4040, the measure of $\angle B$ is closer to 23° than to 22°. So $m\angle B \approx 23°$.

CLASS ACTIVITY

Find $\tan B$. If necessary, use a calculator or the table of trigonometric ratios on page 563.

1.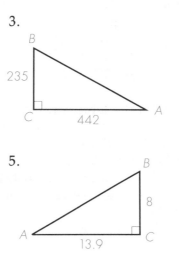

2.

Find $m\angle A$. Use a calculator or the table of trigonometric ratios on page 563.

3.

4.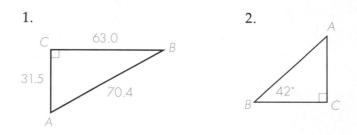

5.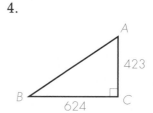

6.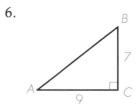

USING TRIGONOMETRIC RATIOS

You can often use one of the three trigonometric ratios to find a missing part of a right triangle. Depending on which part is missing and which parts are known, you can choose either the sine, cosine, or tangent. Often there is more than one approach that works.

Examples:

1. In right triangle ABC, $m\angle A = 40°$ and $b = 10$ yd. Find a.

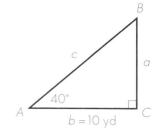

The part you need to find is *opposite* $\angle A$. The known part is *adjacent* to $\angle A$. This suggests the ratio $\frac{\text{opp}}{\text{adj}}$. So, use the *tangent* ratio.

$$\tan A = \frac{\text{opp}}{\text{adj}}$$

$$\tan 40° = \frac{a}{10}$$

From the table, $\tan 40° = 0.8391$.

$$0.8391 = \frac{a}{10}$$
$$10 \times 0.8391 = a$$
$$a = 10 \times 0.8391$$
$$= 8.391, \text{ or about } 8.4 \text{ yd}$$

2. In right triangle XYZ, $m\angle X = 25°$ and $z = 30$ mm. Find x.

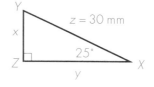

The part you need to find is *opposite* $\angle X$. The known part is the *hypotenuse*. This suggests the ratio $\frac{\text{opp}}{\text{hyp}}$. So, use the *sine* ratio.

$$\sin X = \frac{\text{opp}}{\text{hyp}}$$

$$\sin 25° = \frac{x}{30}$$

From the table, $\sin 25° = 0.4226$.

$$0.4226 = \frac{x}{30}$$
$$30 \times 0.4226 = x$$
$$x = 30 \times 0.4226$$
$$= 12.678, \text{ or about } 12.7 \text{ mm}$$

3. When a low-flying plane is directly over point *R*, the *angle of elevation* from a ship at point *Q* is 50°. The distance *QR* is 3000 ft. What is the direct-line distance from the plane to the ship?

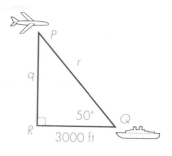

The known part is *adjacent to* ∠*Q*. The part you need to find is the *hypotenuse*. This suggests the ratio $\frac{adj}{hyp}$. So, use the *cosine* ratio.

$$\cos Q = \frac{adj}{hyp}$$

$$\cos 50° = \frac{3000}{r}$$

From the table, cos 50° = 0.6428.

$$0.6428 = \frac{3000}{r}$$

$$r \times 0.6428 = 3000$$

Divide each side of the equation by 0.6428.

$$r = \frac{3000}{0.6428}$$

$$= 4667, \text{ or about 4670 ft from the ship}$$

CLASS ACTIVITY

Tell which trigonometric ratio you could use to find the value of *a*.

1. *B* — *c* = 10, *a*, 26°, *C*, *b*, *A*

2. *A* — *b* = 7.2 — *C*, 55°, *a*, *c*, *B*

In Exercises 3 and 4, use a calculator or the table on page 563. Give answers to the nearest tenth.

3. *m*∠*F* = 38° and *h* = 18 m
Find *g*.

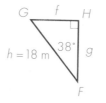

4. *m*∠*S* = 63° and *s* = 12 in.
Find *r*.

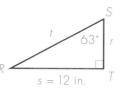

HOME ACTIVITY

In the Home Activity exercises, use a calculator or the table on page 563. Give lengths and distances to the nearest tenth.

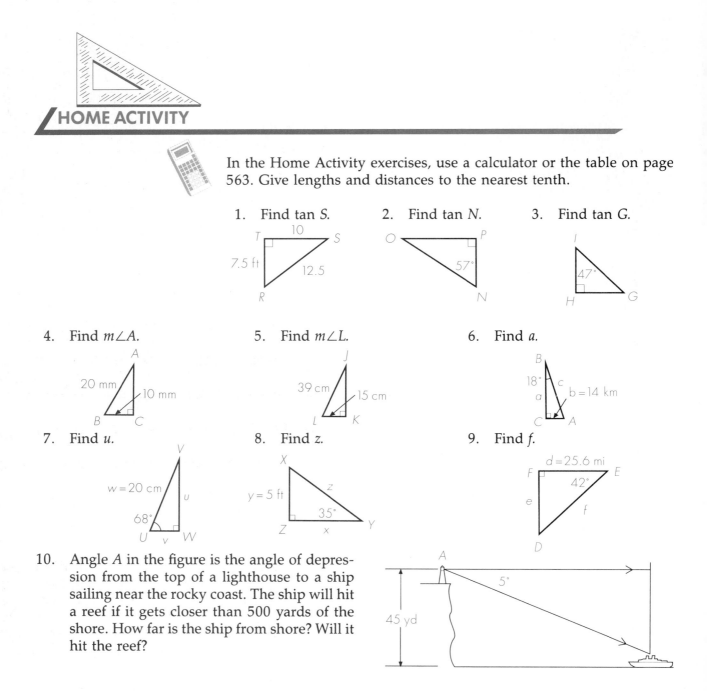

1. Find tan *S*.

2. Find tan *N*.

3. Find tan *G*.

4. Find *m∠A*.

5. Find *m∠L*.

6. Find *a*.

7. Find *u*.

8. Find *z*.

9. Find *f*.

10. Angle *A* in the figure is the angle of depression from the top of a lighthouse to a ship sailing near the rocky coast. The ship will hit a reef if it gets closer than 500 yards of the shore. How far is the ship from shore? Will it hit the reef?

11. The angle of elevation from a sailboat to the top of an oil rig is 32°. The oil rig is 130 m high. Draw a diagram. Then find how far the sailboat is from the base of the oil rig.

CRITICAL THINKING

12. Without using tables or a calculator, show that sin (30° + 30°) does not equal sin 30° + sin 30°. (Hint: Review Section 7.3.)

SECTION 14.3

The Sine Function

After completing this section, you should understand
▲ what an angle of rotation is.
▲ how the sine and cosine functions can be extended.
▲ how to graph the sine function.

In earlier sections we defined the sine, cosine, and tangent ratios in terms of right triangles. In this section you will begin to extend your definition of sine and cosine.

DISCOVERY ACTIVITY

1. On graph paper, prepare coordinate axes like the one at the right.
2. Copy the partly filled-in table.
3. Complete the table using either the table of trigonometric ratios on page 563 or a scientific calculator. Round the values of the sine ratio to the nearest hundredth.
4. Using your graph paper, plot the points defined by your table. Locate each point as precisely as you can.
5. Carefully connect the seven points with a smooth curve.
6. Is the curve you drew a line segment?
7. If you were to extend the curve at both ends, what do you think would be the value of y for x = 0°? for x = 90°?

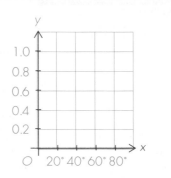

$y = \sin x$

x	y
5°	0.09
20°	
40°	
50°	0.77
60°	
70°	
85°	0.99

The graph you drew should look like this:

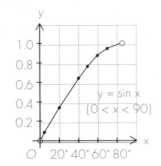

Recall that a **function** is a set of ordered pairs of real numbers such that each x-value has only one y-value. The table on page 529 is for part of the **sine function.** The graph shown above is for the equation $y = \sin x$ for values of x between $0°$ and $90°$.

The sine ratio has been defined only for *acute* angles. Neither $0°$ nor $90°$ is the measure of an acute angle. For this reason, there are not points on the graph at $x = 0°$ or $x = 90°$. However, it is possible to remove the restriction that limits the sine function to acute angles. If this is done, the graph can be extended at both ends.

EXTENDING THE SINE FUNCTION

To extend the sine function you must first extend the angle measures that are allowed for the sine. This is done by beginning with a new kind of angle, the **angle of rotation,** which is shown below.

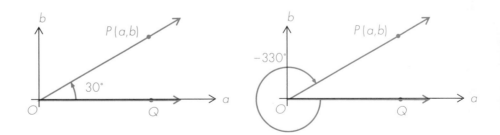

In the diagram, the a-axis and b-axis define a coordinate plane. The angle of rotation is formed by two rays meeting at vertex O. The first ray, $\overrightarrow{OQ}$, is called the **initial side** and lies on the positive horizontal axis. The second ray, $\overrightarrow{OP}$, is the **terminal side** and may lie in any of the four quadrants or along the vertical or horizontal axes.

You can rotate the initial side about the vertex O in a *counterclockwise* direction in order to reach the terminal side. In this case the angle of rotation is defined to be *positive*. If you rotate the initial side in a *clockwise* direction to reach the terminal side, the angle of rotation is defined to be *negative*. In the diagram at the bottom of page 530, the terminal side $\overrightarrow{OP}$ is shown with two measures, 30° and −330°. (Recall that there are 360° in a full revolution.)

Examples: Find the measure of another angle of rotation so that the sides are in the same position.

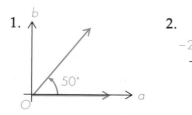

1.

Find a clock-wise angle. The measure will be negative: 360 − 50 = 310. A second measure is −310°.

2.

Find a counter-clockwise angle. The measure will be positive: 360 − 240 = 120. A second measure is 120°.

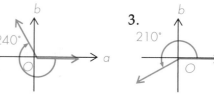

3.

Find a clock-wise angle. The measure will be negative: 360 − 210 = 150. A second measure is −150°.

CLASS ACTIVITY

Find a second measure for each angle shown at the right.

1. $\angle QOP_1$

2. $\angle QOP_2$

3. $\angle QOP_3$

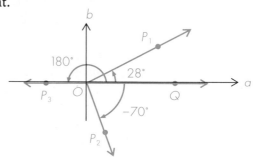

A NEW DEFINITION FOR THE SINE

In the figure at the right, line $\overleftrightarrow{P_1OP_2}$ includes the terminal sides of two angles, $\angle QOP_1$ and $\angle QOP_2$, with measures 37° and 217°, respectively. So far, only one of these angles, the 37° angle, has a sine:

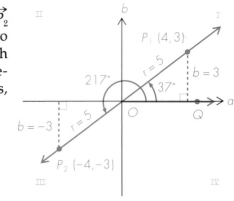

$$\sin 37° = \frac{\text{leg opp. 37° angle}}{\text{hypotenuse}}$$
$$= \frac{b}{r}$$
$$= \frac{3}{5}, \text{ or } 0.6000$$

To get a new definition of the sine ratio of 37°, replace
   "leg opp. 37° angle" by "vertical coordinate of point P_1" and replace
   "hypotenuse" by "distance r from the origin."
Then with this new definition, you will have:

$$\sin 37° = \frac{\text{vertical coordinate}}{\text{distance from origin}} = \frac{b}{r} = \frac{3}{5} = 0.6000$$

For 37°, in quadrant I, either definition will work.

For 217°, in quadrant III, only the new definition will work.

$$\sin 217° = \frac{\text{vertical coordinate}}{\text{distance from origin}}$$
$$= \frac{b}{r}$$
$$= \frac{-3}{5}, \text{ or } -0.6000$$

In quadrant I, either definition of the sine ratio will work. In quadrants II, III, and IV, only the new definition of the sine ratio will work.

Example: Use the diagram and the table on page 563 to find the value of sin (−30°).
In quadrant I, $\sin 30° = \frac{b}{r}$.
From the table, $\sin 30° = 0.5000$.
So, $\frac{b}{r} = 0.5000$.
In quadrant IV,
$\sin (-30) = \frac{-b}{r}$
$\qquad = -\frac{b}{r} = -0.5000$

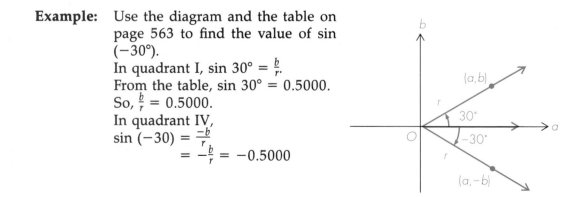

If you have a scientific calculator, then you can find the sine of any angle by using the SIN function button as shown below for sin (−30°). First, be sure that the calculator is in "degree" mode. Then use the following keystrokes.

$$30 \boxed{\pm} \boxed{\text{SIN}} = -0.5$$

CLASS ACTIVITY

Find the sine of each angle.

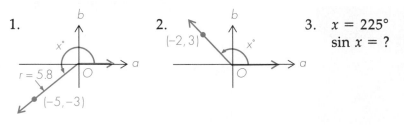

1.

sin x = ?

2.

(−2, 3)

sin x = ?
Hint: To find r use the
distance formula (Section 7.5).

3. x = 225°
   sin x = ?

**GRAPHING THE
SINE FUNCTION**

On page 530 you saw the graph of the sine function for angle measures between 0° and 90°, that is, for $0° < x° < 90°$. With the new definition of the sine ratio, you can graph the sine function for any values of x.

CLASS ACTIVITY

Copy and complete the partially filled-in table below. Use a scientific calculator or the table on page 563. If you use the table on page 563 for angle measures greater than 90°, then you will have to find the sine values by using the method in the example at the bottom of page 532.

x°	sin x°	x°	sin x°	x°	sin x°
0°	0.00	120°	0.87	240°	
5°	0.09	130°		250°	
20°	0.34	140°		270°	
40°	0.64	160°	0.34	280°	−0.98
50°	0.77	180°	0.00	290°	
60°	0.87	190°	−0.17	300°	
70°	0.94	200°	−0.34	310°	−0.77
85°	0.996	210°	−0.50	320°	−0.64
90°	1.00	220°		340°	
95°	0.996	230°		360°	0.00

The complete graph for $0° \leq x° \leq 360°$ appears at the top of page 534.

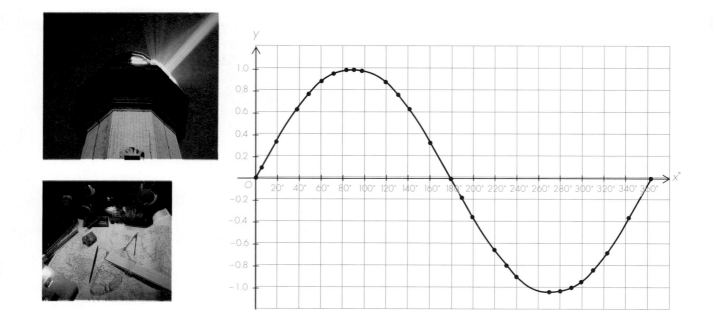

The cosine function's definition can be extended in much the same way. The main ideas of this section can be summarized in the following definition.

NOTEBOOK

DEFINITION 14-3-1

Suppose that an angle with measure $x°$ has its vertex at the origin of a coordinate plane with one ray on the positive horizontal axis and a point $P(a, b)$ on the other ray r units from the origin. Then the SINE and COSINE of the angle are defined to be

$$\sin x° = \frac{b}{r}$$

$$\cos x° = \frac{a}{r}$$

HOME ACTIVITY

In this activity, you may use a calculator or the tables on pages 562 and 563. Find a second measure for each angle.

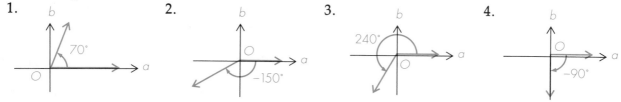

1. 70°

2. −150°

3. 240°

4. −90°

Find the sine of each angle. Express answers to the nearest hundredth.

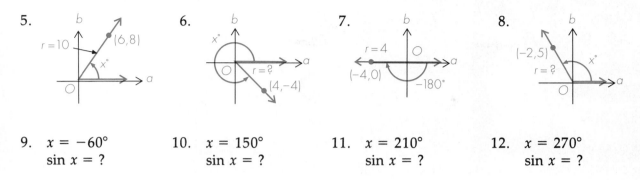

5.   6.   7.   8.

9. $x = -60°$
    $\sin x = ?$

10. $x = 150°$
    $\sin x = ?$

11. $x = 210°$
    $\sin x = ?$

12. $x = 270°$
    $\sin x = ?$

Use the Definition 14-3-1 to find r and $\cos x°$. Express answers to the nearest hundredth.

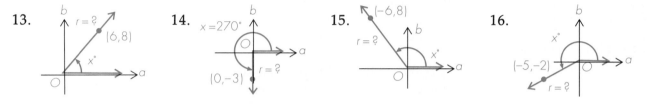

13.   14.   15.   16.

17. Copy and complete the following table for $\cos x°$, where $0° [\leq] x° [\leq] 360°$. Then graph the cosine function in the coordinate plane.

$x°$	$\cos x°$		$x°$	$\cos x°$		$x°$	$\cos x°$
0°			130°			250°	
15°			140°			260°	
30°			150°	−0.87		270°	
40°			160°			280°	
50°			170°			290°	
60°			180°			300°	0.50
70°			190°	−0.98		315°	
80°			200°			330°	
90°			215°			340°	0.94
100°	−0.17		230°			350°	
115°			240°			360°	

CRITICAL THINKING

18. Refer to Definition 14-3-1. The old definition for the tangent ratio was $\frac{\text{opp}}{\text{hyp}}$. In terms of the numbers a, b, and r of Definition 14-3-1, what would be a logical way to give a new definition of $\tan x°$? Will there be a value for $\tan x°$ for all values of $x°$ such that $0 \leq x° \leq 360°$? Explain your answer.

SECTION 14.4

Solving Non-right Triangles

After completing this section, you should understand
▲ how to use the law of sines.
▲ how to use the law of cosines.

In △*ABC* at the right, you can see that ∠*B* is greater than ∠*A*. What can you say about the relationship between the sides opposite these two angles? From Theorem 7-4-1 you know that if a triangle has one angle greater than another angle, then the side opposite the greater angle is longer than the side opposite the other angle.

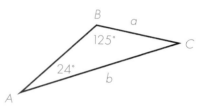

Applying this theorem to *ABC*, you can conclude that since $m\angle B > m\angle A$, the inequality $b > a$ is true. You can use trigonometry to show exactly how the two angles and two sides are related.

First, you can use the ideas from the previous section to find sin 125°. Use the figure at the right:

$$\sin 125° = \frac{\text{vertical coordinate}}{\text{distance of } P \text{ from } 0} = \frac{b}{r}$$

Also, for △*OMP* you have the following:

$$\sin 55° = \frac{\text{opp}}{\text{hyp}} = \frac{b}{r}$$

Therefore, sin 125° = sin 55°.

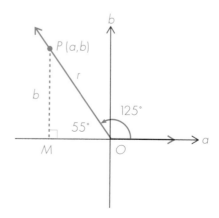

DISCOVERY ACTIVITY

1. Refer to △ABC at the right. Use a table or a scientific calculator to express the following ratios as decimals:

   $\frac{a}{b} = ?$ $\frac{A}{B} = ?$ $\frac{\sin A}{\sin B} = ?$

   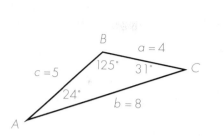

2. Which of the above ratios, if any, have the same value?

3. Based on your answers to questions 1 and 2, write a true proportion.

As a result of your work in the Discovery Activity, you should have concluded that two sides of a triangle are proportional not to the opposite angles, as you might have expected, but to the *sines* of those angles. The proportion can be written in several ways:

$$\frac{a}{b} = \frac{\sin A}{\sin B} \quad \text{or} \quad \frac{\sin A}{a} = \frac{\sin B}{b}$$

The second of these ways leads to a theorem called the **Law of Sines.** The theorem holds for any two angles of a triangle and their opposite sides.

THEOREM 14-4-1: **(Law of Sines)** In any triangle *ABC*,

$$\frac{\sin A}{a} = \frac{\sin B}{b} = \frac{\sin C}{c}.$$

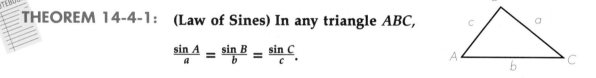

You can use the Law of Sines to find measurements of sides or angles of a triangle, even if the triangle is not a right triangle. Finding all the missing measurements is called **solving the triangle.**

Example: Find *c* for the triangle shown in the diagram.

$$\frac{\sin B}{b} = \frac{\sin C}{c}$$

$$\frac{\sin 40°}{100} = \frac{\sin 75°}{c}$$

Now solve the proportion.

$$\sin 40° \times c = 100 \times \sin 75°$$

$$c = \frac{100 \times \sin 75°}{\sin 40°} = \frac{100 \times 0.9659}{0.6428} \approx 150.3 \text{ yd}$$

CLASS ACTIVITY

Find the indicated length. Use a calculator or the table on page 563.

1.

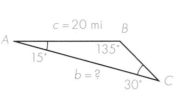

2.

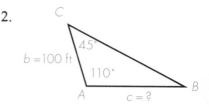

Hint: First find $m\angle B$.

ANOTHER LAW FOR SOLVING TRIANGLES

In some cases you may need more information than you are given in order to solve the triangle. The Discovery Activity below illustrates this.

DISCOVERY ACTIVITY

Study the three triangles shown below. One of the triangles can be solved using the Law of Sines. Another cannot be solved by any method.

1. Which triangle can be solved using the law of sines?

2. Which triangle cannot be solved by any method? Why?

In the Discovery Activity you probably noticed that the triangle on the left can be solved by the Law of Sines. You can find $m\angle A$ by solving the equation $m\angle A + 68° + 85° = 180°$. Then you can use the Law of Sines to find a and b. The triangle on the right cannot be solved by any method. To solve a triangle you must know the length of at least one side.

What about the triangle in the middle? It cannot be solved by using the law of sines but it can be solved by using another law, the Law of Cosines.

NOTEBOOK

THEOREM 14-4-2

(Law of Cosines) In any triangle ABC,
$$a^2 = b^2 + c^2 - 2bc \cos A$$
$$b^2 = a^2 + c^2 - 2ac \cos B$$
and $c^2 = a^2 + b^2 - 2ab \cos C$.

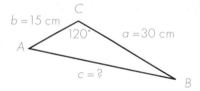

Examples:

1. Find b for the triangle in the diagram. Since $m\angle B$ is known, use the middle formula in the Law of Cosines.
$$b^2 = a^2 + c^2 - 2ac \cos B$$
$$= 9.8^2 + 20^2 - 2 \times 9.8 \times 20 \times \cos 85°$$
$$= 96.04 + 400 - 392 \times 0.0872$$
$$\approx 496.04 - 34.18, \text{ or } 461.86$$

Use a calculator or the table of squares and square roots on page 562 to find b. $b \approx 21.5$ ft

2. Find c in $\triangle ABC$. Since 120° is not an acute angle, you must use the definition of the cosine ratio on page 535.

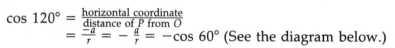

$$\cos 120° = \frac{\text{horizontal coordinate}}{\text{distance of } P \text{ from } O}$$
$$= \frac{-a}{r} = -\frac{a}{r} = -\cos 60° \text{ (See the diagram below.)}$$

Use the last equation from the Law of Cosines.

$$c^2 = a^2 + b^2 - 2ab \cos C$$
$$= 30^2 + 15^2 - 2 \times 30 \times 15 \times \cos 120°$$
$$= 900 + 225 - 900 \times (-\cos 60°)$$
$$= 1125 - 900 \times (-0.5000)$$
$$= 1125 + 450, \text{ or } 1575$$
$$c \approx 39.7 \text{ cm}$$

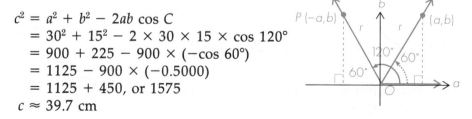

If you want to solve a triangle, you can decide whether to use the law of sines or the law of cosines by referring to the chart below.

Given information	Law to use
Two angles and any side given (AAS or ASA)	Law of Sines
Two sides and the included angle given (SAS)	Law of Cosines
Three sides given (SSS)	Law of Cosines

Example: Find $m\angle A$ in $\triangle ABC$.
The pattern is *SSS*.
Use the Law of Cosines.
$a^2 = b^2 + c^2 - 2bc \cos A$
$15^2 = 10^2 + 19^2 - 2 \times 10 \times 19 \times \cos A$
Simplify and solve for $\cos A$.
$\cos A = 0.6211$
$m\angle A = 52°$, to the nearest degree

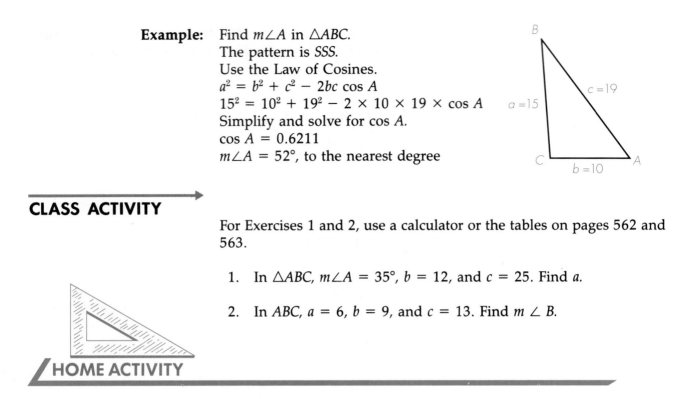

CLASS ACTIVITY

For Exercises 1 and 2, use a calculator or the tables on pages 562 and 563.

1. In $\triangle ABC$, $m\angle A = 35°$, $b = 12$, and $c = 25$. Find a.

2. In ABC, $a = 6$, $b = 9$, and $c = 13$. Find $m\angle B$.

HOME ACTIVITY

In this activity, you may use a calculator or the tables on pages 562 and 563. Use the law of sines to find the indicated lengths to the nearest tenth.

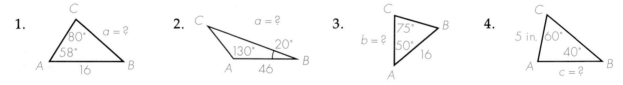

1. 2. 3. 4.

Use the laws of cosines to find the indicated measure. Give lengths to the nearest tenth and angle measures to the nearest degree.

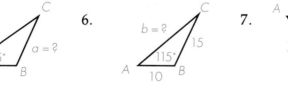
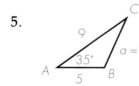

5. 6. 7. 8.

9. Two lighthouses are located at points A and B. A ship is at point C. (See diagram.) How far is the ship from the lighthouse at point A?

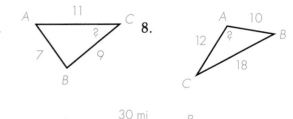

CRITICAL THINKING

10. Show that for $\triangle ABC$, $\frac{a-b}{b} = \frac{\sin A - \sin B}{\sin B}$.

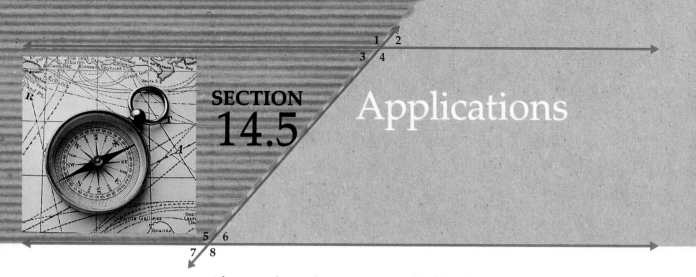

Applications

After completing this section, you should understand
▲ how to use trigonometric functions to solve practical problems.

You have seen how to measure distances indirectly using the sine, cosine, and tangent ratios. Here is a review of those ratios for a positive acute angle A.

$\sin X = \frac{\text{opp}}{\text{hyp}}$

$\cos X = \frac{\text{adj}}{\text{hyp}}$

$\tan X = \frac{\text{opp}}{\text{adj}}$

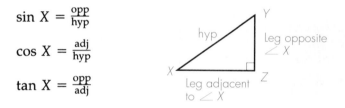

You can use the above definitions to solve a *right* triangle. To solve other kinds of triangles, you need to extend the functions with the following definitions:

$$\sin x = \frac{\text{vertical coordinate}}{\text{distance of } P \text{ from origin}} = \frac{b}{r}$$

$$\cos x = \frac{\text{horizontal coordinate}}{\text{distance of } P \text{ from origin}} = \frac{a}{r}$$

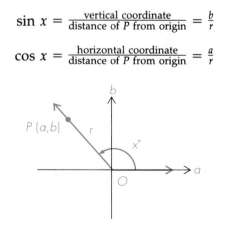

In this section, you will see how to apply these definitions to solving practical problems.

ANGLES OF ELEVATION AND DEPRESSION

When you wish to measure a vertical distance indirectly, you often need to know the angle of elevation or angle of depression. These are illustrated below.

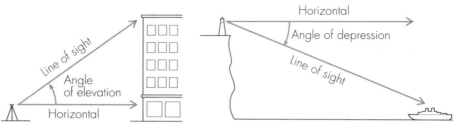

In both cases, the angle of elevation and angle of depression are formed by a horizontal ray and a second ray called the **line of sight.**

Examples:

1. The angle of elevation from a transit to the top of a flagpole is 70°. The transit is 1.5 m high and is 6 m from the flagpole. Find the height of the flagpole and the line-of-sight distance from the observer to the top of the flagpole.

Refer to the diagram. Use trigonometry to find YZ. You know XY and $m\angle X$, so use the tangent ratio.

$$\tan X = \frac{\text{opp}}{\text{adj}}$$

$$\tan 70° = \frac{YZ}{6}$$

$$6 \tan 70° = YZ$$

$$YZ = 6 \tan 70°$$

$$= 6 \times 2.747, \text{ or } 16.48$$

$$BY = \text{height of transit} + YZ$$
$$= 1.5 + 16.48 = 17.98, \text{ or about 18 m}$$

Next, use the cosine ratio to find XZ.

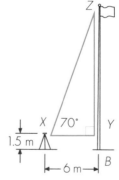

$$\cos X = \frac{\text{adj}}{\text{hyp}}$$

$$\cos 70° = \frac{6}{XZ}$$

$$XZ \cos 70° = 6$$

$$XZ = \frac{6}{\cos 70°} = \frac{6}{0.3420} = 17.54, \text{ or about 17.5 m}$$

2. The string of the kite pictured in the diagram makes an angle of 34° with the horizontal. The length of the string is 170 yd. Find the altitude of the kite. Assume that the kite string is held from a point 1 yd above the ground.

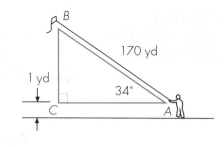

Use the sine ratio to find *BC*. $\sin A = \frac{\text{opp}}{\text{hyp}}$

$$\sin 34° = \frac{BC}{170}$$

$$170 \sin 34° = BC$$
$$BC = 170 \sin 34°$$
$$= 170 \times 0.5592 \approx 95.06, \text{ or about 95 yd}$$
Altitude of kite = 1 yd + 95 yd = 96 yd

3. In Example 2, find the distance from the person flying the kite to the point on the ground directly below the kite.

Use the cosine ratio to find *AC*. $\cos A = \frac{\text{adj}}{\text{hyp}}$

$$\cos 34° = \frac{AC}{170}$$

$$170 \cos 34° = AC$$
$$AC = 170 \cos 34°$$
$$= 170 \times 0.8290 \approx 140.9, \text{ or about 141 yd}$$

CLASS ACTIVITY

The angle of elevation from a control tower to an airplane is 6°. The tower is 40 ft above the ground. From a point directly under the plane, the distance to the tower is 3000 ft.

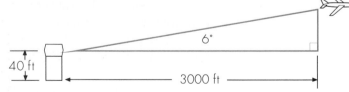

Use a calculator or the table on page 563 to find the indicated distances.

1. Find the altitude of the plane.

2. Find the line-of-sight distance from the tower to the plane.

SOLVING NON-RIGHT TRIANGLES

To solve non-right triangles, use the following two laws.

Law of Sines

$$\frac{\sin A}{a} = \frac{\sin B}{b} = \frac{\sin C}{c}$$

Law of Cosines

$$a^2 = b^2 + c^2 - 2bc \cos A$$
$$b^2 = a^2 + c^2 - 2ac \cos B$$
$$c^2 = a^2 + b^2 - 2ab \cos C$$

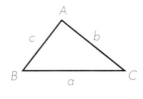

Example: Two surveyors need to find the distance between two points A and B on opposite sides of a lake. They measure $\angle A$ and $\angle C$. They also measure the distance BC. Find the distance across the lake.

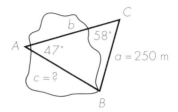

Two angles and one side are known. The pattern is AAS. (See the chart on page 539.) Use the Law of Sines.

$$\frac{\sin A}{a} = \frac{\sin C}{c}$$

$$\frac{\sin 47°}{250} = \frac{\sin 58°}{c}$$

$$c \sin 47° = 250 \sin 58°$$

$$c = \frac{250 \sin 58°}{\sin 47°}$$

$$= \frac{250 \times 0.8480}{0.7314} \approx 289.9, \text{ or about 290 m}$$

If you use a scientific calculator to find the sine and cosine ratios, you will immediately know whether the ratio is positive or negative. For example, on a calculator with an 8-digit display,

$$\sin 110° = 0.9396926 \approx 0.94$$
and $\cos 110° = -0.3420201 \approx -0.34.$

If you use the table on page 563 to find the ratios, draw a diagram such as the one at the right to see how your angle is related to the angles in the table. In this case,

$$\sin 110° = \sin 70°$$
and $\cos 110° = -\cos 70°$

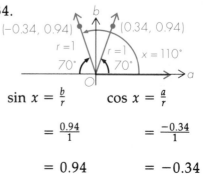

$$\sin x = \frac{b}{r} \qquad \cos x = \frac{a}{r}$$

$$= \frac{0.94}{1} \qquad = \frac{-0.34}{1}$$

$$= 0.94 \qquad = -0.34$$

Examples:

1. The navigators of two ships at sea can see one another's ships as well as a lighthouse on shore. The relative positions of the ships are as shown in the diagram. How far apart are the ships?

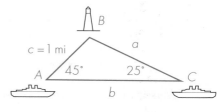

Two angles and a side are known. Use the law of sines. Since b is asked for, $m\angle B$ must also be found.

$$m\angle B + 45° + 25° = 180°$$
$$m\angle B = 180° - 45° - 25°, \text{ or } 110°$$

$$\frac{\sin B}{b} = \frac{\sin C}{c}$$

$$\frac{\sin 110°}{b} = \frac{\sin 25°}{1}$$

$$\sin 110° = b \sin 25°$$

$$b = \frac{\sin 110°}{\sin 25°} = \frac{0.9397}{0.4226} \approx 2.224, \text{ or about 2.2 mi}$$

2. A ship is 100 yd from point A on shore and 80 yd from point C on shore. The measure of $\angle B$ is 95°. How far apart are A and C? The pattern is ASA. Use the law of cosines.

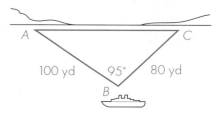

$$b^2 = a^2 + c^2 - 2ac \cos B$$
$$= 80^2 + 100^2 - 2 \times 80 \times 100 \times \cos 95°$$
$$= 6400 + 10{,}000 - 16{,}000 \times (-0.0872)$$
$$= 17{,}795$$
$$b = 133.4, \text{ or about 130 yd}$$

CLASS ACTIVITY

In this activity, use a calculator or the tables on pages 562 and 563. Give answers to the nearest tenth of a unit.

1. A saltbox house has a back and front roof constructed with the angles shown. Find the length of the longer roof.

2. Points A and B are at the ends of a rail tunnel. How long is the tunnel?

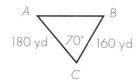

HOME ACTIVITY

In this activity, you may use a calculator or the tables on pages 562 and 563. In your answers, give lengths to the nearest tenth and angle measures to the nearest degree.

1. A lifeguard observes a boat that she thinks is too close to shore. How far is the boat from the base of the life guard's platform?

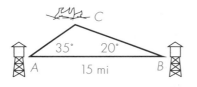

2. An airplane is climbing at an angle of 13°. The distance it has covered from A to B is 2 mi. Find its altitude.

3. Two ranger stations 15 mi apart notice a brush fire at the angles indicated. Which station is closer to the fire? How far away from the fire is it?

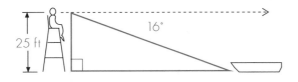

4. Two ships are located off shore at the indicated positions. Ship A is 3 mi from a lighthouse at C. How far from the lighthouse is ship B?

Find the missing lengths and angle measures indicated in the diagrams.

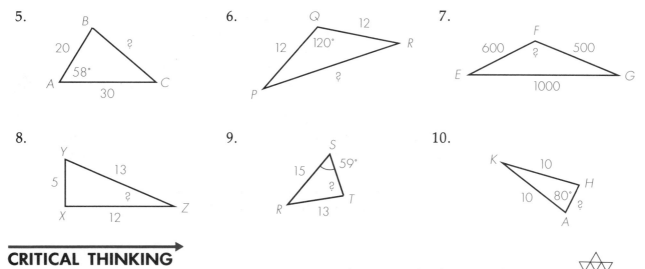

5.

6.

7.

8.

9.

10.

CRITICAL THINKING

11. How many triangles are in the figure at the right?

SECTION
14.6

Trigonometric Identities

After completing this section, you should understand
▲ some basic identities for the sine, cosine, and tangent.
▲ how to recognize a trigonometric identity.

In algebra there is a difference between an equation such as
$$3x + 15 = 3$$
and an equation such as
$$3x + 15 = 3(x + 5).$$
The first of these equations can be solved to obtain just one solution:
$$3x + 15 = 3$$
$$3x = 3 - 15$$
$$3x = -12$$
$$x = -4$$

The second equation, $3x + 15 = 3(x + 5)$, cannot be solved to obtain just one solution. Notice what happens after the first step (removing parentheses on the right side).
$$3x + 15 = 3(x + 5)$$
$$3x + 15 = 3 \cdot x + 3 \cdot 5$$
(Algebraic property: $a(b + c) = ab + ac$)
The result is the equation
$$3x + 15 = 3x + 15$$
which is true for all values of x. The equation is an *algebraic identity*.

DEFINITION 14-6-1   **An equation is an identity** if the equation is true for all values of the variables for which the expressions in the equation have meaning.

Trigonometry also has equations that are identities. These are used in advanced mathematics courses such as calculus.

DISCOVERY ACTIVITY

1. Copy and complete the table below. Use a scientific calculator if you can. Some entries have been filled in for you.

x	$\sin x$	$\cos x$	$(\sin x)^2$	$(\cos x)^2$
30°	0.5	_____	0.25	_____
40°	_____	0.7660	_____	0.5868
70°	_____	_____	_____	_____
90°	_____	_____	_____	_____
135°	_____	_____	_____	_____
155°	_____	_____	_____	_____

2. Examine the last two numbers in each row. What pattern do you observe?

In the Discovery Activity, it appears that the sum of the last two numbers in each row is 1. If this is true for all possible values of x, then the following equation is **trigonometric identity:**

$$(\sin x)^2 + (\cos x)^2 = 1$$

You know that this equation is true for *some* values of x, namely 30°, 40°, 70°, 90°, 135°, and 155°. Here is a proof that the equation is an identity and hence true for *all* values of x.

From the diagram at the right, you know that

$$\sin x = \frac{b}{r} \quad \text{and} \quad \cos x = \frac{a}{r}.$$

$$\text{So } (\sin x)^2 = \frac{b^2}{r^2} \quad \text{and} \quad (\cos x)^2 = \frac{a^2}{r^2}.$$

Next, add the last two equations.

$$(\sin x)^2 + (\cos x)^2 = \frac{b^2}{r^2} + \frac{a^2}{r^2}$$

$$= \frac{b^2 + a^2}{r^2}, \text{ or } \frac{a^2 + b^2}{r^2}$$

By the Pythagorean Theorem you know that
$$r^2 = a^2 + b^2$$

Therefore $$(\sin x)^2 + (\cos x)^2 = \frac{r^2}{r^2}$$

or $(\sin x)^2 + (\cos x)^2 = 1.$

The identity $(\sin x)^2 + (\cos x)^2 = 1$ is one of the basic identities of trigonometry. In order to avoid parentheses, the left side of the identity is usually written as $\sin^2 x + \cos^2 x$. Because of its similarity to the Pythagorean formula $a^2 + b^2 = c^2$, the identity $\sin^2 x + \cos^2 x = 1$ is referred to as a Pythagorean identity.

In a full course on trigonometry, you will learn more about Pythagorean identities. You will also learn about other basic kinds of identities, such as quotient identities and double-angle identities.

In the next Discovery Activity you will learn about one of the double-angle identities.

DISCOVERY ACTIVITY

1. Copy and complete the table below. Use a scientific calculator if you can. Some entries have been filled in for you.

x	$2x$	$\cos 2x$	$\cos x$	$\sin x$	$\cos^2 x$	$\sin^2 x$
30°	60°	0.5	0.8660	0.5	0.75	0.25
40°	80°	0.1736	_____	0.6428	_____	0.4132
70°	_____	_____	_____	_____	_____	_____
90°	_____	_____	_____	_____	_____	_____
135°	_____	0	−0.7071	0.7071	_____	_____
155°	_____	_____	_____	_____	_____	_____

2. Examine the "$\cos^2 x$" column and "$\sin^2 x$" column. How are they related to the "$\cos 2x$" column?
3. Based upon your investigation, complete the following to show an identity: $\cos 2x$ = _____

In the Discovery Activity, you may have noticed that when you subtract the number in the "$\sin^2 x$" column from the number in the "$\cos^2 x$" column, you obtain the number in the "$\cos 2x$" column. This fact can be written as

$$\cos 2x = \cos^2 x - \sin^2 x$$

This is the double-angle identity for the cosine.

You have seen examples of two kinds of basic identities:
Pythagorean identity: $\sin^2 x + \cos^2 x = 1$
Double-angle identity: $\cos 2x = \cos^2 x - \sin^2 x$
An example of a **quotient identity** is the following:

$$\tan x = \frac{\sin x}{\cos x}$$

You can use these basic identities to discover other identities.

Examples: 1.   Show that the following equation is an identity:
$$\cos 2x = 2 \cos^2 x - 1$$

Change one side of the equation by using the basic identities you already know. Reduce one side of the equation to the other.

Since $\sin^2 x + \cos^2 x = 1$ is an identity, substitute $\sin^2 x + \cos^2 x$ for 1 in the equation.

$$\cos 2x = 2 \cos^2 x - (\sin^2 x + \cos^2 x)$$
$$2 \cos^2 x - \sin^2 x - \cos^2 x$$
$$2 \cos^2 x - 1 \cos^2 x - \sin^2 x$$
$$\cos^2 x - \sin^2 x \quad \text{(Collecting terms)}$$
$$\cos 2x = \cos 2x \quad \text{(Substitute } \cos 2x \text{ for } \cos^2 x - \sin^2 x.)$$

The two sides of the final equation are equal. Therefore, the original equation is an identity.

2.   Show that $\sin x - \cos x = 1$ is *not* an identity. If the equation is false for any value of x, then it is not an identity. Try $x = 0$.
$$\sin 0° - \cos 0° = 0 - 1$$
The statement "$0 - 1 = 1$" is false. Thus, the equation is not an identity.

PROJECT 14-6-1   You can use a graphing calculator to help you determine whether an equation may be an identity. For example, to see whether sin $(90° - x) = \cos x$ might be an identity, you can use a graphing calculator to graph these two equations on the same set of axes.

$$y = \sin (90° - x) \text{ and } y = \cos x$$

If the graphs displayed are the same, then the original equation *may* be an identity. If the graphs are different, you can be sure that the original equation is *not* an identity.

Investigate the following equations. Use a graphing calculator to decide which equations may be identities and which are definitely not identities.

1. $\sin (90° - x) = \cos x$ 2. $1/\sin x = \sin x \tan x$

3. $\tan 2x = \frac{2 \tan x}{1 - \tan^2 x}$ 4. $\tan^2 x - \sin^2 x = \tan^2 x \sin^2 x$

5. $\sin x \cos x \tan x = 1$ 6. $\sin (x - 90°) = \cos x$

The basic identities of this section can be stated as theorems.

NOTEBOOK

THEOREM 14-6-1 If x is the measure of any angle, then $\sin^2 x + \cos^2 x = 1$.

NOTEBOOK

THEOREM 14-6-2 If x is the measure of any angle, then $\cos 2x = \cos^2 x - \sin^2 x$.

NOTEBOOK

THEOREM 14-6-3 If x is the measure of any angle such that $\cos x \neq 0$, then

$$\tan x = \frac{\sin x}{\cos x}.$$

CLASS ACTIVITY

If necessary, use a calculator for Exercise 1.

1. Show that the following equation is not an identity:
   $\tan x = \sin x + 1$

2. Show that the following equation is an identity:
   $\sin x = \cos x \cdot \tan x$

HOME ACTIVITY

In the Home Activity exercises, you may use a scientific calculator.

Show that the following equations are *not* identities.

1. $\sin x + \cos x = 0$
2. $\sin x \cdot \cos x = 1$
3. $\tan x \cdot \sin x = \cos x$
4. $\sin 2x = 2 \sin x$
5. $\sin x = x$
6. $\tan^2 x = \tan x^2$

Complete the missing steps in the following proofs of identities.

7. $\tan^2 x = \frac{1 - \cos^2 x}{\cos^2 x}$

   $$\frac{?}{?}$$

   $$\frac{\sin^2 x}{\cos^2 x}$$

   $\tan^2 x = \tan^2 x$

8. $\sin^2 x - \cos^2 x = 1 - 2 \cos^2 x$
   $$(\cos^2 x + \sin^2 x) - 2 \cos^2 x$$

   $$\sin^2 x - \cos^2 x = \frac{?}{?}$$

Prove the following identities.

9. $\frac{\sin^2 x + \cos^2 x}{\cos x} = \frac{\tan x}{\sin x}$

10. $\cos 2x = 1 - 2 \sin^2 x$

CRITICAL THINKING

11. Prove the following identity: $\cos 2x = \cos^4 x - \sin^4 x$.

14.1 Refer to the figure. The sine and cosine of a positive acute angle are defined by the following equations:

$$\sin A = \frac{\text{opp}}{\text{hyp}} \qquad \cos A = \frac{\text{adj}}{\text{hyp}}$$

If $AC = 4$, $BC = 3$, and $AB = 5$,

then $\sin A = \frac{3}{5}$ and $\cos A = \frac{4}{5}$.

14.2 Refer to the figure. The tangent of a positive acute angle is defined by the following equation:

$$\tan A = \frac{\text{opp}}{\text{adj}}$$

If $AC = 4$ and $BC = 3$,

then $\tan A = \frac{3}{4}$.

14.3 Refer to the figure. The sine and cosine of an angle of rotation are defined by the following equations:

$$\sin x° = \frac{b}{r} \qquad \cos x° = \frac{a}{r}$$

For example, if $x° = 120°$ and $P(a, b) =$

$(-1, \sqrt{3})$, then $\sin 120° = \frac{\sqrt{3}}{2}$

and $\cos 120° = -\frac{1}{2}$.

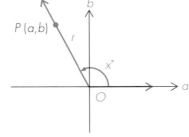

The figure below is the graph of the sine function for values of x such that $0° \leq x° \leq 360°$.

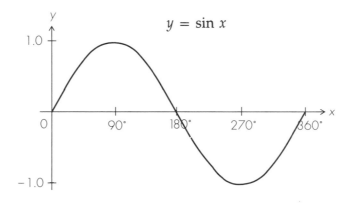

14.4 To solve a non-right triangle where the given information has the pattern *ASA* or *AAS,* use the Law of Sines.

$$\frac{\text{sine } A}{a} = \frac{\text{sine } B}{b} = \frac{\text{sine } C}{c}$$

For $\triangle ABC$, you can find b by

solving $\frac{\sin 25°}{b} = \frac{\sin 45°}{12}$.

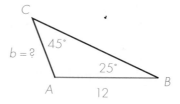

Refer to the figure. To solve a non-right triangle where the given information has the pattern *SAS* or *SSS,* use the Law of Cosines.

$$a^2 = b^2 + c^2 - 2bc \cos A$$

$$b^2 = a^2 + c^2 - 2ac \cos B$$

$$c^2 = a^2 + b^2 - 2ab \cos C$$

In $\triangle ABC$, to find BC you solve

$$a^2 = 9^2 + 12^2 - 2(9)(12) \cos 120°$$

In $\triangle A_1B_1C_1$ to find $m\,A$, use

$$18^2 = 9^2 + 12^2 - 2(9)(12)\cos A.$$

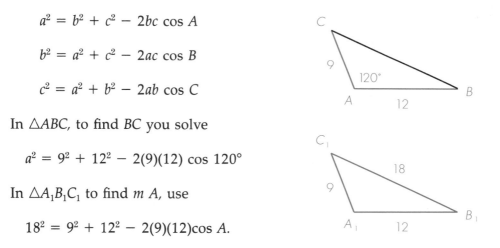

14.5 Refer to the figure. Angles of elevation and depression are used with the trigonometric ratios to measure distances indirectly.

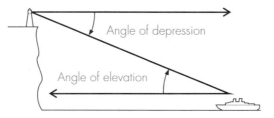

14.6 A trigonometric identity is an equation that is true for all values of the variables for which the expressions in the equation have meaning. Examples of trigonometric identities are:

$$\sin^2 x + \cos^2 x = 1 \text{ and } \frac{\sin x}{\cos x} = \tan x.$$

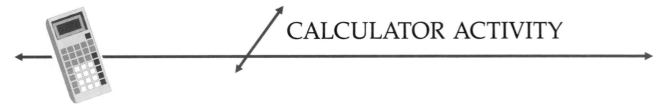

CALCULATOR ACTIVITY

Recall that to use a calculator to find the sine of an angle, you must first enter the degree measure of the angle, press the $\boxed{+/-}$ key if the degree measure of the angle is negative, and then press the $\boxed{SIN}$ key. The same procedure can be followed to find the cosine of an angle if you press the $\boxed{COS}$ key instead of the $\boxed{SIN}$ key.

1. Use a calculator. Copy and complete the chart below.

x	$\sin x$	$\cos x$
65		
425		
90		
450		
235		
595		
280		
1000		
120		
1200		
−50		
−410		
−155		
−875		
−140		
−2300		

2. What do you notice about the values of the sine function?

3. Use your calculator to subtract the angle measures that produce the same values for the sine function. What do you notice about the differences?

4. What do you notice about the values of the cosine function?

5. Use your calculator to subtract the angle measures that produce the same values for the cosine function. What do you notice about the differences?

You have learned that there are 360° in a complete revolution of the initial side of an angle of rotation. In Exercises 1 through 5, you discovered that the sine and cosine functions repeat their values every 360°. Thus, these functions are said to be **periodic,** with a period of 360°.

6. Use a calculator. Copy and complete the chart below.

x	tan x
65	
245	
110	
290	
90	
450	
85	
625	
135	
1035	
−155	
−335	
−50	
−410	
−140	
−680	

7. What do you notice about the values of the tangent function?

8. Use your calculator to subtract the angle measures that produce the same values for the tangent function. What do you notice about the differences?

9. Is the tangent a periodic function? If it is periodic, what is the period?

Use a calculator or the table on page 563.

In Exercises 1 through 4, express answers to the nearest ten-thousandth. Find sin A, cos A, and tan A.

1.

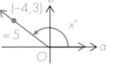

2.

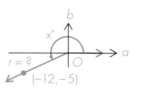

Find the sine and cosine of each angle.

3.

4.

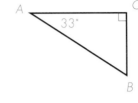

In Exercises 5 through 8, express lengths to the nearest tenth and angle measures to the nearest whole degree. Find each indicated measure.

5.

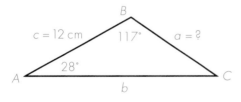

6.

7. Two boats are positioned as shown with respect to a buoy. The boat at A is 100 m from the buoy. How far apart are the two boats?

8. The lengths of two adjacent sides of parallelogram $ABCD$ are 6 and 10 and the measure of the angle between them is 75°. Find the length of the longer diagonal and the measure of the angle formed by that diagonal and $\overline{AB}$.

Show that the equations are not identities.

9. $\sin^2 x + 1 = \cos x$

10. $\frac{\cos x}{\sin x} = \tan x$

Use a calculator or the table on page 563.

Find sin *A*, cos *A*, and tan *A*.

1.

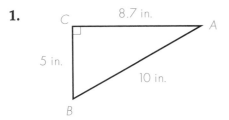

2.

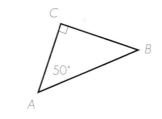

Find the sine and cosine of each angle.

3.

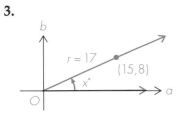

4.

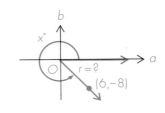

Find each indicated measure. Answer to the nearest tenth.

5.

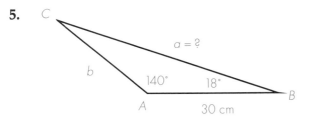

6.

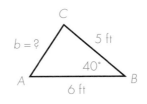

7. Two surveyors need to find the distance between two points *B* and *C* on opposite sides of a river. What is that distance?

8. The lengths of two adjacent sides of parallelogram *ABCD* are 7 and 11. The measure of the angle between the sides is 55°. Find the length of the shorter diagonal.

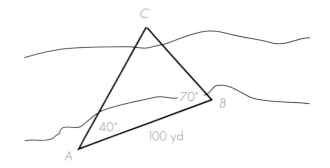

Tell whether each equation is an identity. If not, give an example to support your answer.

9. $\sin^2 x = 1 + \cos^2 x$

10. $\cos^2 x - \sin^2 x = \cos 2x$

Use the diagram. Identify the following.

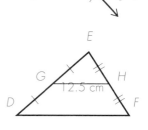

1. vertical angles

2. corresponding angles

3. alternate interior angles

4. alternate exterior angles

5. The vertices of a triangle in the x-y plane are $A(2, 0)$, $B(-1, 0)$, and $C(-1, -3)$. What type of triangle is $\triangle ABC$?

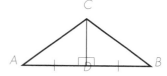

6. What is the length of $\overline{DF}$ in $\triangle DEF$?

Refer to $\triangle ABC$.

7. Is $\triangle ACD \sim \triangle BCD$? If so, why?

8. Is $\triangle ACD \cong \triangle BCD$? If so, why?

The members of the school baseball team were asked what professional baseball teams they root for.
  15 root for the Dodgers (D).
  6 root for both D and A.
  10 root for the Athletics (A).
  5 root for neither team.

9. Draw a Venn diagram illustrating this information.

10. How many members does the baseball team have?

11. Draw an acute angle and construct its bisector.

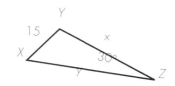

Find the missing lengths to the nearest tenth. Use a calculator or the table of squares and square roots on page 562.

12.

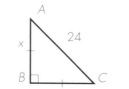

13.

14. Find the length of the line segment that connects *P* and *Q*.

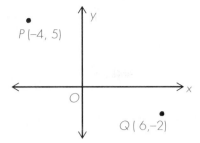

Is the figure a parallelogram? Give a reason for your answer.

15.

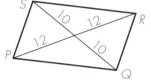

16.

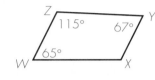

17. Find the indicated angle measures for *□ABCD*. Is *m∠DCA* + *m∠ABD* equal to *m∠AED*? Explain.

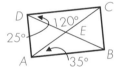

18. *PQRS* is an isosceles trapezoid. *PQ* = 25 and *SR* = 18. *T* and *V* are midpoints. What is the length of *TU*?

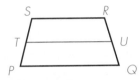

Find the perimeter and area of each trapezoid. Answer to the nearest tenth.

19.

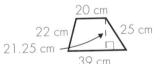

20.

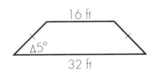

21. The sum of the angle measures for a regular polygon is 7740°. How many sides does the polygon have?

Find the circumference and area of each circle. Use *π* ≈ 3.14.

22.

23.

In circle O, $m\overarc{AC} = 90°$, $m\angle EOC = 60°$, $m\angle FCB = 25°$, and $\overline{BC}$ is a diameter. $\overline{ED}$ is tangent to the circle. Find these measures.

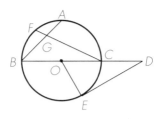

24. $m\angle ABC$ **25.** $m\angle FGB$ **26.** $m\angle OED$

27. $m\angle BDE$ **28.** $m\overarc{FA}$ **29.** $m\angle FGA$

30. In the circle, two chords intersect as shown. Find the value of x.

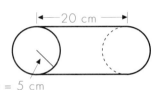

Find the total area and volume for each prism or right circular cylinder. Use $\pi \approx 3.14$.

31.

6 ft

8 ft

15 ft

32.

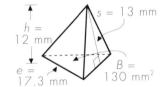

20 cm

$y = 5$ cm

Find the total area and volume of each regular pyramid or right circular cone. Answer to the nearest whole number. Use $\pi \approx 3.14$.

33.

$s = 15$ in.

$y = 9$ in.

34.

$h = 12$ mm

$e = 17.3$ mm

$s = 13$ mm

$B = 130$ mm^2

35. Find the volume and area of a sphere with a radius of 20 cm. Answer to the nearest whole number. Use $\pi \approx 3.14$.

36. In a plane, what is the locus of all points equidistant from a circle and the center of the circle?

Does the figure have line symmetry? If so, trace the figure and draw the line(s) of symmetry.

37.

38.

39.

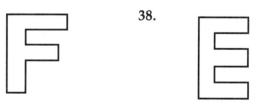

40. Write the converse, inverse, and contrapositive for this statement: If a creature can fly, then it is a bird.

Tell whether the lines are parallel, perpendicular, or neither.

41. $y = \frac{1}{2}x - 2$
$y = -2x - 2$

42. $y = -2x + 5$
$y = -2x - 5$

43. $y - 2x = 5$
$y = 2x - \frac{1}{5}$

Tell whether the graph of the equation is a circle or a parabola. If it is a circle, give its center and radius. If it is a parabola, tell whether it opens up or down.

44. $x^2 + y^2 = 36$

45. $y = 4x^2$

46. $x^2 + y^2 = \frac{1}{4}$

47. Refer to the figure. Use coordinate geometry to prove that $\overline{AC}$ and $\overline{BD}$ bisect each other.

For items 48–53 you may use a calculator or the table on page 563. Find sin A, cos A, and tan A. Give answers to the nearest ten-thousandth.

48.

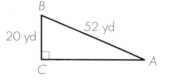

49.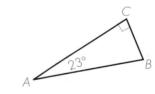

Find the sine and cosine of each angle. Answer to the nearest tenth.

50.

51.

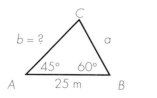

Find each indicated measure. Answer to the nearest tenth.

52.

53.

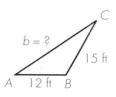

Squares and Approximate Square Roots

n	n²	√n	n	n²	√n
1	1	1.000	51	2601	7.141
2	4	1.414	52	2704	7.211
3	9	1.732	53	2809	7.280
4	16	2.000	54	2916	7.348
5	25	2.236	55	3025	7.416
6	36	2.449	56	3136	7.483
7	49	2.646	57	3249	7.550
8	64	2.828	58	3364	7.616
9	81	3.000	59	3481	7.681
10	100	3.162	60	3600	7.746
11	121	3.317	61	3721	7.810
12	144	3.464	62	3844	7.874
13	169	3.606	63	3969	7.937
14	196	3.742	64	4096	8.000
15	225	3.873	65	4225	8.062
16	256	4.000	66	4356	8.124
17	289	4.123	67	4489	8.185
18	324	4.243	68	4624	8.246
19	361	4.359	69	4761	8.307
20	400	4.472	70	4900	8.367
21	441	4.583	71	5041	8.426
22	484	4.690	72	5184	8.485
23	529	4.796	73	5329	8.544
24	576	4.899	74	5476	8.602
25	625	5.000	75	5625	8.660
26	676	5.099	76	5776	8.718
27	729	5.196	77	5929	8.775
28	784	5.292	78	6084	8.832
29	841	5.385	79	6241	8.888
30	900	5.477	80	6400	8.944
31	961	5.568	81	6561	9.000
32	1024	5.657	82	6724	9.055
33	1089	5.745	83	6889	9.110
34	1156	5.831	84	7056	9.165
35	1225	5.916	85	7225	9.220
36	1296	6.000	86	7396	9.274
37	1369	6.083	87	7569	9.327
38	1444	6.164	88	7744	9.381
39	1521	6.245	89	7921	9.434
40	1600	6.325	90	8100	9.487
41	1681	6.403	91	8281	9.539
42	1764	6.481	92	8464	9.592
43	1849	6.557	93	8649	9.644
44	1936	6.633	94	8836	9.695
45	2025	6.708	95	9025	9.747
46	2116	6.782	96	9216	9.798
47	2209	6.856	97	9409	9.849
48	2304	6.928	98	9604	9.899
49	2401	7.000	99	9801	9.950
50	2500	7.071	100	10000	10.000

Trigonometric Ratios

Angle	sin	cos	tan	Angle	sin	cos	tan
0°	0.0000	1.0000	0.0000	45°	0.7071	0.7071	1.0000
1°	0.0175	0.9998	0.0175	46°	0.7193	0.6947	1.0355
2°	0.0349	0.9994	0.0349	47°	0.7314	0.6820	1.0724
3°	0.0523	0.9986	0.0524	48°	0.7431	0.6691	1.1106
4°	0.0698	0.9976	0.0699	49°	0.7547	0.6561	1.1504
5°	0.0872	0.9962	0.0875	50°	0.7660	0.6428	1.1918
6°	0.1045	0.9945	0.1051	51°	0.7771	0.6293	1.2349
7°	0.1219	0.9925	0.1228	52°	0.7880	0.6157	1.2799
8°	0.1392	0.9903	0.1405	53°	0.7986	0.6018	1.3270
9°	0.1564	0.9877	0.1584	54°	0.8090	0.5878	1.3764
10°	0.1736	0.9848	0.1763	55°	0.8192	0.5736	1.4281
11°	0.1908	0.9816	0.1944	56°	0.8290	0.5592	1.4826
12°	0.2079	0.9781	0.2126	57°	0.8387	0.5446	1.5399
13°	0.2250	0.9744	0.2309	58°	0.8480	0.5299	1.6003
14°	0.2419	0.9703	0.2493	59°	0.8572	0.5150	1.6643
15°	0.2588	0.9659	0.2679	60°	0.8660	0.5000	1.7321
16°	0.2756	0.9613	0.2867	61°	0.8746	0.4848	1.8040
17°	0.2924	0.9563	0.3057	62°	0.8829	0.4695	1.8807
18°	0.3090	0.9511	0.3249	63°	0.8910	0.4540	1.9626
19°	0.3256	0.9455	0.3443	64°	0.8988	0.4384	2.0503
20°	0.3420	0.9397	0.3640	65°	0.9063	0.4226	2.1445
21°	0.3584	0.9336	0.3839	66°	0.9135	0.4067	2.2460
22°	0.3746	0.9272	0.4040	67°	0.9205	0.3907	2.3559
23°	0.3907	0.9205	0.4245	68°	0.9272	0.3746	2.4751
24°	0.4067	0.9135	0.4452	69°	0.9936	0.3584	2.6051
25°	0.4226	0.9063	0.4663	70°	0.9397	0.3420	2.7475
26°	0.4384	0.8988	0.4877	71°	0.9455	0.3256	2.9042
27°	0.4540	0.8910	0.5095	72°	0.9511	0.3090	3.0777
28°	0.4695	0.8829	0.5317	73°	0.9563	0.2924	3.2709
29°	0.4848	0.8746	0.5543	74°	0.9613	0.2756	3.4874
30°	0.5000	0.8660	0.5774	75°	0.9659	0.2588	3.7321
31°	0.5150	0.8572	0.6009	76°	0.9703	0.2419	4.0108
32°	0.5299	0.8480	0.6249	77°	0.9744	0.2250	4.3315
33°	0.5446	0.8387	0.6494	78°	0.9781	0.2079	4.7046
34°	0.5592	0.8290	0.6745	79°	0.9816	0.1908	5.1446
35°	0.5736	0.8192	0.7002	80°	0.9848	0.1736	5.6713
36°	0.5878	0.8090	0.7265	81°	0.9877	0.1564	6.3138
37°	0.6018	0.7986	0.7536	82°	0.9903	0.1392	7.1154
38°	0.6157	0.7880	0.7813	83°	0.9925	0.1219	8.1443
39°	0.6293	0.7771	0.8098	84°	0.9945	0.1045	9.5144
40°	0.6428	0.7660	0.8391	85°	0.9962	0.0872	11.4301
41°	0.6561	0.7547	0.8693	86°	0.9976	0.0698	14.3007
42°	0.6691	0.7431	0.9004	87°	0.9986	0.0523	19.0811
43°	0.6820	0.7314	0.9325	88°	0.9994	0.0349	28.6363
44°	0.6947	0.7193	0.9657	89°	0.9998	0.0175	57.2900
45°	0.7071	0.7071	1.0000	90°	1.0000	0.0000	∞

Metric and Customary Measures

Abbreviations

Metric

centimeter (cm)   liter (L)
decimeter (dm)   meter (m)
gram (g)   milliliter (mL)
kilogram (kg)   millimeter (mm)

Customary

cup (c)   mile (mi)
foot (ft)   pint (pt)
gallon (gal)   quart (qt)
inch (in.)   yard (yd)

LENGTH

Metric

1 km = 1000 m
1 m = 100 cm = 1000 mm
1 cm = 10 mm
1 mm = 0.1 cm = 0.001 m
1 cm = 0.01 m

Customary

1 mi = 1760 yd
1 mi = 5280 ft
1 yd = 3 ft = 36 in.
1 ft = 12 in.

Metric/Customary

1 mi = 1.609 km
1 in. = 2.54 cm
1 km = 0.621 mi
1 m = 39.37 in.

AREA

$1 \text{ cm}^2 = 100 \text{ mm}^2$
$1 \text{ m}^2 = 10{,}000 \text{ cm}^2$
$1 \text{ hectare} = 10{,}000 \text{ m}^2$

$1 \text{ yd}^2 = 9 \text{ ft}^2$
$1 \text{ ft}^2 = 144 \text{ in}^2$
$1 \text{ acre} = 4840 \text{ yd}^2$

$1 \text{ mi}^2 = 2.59 \text{ km}^2$
$1 \text{ hectare} = 2.471 \text{ acres}$
$1 \text{ m}^2 = 1.196 \text{ yd}^2$
$1 \text{ cm}^2 = 0.155 \text{ in}^2$

VOLUME

$1 \text{ dm}^3 = 1000 \text{ cm}^3$
$1 \text{ cm}^3 = 1000 \text{ mm}^3$

$1 \text{ yd}^3 = 27 \text{ ft}^3$
$1 \text{ ft}^3 = 1728 \text{ in}^3$

$1 \text{ yd}^3 = 0.765 \text{ m}^3$
$1 \text{ ft}^3 = 0.028 \text{ m}^3$
$1 \text{ in.}^3 = 16.387 \text{ cm}^3$

LIQUID

1 L = 1000 mL

1 gal = 4 qt
1 qt = 2 pt
1 pt = 2 c

1 gal = 3.785 L
1 qt = 0.946 L
1 L = 1.057 qt

WEIGHT

1 kg = 1000 g

1 ton = 2000 lb
1 lb = 16 oz

1 lb = 0.454 kg
1 kg = 2.205 lb

POSTULATES AND THEOREMS

CHAPTER 1

Postulate 1-1-1: Two points determine exactly one line. [p. 3]

Postulate 1-2-1: There is a one-to-one matching between the points on a line and the real numbers. The real number assigned to each point is its coordinate. The distance between two points is the positive difference of their coordinates. If A and B are two points with coordinates a and b such that $a > b$, then $AB = a - b$. [p. 8]

Postulate 1-2-2: Let O be a point on $\overleftrightarrow{XY}$ such that X is on one side of O and Y is on the other side of O. Real numbers from 0 through 180 can be matched with $\overrightarrow{OX}$, $\overrightarrow{OY}$, and all rays that lie on one side of $\overleftrightarrow{XY}$ so that each of the following is true:

(1) 0 is the number matched with $\overrightarrow{OX}$.

(2) 180 is the number matched with $\overrightarrow{OY}$.

(3) If $\overrightarrow{OA}$ is matched with a and $\overrightarrow{OB}$ is matched with b and $a > b$, then the number matched with $\angle AOB$ is $a - b$. [p. 9]

CHAPTER 2

Postulate 2-1-1: Three noncollinear points determine a plane. [p. 43]

Postulate 2-1-2: If two lines intersect, then they intersect at exactly one point. [p. 44]

Postulate 2-2-1: If two parallel lines are intersected by a transversal, then the corresponding angles are equal in measure. [p. 49]

Theorem 2-3-1: If two lines intersect, then the vertical angles are equal. [p. 55]

Theorem 2-3-2: If two parallel lines are intersected by a transversal, then the alternate interior angles are equal. [p. 57]

Theorem 2-5-1: If two parallel lines are intersected by a transversal, then the alternate exterior angles are equal. [p. 67]

CHAPTER 3

Postulate 3-1-1: Through a point not on a line there is only one line parallel to the given line. [p. 84]

Theorem 3-1-1: If the figure is a triangle, then the measure of the sum of the angles is 180°. [p. 83]

Theorem 3-2-1: The sum of the lengths of two sides of a triangle is greater than the length of the third side. [p. 89]

Theorem 3-2-2: If given a triangle, then any side is greater than the absolute value of the difference of the other two sides. [p. 90]

Theorem 3-5-1: A measure of an exterior angle of a triangle is equal to the sum of the measures of the two non-adjacent, or remote, interior angles. [p. 107]

CHAPTER 4

Theorem 4-5-1: If a segment connects the midpoints of two sides of a triangle, then the length of the segment is equal to $\frac{1}{2}$ the length of the third side. [p. 147]

Theorem 4-5-2: If a line is parallel to one side of a triangle and intersects the other sides at any points except the vertex, then the line divides the sides proportionally. [p. 148]

CHAPTER 5

Theorem 5-1-1: The empty set is a subset of every set. The symbol is either $\emptyset$ or { }. [p. 169]

Theorem 5-2-1: If a triangle has two equal sides and the included angle is bisected, then the triangle is divided into two congruent triangles. [p. 174]

Theorem 5-3-1: If a triangle has two equal angles, then the sides opposite the equal angles are equal. [p. 179]

Theorem 5-3-2: If a triangle is equiangular, then it is equilateral. [p. 179]

Theorem 5-4-1: If the triangle is isosceles, then the angles opposite the equal sides are equal. [p. 186]

CHAPTER 6

Theorem 6-3-1: If two lines are intersected by a transversal so that the alternate interior angles are equal, then the lines are parallel. [p. 220]

Theorem 6-6-1: If the midpoints of two sides of a triangle are joined, then the line segment determined or formed is parallel to the third side and is equal to $\frac{1}{2}$ its length. [p. 235]

Theorem 6-6-2: If a point is on the perpendicular bisector, then it is equidistant from the endpoints of the segment. [p. 237]

Theorem 6-6-3: If a point is on the angle bisector, then the point is equidistant from the sides of the angle. [p. 238]

CHAPTER 7

Theorem 7-1-1: (Pythagorean Theorem) If a right triangle has sides of lengths a, b, and c, where c is the hypotenuse, then $a^2 + b^2 = c^2$. [p. 251]

Theorem 7-1-2: (Converse of the Pythagorean Theorem) If a triangle has three sides of lengths a, b, and c, such that $a^2 + b^2 = c^2$, then the triangle is a right triangle. [p. 251]

Theorem 7-2-1: If the triangle is an isosceles right triangle, then the acute angles are each 45°. [p. 254]

Theorem 7-2-2: The length of the hypotenuse of an isosceles right triangle is $\sqrt{2}$ (about 1.414) times the length of either leg. In symbols this is written as $c = s\sqrt{2}$, or $c \approx 1.414s$. [p. 255]

Theorem 7-3-1: If a triangle is a 30°-60° right triangle, then the length of the leg opposite the 30° angle is half the length of the hypotenuse. [p. 261]

Theorem 7-3-2: If a triangle is a 30°-60° right triangle, then the length of the leg opposite the 60° angle is $\frac{\sqrt{3}}{2}$ times the length of the hypotenuse. [p. 262]

Theorem 7-4-1: If a triangle has one angle greater than another angle, then the side opposite the greater angle is longer than the side opposite the other angle. [p. 266]

Theorem 7-4-2: If a triangle has one side longer than another side, the angle opposite the longer side is greater than the angle opposite the other side. [p. 268]

Theorem 7-5-1: (The Distance Formula) The distance between points $A(x_1, y_1)$ and $B(x_2, y_2)$ is $d = \sqrt{(x_2 - x_1)^2 + (y_2 - y_1)^2}$. [p. 273]

CHAPTER 8

Theorem 8-1-1: The sum of the measures of the interior angles of a quadrilateral is 360°. [p. 285]

Theorem 8-1-2: A triangle is a rigid or non-flexible figure.

Theorem 8-2-1: In an isosceles trapezoid, the measures of each pair of base angles are equal. [p. 289]

Theorem 8-2-2: The median of a trapezoid is parallel to the bases and equal to one-half the sum of their lengths. [p. 290]

Theorem 8-3-1: If a quadrilateral is a parallelogram, then the opposite sides are equal and the opposite angles are equal. [p. 294]

Theorem 8-3-2: If the quadrilateral is a parallelogram, then the diagonals bisect each other. [p. 296]

Theorem 8-4-1: If a quadrilateral is a rectangle, then the diagonals are equal and all four angles are right angles. [p. 301]

Theorem 8-5-1: If a parallelogram is a rhombus, then the diagonals are perpendicular. [p. 306]

Theorem 8-5-2: If a figure is a rhombus, then opposite sides are parallel, opposite sides are equal, opposite angles are equal in length, opposite angles have equal measures, the diagonals bisect each other, and the diagonals are perpendicular. [p. 306]

Theorem 8-5-3: If a figure is a square, then opposite sides are parallel, opposite sides are equal in length, there are four right angles, the diagonals are equal in length, the diagonals bisect each other, and the diagonals are perpendicular. [p. 307]

CHAPTER 9

Postulate 9-1-1: If the figure is a rectangle, then the number assigned for the area is the length times the width. [p. 318]

Postulate 9-5-1: If two similar figures have a ratio of a/b for their corresponding sides, then:
a) The ratio of their perimeters is a/b, and
b) The ratio of their areas is $(a/b)^2$.

Theorem 9-2-1: If the figure is a triangle, then the number assigned to the area is $\frac{1}{2}$ times the base times the height. [p. 322]

Theorem 9-3-1: If the figure is a parallelogram, then the area is the length of the base times the height. [p. 329]

Theorem 9-3-2: If the figure is a parallelogram, then the perimeter is twice the sum of two adjacent sides. [p. 331]

Theorem 9-4-1: If the figure is a trapezoid, then the area is $\frac{1}{2}$ the height times the sum of the bases. [p. 337]

CHAPTER 10

Postulate 10-3-1: The circumference of a circle is π times the diameter (d). [p. 368]

Theorem 10-1-1: Each angle of a regular N-gon is equal to ($n - 2$) times 180° divided by n, where n is the number of sides in the N-gon. [p. 358]

Theorem 10-2-1: The area of a regular polygon is $\frac{1}{2}Ph$, where P is the perimeter of the polygon and h is the length of the perpendicular from the center of the polygon to each of its sides. [p. 362]

Theorem 10-3-1: The area of a circle is the radius squared times the number π. [p. 370]

Theorem 10-4-1: The measure of an inscribed angle is $\frac{1}{2}$ the number of degrees in the intersected arc. [p. 375]

Theorem 10-4-2: If the vertex of the angle is within the circle, then the measure of the angle is $\frac{1}{2}$ the sum of the degrees in the two intersected arcs. [p. 375]

Theorem 10-4-3: If the vertex of the angle is outside the circle and the rays intersect the circle, then the measure of the angle is $\frac{1}{2}$ the difference of the measure in degrees of the two intersected arcs. [p. 376]

Theorem 10-4-4: A tangent to a circle forms a 90° angle with the radius of the circle at the point of intersection. [p. 377]

Theorem 10-5-1: If two chords intersect, then the product of the segment lengths of one chord is equal to the product of the segment lengths of the other chord. [p. 380]

Theorem 10-5-2: The perpendicular bisector of a chord of a circle passes through the center of the circle. [p. 381]

Theorem 10-5-3: Given a tangent and a secant from a point, then the length of the tangent squared is equal to the product of the lengths of the secant and its external segment. [p. 382]

CHAPTER 11

Postulate 11-2-1: The volume of a right rectangular prism is the product of its length, width, and height: $V = lwh$.

Postulate 11-2-2: (Cavalieri's Principle) If two solid figures have equal heights and bases of the same area, and if every plane parallel to the bases always cuts off two cross-sections of equal area, then the two solids have equal volume.

Theorem 11-1-1: If the height of a prism is h and each base has perimeter P and area B, then the formula for the total surface area TA is: $TA = 2B + Ph$. [p. 405]

Theorem 11-1-2: If the height of a right circular cylinder is h and the radius of each circular base is r, then the formula for the total surface area is Total area $= 2\pi r^2 + 2\pi rh$. [p. 405]

Theorem 11-2-1: The volume of a prism is the product of the area B of one of its bases and the height h of the prism: $V = Bh$. [p. 410]

Theorem 11-2-2: The volume of a circular cylinder is the product of the area B of one of its bases and the height h of the cylinder: $V = Bh$. [p. 411]

Theorem 11-3-1: If p is the perimeter of the base of a regular pyramid and s is the slant height, the lateral area LA is given by the formula $LA = \frac{1}{2} ps$. [p. 415]

Theorem 11-3-2: If B is the area of the base of a regular pyramid, p is the perimeter of the base, and s is the slant height of the pyramid, then the total area TA is given by the formula $TA = B + \frac{1}{2} ps$. [p. 415]

Theorem 11-3-3: The lateral area LA of a right circular cone with a base of radius r and slant height s is given by $LA = \pi rs$. [p. 417]

Theorem 11-3-4: The total area TA of a right circular cone is the sum of the area of the base and the lateral area. If r is the radius of the base and s is the slant height, the total area is given by the formula $TA = \pi r^2 + \pi rs$. [p. 417]

Theorem 11-4-1: The volume of a triangular pyramid is equal to one third of the product of the area of the base and the height. If B is the area of the base and h is the height, then the volume is given by the formula $V = \frac{1}{3}(B)h$. [p. 421]

Theorem 11-4-2: The volume of any pyramid is equal to one-third the product of the area of the base and the height. The formula is $V = \frac{1}{3} Bh$. [p. 422]

Theorem 11-4-3: The volume of a right circular cone is equal to one-third the product of the area of the base and the height of the cone: $V = \frac{1}{3} Bh = \frac{1}{3}\pi r^2 h$. [p. 423]

Theorem 11-6-1: The formula for the volume V of a sphere is $V = \frac{4}{3}\pi r^3$, where r is the radius of the sphere. [p. 432]

Theorem 11-6-2: The formula for the surface area A of a sphere of radius r is $A = 4\pi r^2$. [p. 434]

CHAPTER 12

Theorem 12-1-1: The locus of points equidistant from the sides of an angle is the angle bisector. [p. 446]

Theorem 12-1-2: The locus of points equidistant from the endpoints of a line segment is the perpendicular bisector of the line segment. [p. 447]

Theorem 12-1-3: The locus of points in a plane that are a given distance from a line on the plane is two parallel lines. [p. 447]

Theorem 12-1-4: The locus of points a given distance from a point on a plane is a circle. [p. 447]

Theorem 12-1-5: The locus of the vertex (C) of all right triangles with given hypotenuse (AB) is a circle with diameter AB minus the endpoints A and B of the hypotenuse. [p. 447]

CHAPTER 13

Theorem 13-2-1: If (x_1, y_1) and (x_2, y_2) are two points on the graph of $y = mx + b$, then the slope m of the graph is equal to $\frac{y_2 - y_1}{x_2 - x_1}$ [p. 484]

Theorem 13-2-2: For any real numbers m and b, the y-intercept of the graph line of $y = mx + b$ is b, which means that the line crosses the y-axis at $(0, b)$. [p. 487]

Theorem 13-3-1: If two nonvertical lines in the coordinate plane are parallel, then their slopes are equal. [p. 490]

Theorem 13-3-2: If l_1 and l_2 are perpendicular lines with slopes m_1 and m_2 respectively, then $m_1 m_2 = -1$. [p. 492]

Theorem 13-4-1: The points $P(x, y)$ whose coordinates satisfy the equation $x^2 + y^2 = r^2$, where r is a positive real number, form a circle of radius r whose center is $(0, 0)$. [p. 496]

Theorem 13-5-1: An equation for the parabola determined by $(0, a)$ and the line $y = -a$, where $a > 0$, is $y = \frac{1}{4}a \cdot x^2$. [p. 502]

Theorem 13-6-1: (The Midpoint Theorem) If $P(a, b)$ and $Q(c, d)$ are any two points in the coordinate plane, then the midpoint of PQ is $M\left(\left(a + \frac{c}{2}\right), \left(b + \frac{d}{2}\right)\right)$. [p. 507]

CHAPTER 14

Theorem 14-4-1:   (Law of Sines) In any triangle ABC, $\frac{\sin A}{a} = \frac{\sin B}{b} = \frac{\sin C}{c}$. [p. 537]

Theorem 14-4-2:   (Law of Cosines) In any triangle are ABC,

$$a^2 = b^2 + c^2 - 2bc \cos A$$

$$b^2 = a^2 + c^2 - 2ac \cos B \text{ and}$$

$$c^2 = a^2 + b^2 - 2ab \cos C. \text{ [p. 539]}$$

Theorem 14-6-1:   If x is the measure of any angle, then $\sin^2 x + \cos^2 x = 1$. [p. 551]

Theorem 14-6-2:   If x is the measure of any angle, then $\cos 2x = \cos^2 x - \sin^2 x$. [p. 551]

Theorem 14-6-3:   If x is the measure of any angle such that $\cos x \neq 0$, then $\tan x = \sin x / \cos x$. [p. 551]

PHOTO ACKNOWLEDGMENTS

All photography by The Image Bank, Chicago, unless marked with asterisk (*)

Cover Comstock: Mike & Carol Werner*

Preface p. viii: Jeff Hunter; Joseph Devenney; p. ix: Gary Gladstone, 1989; Dan Rest*; p. x: Mel Di Giacomo and Bob Masini; Gary Gladstone; p. xi: Murray Alcosser; Murray Alcosser; Michael Salas, 1976; p. xii: Kay Chernush; Hank Delespinasse; p. xiii: Peter Miller; Cesar Lucas; Joseph Devenney; Harald Sund; p. xix: Jacques Cochin, 1988.

Chapter 1 p. 1: Larry Keenan Associates; p. 2: Eric Meola; Sobel/Klonsky, 1988; B. Lindhout; Guido Alberto Rossi; p. 3: Michael Melford; p. 7: Michael Salas; Marc Romanelli; p. 9: Steve Krongard, 1990; p. 13: William Rivelli; Lou Jones; p. 19: Alvis Upitis, 1989; p. 24: D.W. Productions*; Michael Salas; p. 25: Alfredo Tessi; p. 28: John P. Kelly; p. 30: Schneps; Steve Dunwell.

Chapter 2 p. 41: Grant V. Faint; p. 42: Pete Turner; p. 44: Mitchell Funk; Eddie Hironaka; p. 47: Gary Cralle; David Brownell; p. 53: Gerard Champlong; p. 55: Mitchell Funk; p. 59: Steve Proehl; Harald Sund; Gary Gladstone; p. 65: Ted Russell; Marc Romanelli; Walter Bibikow, 1988; p. 71: Gerard Champlong; p. 72: Jake Rajs.

Chapter 3 p. 81: L. Mason; p. 82: Geoff Gove; Bernard Roussel; p. 83: Benn Mitchell; p. 84: Ron Kadrmas*; p. 87: Jürgen Vogt; Ron Kadrmas*; p. 92: Richard Pan, Stockphotos, Inc.*; p. 93: Ulli Seer; p. 98: Ron Kadrmas*; p. 99: Gary Cralle; p. 100: W. Peisenroth; p. 105: Romilly Lockyer; Paul Silverman, 1984, Stockphotos, Inc.*; p. 110: Tim Bieber; p. 111: P. Runyon; p. 116: Brett Froomer.

Chapter 4 p. 123: Gary Cralle; p. 124: Margarette Mead; p. 129: Al Giddings, 1989; p. 131: James H. Carmichael; Philip A. Harrington; p. 134: Jacky Gucia; p. 135: Weinberg-Clark; p. 140: Arthur Meyerson; p. 141: Jay Freis; p. 144: Stephen Derr; p. 145: Guido Alberto Rossi; Andre Gallant; p. 146: Luis Castañeda, 1989; p. 152: Luis Castañeda; p. 153: Chris Alan Wilton; p. 158: John Kelly; Nicholas Foster.

Chapter 5 p. 165: Charles C. Place; p. 166: Garry Gay; p. 171: Hans Wolf; p. 172: Jürgen Vogt; p. 173: A. M. Rosario; p. 177: Terje Rakke; Guilano Colliva; Jeffrey M. Spielman; p. 183: David W. Hamilton; Eddie Hironaka; p. 189: Steve Dunwell; p. 192: Terje Rakke; Mahaux Photography; Kenneth Redding; p. 193: Jeff Cadge.

Chapter 6 p. 205: Robert Holland; p. 206; Stephen Derr; Don Klumpp; p. 211: Guido Alberto Rossi; p. 213: Stephen Marks; John Ramey; p. 216: Jürgen Vogt, 1988; T. Bieber; p. 217: Color Day Productions*; p. 222: Marc Romanelli; Color Day Photography*; p. 223: Kim Steele; Bernard Roussel p. 230: Joe Azzara; Murray Alcosser; p. 232: Murray Alcosser; p. 235: Bernard Van Berg.

Chapter 7 p. 227: Mark Solomon; p. 248: Merrell Wood; p. 250: Francois Dardelet; Kay Chernush; Barrie Rokeach; p. 253: Patrick Doherty, Stockphotos, Inc.*; Bill Varie; p. 258: Cliff Feulner; p. 259: Benn Mitchell; Peter Miller; p. 265: Peter Miller; Barrie Rokeach, 1988; p. 269: Marc Romanelli; p. 271: Peter Miller; Barrie Rokeach.

Chapter 8 p. 281: Steve Proehl; p. 282: Barrie Rokeach, 1984; p. 283: Bruce Wodder, 1983; p. 285: Gary Bistram, 1989; p. 287: David J. Maenza; p. 288: H. Wendler; p. 293: Harald Sund; p. 294: L. Mason; p. 296: Mel Di Giacomo; p. 297: Patti McConville; p. 299: Jane Sobel, 1980; p. 300: Michael Melford; p. 302: Jay Brousseau; p. 303: Gary Faber; p. 304: Walter Iooss, Jr.; p. 306: Gary Faber; p. 307: Benn Mitchell.

Chapter 9 p. 315: H.G. Kaufmann; p. 316: Harald Sund; p. 317: David W. Hamilton; p. 321: Daniel Hummel; p. 325: Yuri Dojc; p. 327: Larry Dale Gordon; p. 328: Barrie Rokeach, 1989; P. & G. Bowater; p. 333: Steve Proehl; p. 335: Alvis Upitis; p. 336: Jake Rajs, 1987; p. 342: Marvin E. Newman; p. 343: Anthony A. Boccaccio; p. 350: Patti McConville.

Chapter 10 p. 355: David W. Hamilton; p. 356: Jürgen Vogt; p. 361: Tim Bieber; p. 363: David W. Hamilton; p. 365: Gregory Heisler; p. 366: Marc Romanelli; p. 367: Edward Bower; Jerry Yulsman; p. 370: Weinberg-Clark; p. 373: Chuck Kuhn; R. Phillips; p. 379: Andre Gallant; p. 383: Klaus Mitteldorf; p. 385: G. & J. Images*; p. 386: G. V. Faint; p. 388: Chuck Place; p. 389: Michael Tcherevkoff; Gerard Champlong.

Chapter 11 p. 401: Larry Dale Gordon; p. 402: Jeff Hunter; p. 404: Michael Tcherevkoff; p. 405: Alvis Upitis; Tim Bieber; p. 407: Lionel Isy-Schwart; p. 413: Alan Becker; p. 419: Jeff Spielman, Stockphotos, Inc.*; p. 423: Paolo Gori; p. 424: Jake Rajs; p. 425: David Jeffrey; Schmid/Langsfeld; p. 426: Co Rentmeester; p. 427: Harald Sund; p. 430: Garry Gay; p. 431: P. & G. Bowater; Marti Pie; p. 434: Douglas Struthers.

Chapter 12 p. 443: Geoffrey Gove; p. 444: Don Klumpp; p. 449: Joseph Brignolo; p. 451: Lisl Dennis; p. 453: Garry Gay; p. 455: Nicholas Foster; p. 456: Al Satterwhite; p. 459: Marcel Isy-Schwart; p. 461: Gary Gladstone; p. 462: Ulf E. Wallin, 1986, Stockphotos, Inc.*; Gerard Champlong; p. 464: IN FOCUS INTERNATIONAL*; Lou Jones; p. 467: Michael Melford, 1988; p. 470: Nicholas Foster, 1978.

Chapter 13 p. 477: Hans Wolf Studio; p. 478: Eric Schweikardt; p. 480: Geoffrey Gove; p. 483: Albert Normandin; p. 489: Joseph Szkodzinski; p. 490: Weinberg-Clark; Albert Normandin, 1990; p. 494: Steve Proehl; p. 495: John Kelly; p. 497: Brett Froomer; p. 501: Steve Dunwell; p. 503: Guido Alberto Rossi; p. 504: Sobel-Klonsky; Brett Froomer; p. 506: Don King, 1988; p. 507: David W. Hamilton; p. 509: David W. Hamilton.

Chapter 14 p. 517: Trent Swanson; p. 518: Bill Carter; p. 520: ZAO-Longfield; p. 523: Murray Alcosser; p. 526: Nicholas Foster; p. 527: Edward Bower, 1988; p. 529: Tom Mareschal; p. 531: Cesar Lucas; p. 534: Joe Azzara; Denny Tillman; p. 536: Charles C. Place; J. Ramey; p. 541: Jeff Spielman, Stockphotos, Inc.*; p. 542: G. Rossi; Arthur d'Arazien; p. 545: Steve Satushek; p. 547: Michael Melford.

CHAPTER 1

Class Activity, p. 5
1. Point P **3.** $\overrightarrow{GF}$
5. $\overline{KL}$
7. $\angle ADH$, or $\angle HDA$
9. $\angle FGH$, or $\angle HGF$
11.
W   R
13. F ———— G
15.
Z   X
17.
X   Y
19.

H

I   J

Home Activity, p. 6
1. $\overleftrightarrow{SW}$ or $\overleftrightarrow{WS}$ **3.** $\overrightarrow{PT}$
5. $\angle JKL$ or $\angle LKJ$
7.
C   X
9. plane   **11.** ray
13. $\overrightarrow{GA}$, $\overrightarrow{GC}$ **15.** $\overline{GC}$, $\overline{AD}$

Class Activity, p. 8
1. -3, 1, 3, 5   **3.** -2
5. 3 units   **7.** 3 in.
9. 5 cm

Class Activity, p. 11
1. Obtuse   **3.** Right
5.

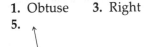

7.

9. 120°; Obtuse
11. 68°; Acute

Home Activity, p. 12
1. -6, -1, 3, and 9
3. 6, 1, 3, and 9 respectively
5. 4   **7.** 8
9. Obtuse   **11.** 75°; Acute
13. $\angle AOB$, 15°, Acute;
  $\angle AOC$, 45°, Acute;
  $\angle BOC$, 30°, Acute

Class Activity, p. 14
1. 3 in., 1.5 in.
3. 8 cm, 4 cm
5. 8 in., 4 in.
7. C

Class Activity, p. 15
1. 120°
3. Acute; 20°; 20°
5. Obtuse; 60°; 60°
7. Straight; 90°; 90°

Class Activity, p. 16
1.

P Q   J

2 3 4 5 6 7 8

3. $2 < 3 < 8$ **5.** 1
7. 6

Class Activity, p. 17
1.

R
    Q

S   T

Home Activity, p. 17
1. 4   **3.** 18
5. 6 in.; 3 in.
7. Acute; 30°; 30°
9. $\angle BEC$ **11.** 50°
13.

C   K

P   A

Class Activity, p. 20
1. Answers may vary.
3. 3°, 5°; 95°
5. $m\angle CEF = 132°$
  $m\angle FEG = 48°$
  $m\angle CED = 132°$
7. 54°; 126°

Class Activity, p. 22
1. 55°
3. *Equation*
  $45 + c = 90$
  $10 + c = 90$
  $80 + c = 90$
  $85 + c = 90$
  $14 + c = 90$
  $29 + c = 90$

$x + 40 + c = 90$
$2x + 20 + c = 90$
Complement
$(90 - x)°$
$(130 - x)°$
$(70 - 2x)°$
5. 59°; 31° **7.** 50°; 40°
9. 72°; 18°

Home Activity, p. 23
1. ∠*EOF* and ∠*FOG*
3. Possible answers: ∠*AOB* and ∠*AOF*, ∠*EOC* and ∠*FOA*; other answers are possible.
5. ∠*EOA*
7. 150° **9.** 18.6°
11. 50° **13.** 83.95°
15. Acute **17.** Obtuse
19. Obtuse **21.** 95°; 85°
23. 30°; 60° **25.** 33°
27. 138°
29. Drawing shows complementary angles of 75° and 15°.

Critical Thinking, p. 24
85 baseballs

Class Activity, p. 26
1. 27; Possible answer: 1 less than number of dots per side times 3—continue the pattern:
$(2 - 1) \times 3 = 3$ — 2 per side;
$(3 - 1) \times 3 = 6$ — 3 per side;
$(4 - 1) \times 3 = 9$ — 4 per side.
(and so on)

Class Activity, p. 28
1. 21, 31, 43
3. 63, 127, 255
5. $0.\overline{36}$; $0.\overline{81}$

Home Activity, p. 29
1. 17, 20, 23
3. 33, 65, 129
5. $\frac{1}{4} + \frac{1}{16} + \frac{1}{32} + \frac{1}{64} = \frac{23}{64}$ shaded
7. 15 **9.** $\frac{1}{2} n (n - 1)$

Class Activity, p. 32
1.

3.

5. Answers may vary.

Home Activity, p. 32
1. Answers may vary.
3. Answers may vary.

Class Activity, p. 34
1.

3. Answers may vary.
5. Answers may vary.

Class Activity, p. 36
1. Answers may vary.
3. Answers may vary.
5. Answers may vary.
7. m∠*ABX* = m∠*XBC*; $\overline{BX}$ is the bisector of ∠*ABC*.
9. Answers may vary.

Chapter 1 Review
1. $\overline{AB}$ or $\overline{BA}$
3. ∠*TAK*, ∠*KAT*, or ∠*A*
5. 7.9 cm **7.** 115°; obtuse
9. −6, −2, 1, 5
11. *L* is between *J* and *K*.
13. $(7x)° = 140°$; $(3x - 20)° = 40°$
15. a) 21, 28, 36
    b) 162, 486, 1458

CHAPTER 2

Class Activity, p. 43
1. b   3. e   5. d
7. True
9. Coplanar
11. Collinear and Coplanar
13. Coplanar

Class Activity, p. 45
1. Intersecting
3. Parallel
5. $\overleftrightarrow{BF}$, $\overleftrightarrow{AC}$ 7. $\overleftrightarrow{BF}$, $\overleftrightarrow{DE}$
9. False

Home Activity, p. 46
1. Yes   3. No
5. Yes   7. Yes
9. $\overleftrightarrow{AB}$, $\overleftrightarrow{EF}$ 11. $\overleftrightarrow{CD}$, $\overleftrightarrow{EF}$
13. Answers may vary.

Critical Thinking, p. 47
16

Class Activity, p. 48
1. a and d
3. Diagram B

Class Activity, p. 50
1. a
3. $\angle 1$, $\angle 5$; $\angle 4$, $\angle 8$; $\angle 2$, $\angle 6$; $\angle 3$, $\angle 7$
5. 8   7. 7
9. 150°   11. 30°
13. All angles have a measure of 90°.

Home Activity, p. 51
1. l and k
3. a and b
5. $\angle 5$; 67°   7. $\angle 7$; 67°
9. 25°   11. 25°   13. 25°
15. 25°   17. 100°; 80°

Class Activity, p. 54
1. H: you live in Utah; C: you live in the United States.

3. H: your pet is a terrier; C: it is a dog.
5. M$\angle PQR > 90°$.
7. Points L, M, and R determine a plane.

Class Activity, p. 56.
Answers for Exercises 1 through 4 may vary.
5. Yes

Class Activity, p. 57
1. t
3. $\angle 1$, $\angle 4$; $\angle 5$, $\angle 8$; $\angle 2$, $\angle 3$; $\angle 6$, $\angle 7$
5. $\angle 3$ and $\angle 6$; $\angle 4$ and $\angle 5$
7. $1 = 85°$; $2 = 95°$
9. Answers may vary.

Home Activity, p. 58
1. H: two angles are vertical angles; C: they are equal.
3. Pete knows the city.
5. The square of 2.001 is greater than 4.
7. $1 = 40°$; $2 = 40°$
9. $m\angle 3 = 93°$; $m\angle 4 = 56°$;
   $m\angle 5 = 87°$; $m\angle 6 = 93°$;
   $m\angle 7 = 37°$; $m\angle 8 = 143°$;
   $m\angle 9 = 143°$; $m\angle 10 = 56°$;
   $m\angle 11 = 124°$

Critical Thinking, p. 58
$m\angle A = m\angle B = 90°$

Class Activity, p. 61
1. Nearest inch: 4 in., error: $\frac{1}{2}$ in; nearest $\frac{1}{8}$ in: $4\frac{1}{8}$ in. error $\frac{1}{16}$ in; nearest $\frac{1}{16}$ in: $4\frac{1}{16}$ or $4\frac{2}{16}$ in, error: $\frac{1}{32}$ in.
3. Nearest cm: 5 cm, error: $\frac{1}{2}$ cm or 5 mm; nearest mm: 52 mm, error: $\frac{1}{2}$ mm
5. Nearest 5 degrees: 70°; error: $2\frac{1}{2}°$; nearest degree: 72°; error: $\frac{1}{2}°$
7. Nearest 5 degrees: 115°; error: $2\frac{1}{2}°$; nearest degree: 116°; error $\frac{1}{2}°$

Class Activity, p. 63
1. 4 ft 10 in.
3. 120 in.; 3.05 m

Home Activity, p. 63
1. Nearest $\frac{1}{4}$ in.: $4\frac{1}{4}$ in.; error: $\frac{1}{8}$ in.;
   nearest $\frac{1}{8}$ in.: $4\frac{2}{8}$ in.; error: $\frac{1}{16}$ in.;
   nearest cm: 11 cm; error: 0.5 cm
3. 32°; error $\frac{1}{2}$°
5. 86°; error $\frac{1}{2}$°
7. $2\frac{3}{4}''$, $1\frac{5}{8}''$, $2\frac{1}{4}''$; about $6\frac{5}{8}$ in.
9. $2\frac{5}{8}''$ long, 1" high; about $7\frac{3}{5}$ in.
11. Answer should be close to 360°.
13. 609 mm   15. 15 yd
17. 13 ft 4 in.
19. 158° 45'   21. 118° 40'
23. 141° 28'   25. 66° 55'

Critical Thinking, p. 64
60°; 180°; 30°

Class Activity, p. 68
1. 2; 4
3. ∠3, ∠6, ∠7
5. ∠2, ∠3, ∠7
7. ∠2, ∠6, ∠7
9. ∠1, ∠4, ∠5
11. 90°   13. 73.4°
15. 106.6°   17. 73.4°
19. ∠1 and ∠11; ∠4 and ∠10
21. ∠3, ∠5, ∠7, ∠9, ∠11, ∠13, ∠15
23. 70°   25. 70°

Home Activity, p. 69
1. ∠1 and ∠8; ∠2 and ∠7
3. ∠3 and ∠6; ∠5 and ∠4
5. 45°   7. 135°   9. 45°
11. 52°   13. 52°   15. 128°
17. 147°   19. 147°   21. 147°
23. 95°   25. 65°   27. 55°

Critical Thinking, p. 70
∠2 and ∠4

Class Activity, p. 73
1.

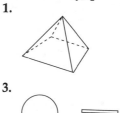

3.

Home Activity, p. 74
1. See students' drawings.
3. Answers may vary.
5. Possible answers:
   • Cubes piled in all 6 corners;
   • 4 cubes (one unseen) piled in corner;
   • Star within a hexagon.
7.

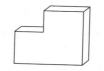

9.

Critical Thinking, p. 74
Both will form a cube.

Chapter 2 Review, p. 78
1. $\overleftrightarrow{AB}$, $\overleftrightarrow{FG}$; $\overleftrightarrow{DH}$, $\overleftrightarrow{AB}$; $\overleftrightarrow{DH}$, $\overleftrightarrow{DC}$; $\overleftrightarrow{BC}$, $\overleftrightarrow{AB}$; $\overleftrightarrow{BC}$, $\overleftrightarrow{DC}$.
3. $\overleftrightarrow{FG}$ and $\overleftrightarrow{AB}$
5. $\overleftrightarrow{FG}$ and $\overleftrightarrow{DC}$, $\overleftrightarrow{DH}$, $\overleftrightarrow{BC}$
7. F, E, G; A, E, B (Other answers are possible.)
9. ∠1 and ∠5, ∠2 and ∠6, ∠3 and ∠7, ∠4 and ∠8
11. ∠1 and ∠8, ∠2 and ∠7
13. 50°   15. 130°   17. 50°
19. 2 in.; error: $\frac{1}{4}$ in.
21. 118°; error: $\frac{1}{2}$°
23. Check students' diagrams.

HAPTER 3

Class Activity, p. 83
1. Yes
3. No; there are 4 sides.

Class Activity, p. 85
1. $x = 80°$ 3. $m = 10°$

Home Activity, p. 86
1. a. $\triangle ABC$
   b. $\overline{AB}, \overline{BC}, \overline{CA}$
   c. A, B, C
   d. $\angle BAC, \angle ABC, \angle ACB$
3. $x = 80°$; A
5. $x = 50°$; A
7. $y = 60°$; $a = 40°$; $x = 80°$; $z = 120°$; $c = 140°$
9. No; the sum of the measures of the angles would be greater than 180°.
11. $m\angle C = 60°$

Class Activity, p. 90
1. $3 < x < 17$
3. $2 < AC < 8$
5. $10 < RS < 40$

Home Activity, p. 91
1. Yes   3. No   5. Yes
7. $2 < x < 8$
9. 2 ft $< x <$ 7 ft
11. 15 km $< x <$ 105 km
13. 16 and 24
15. $4 < HJ < 44$
17. 9 km $< VJ <$ 29 km
19. $150 + 180 > 340$; the triangle is not possible.
21. $16 < x < 36$

Critical Thinking, p. 92
22. The shortcut and sidewalks form a triangle. The shortcut is shorter than the sum of the lengths of the sidewalks.

Class Activity, p. 95
1. Scalene
3. Isosceles
5. $\triangle CED$ or $\triangle ABE$
7. 8   9. Answers may vary.

Home Activity, p. 96
1. Scalene
3. Isosceles
5. Equilateral
7. Scalene
9. Equilateral

11. Isosceles
13. Scalene   15. 4
17. Scalene
19. $m\angle 1 = 35°$; $m\angle 2 = 145°$; $m\angle 3 = 35°$; $m\angle 4 = 145°$; $m\angle 5 = 35°$; $m\angle 6 = 145°$; $m\angle 7 = 35°$; $m\angle 8 = 55°$; $m\angle 9 = 55°$; $m\angle 10 = 90°$
21. They measure 90°.
23. Yes; any two sides chosen are of equal length.

Critical Thinking, p. 98
25. No; 25 km would exceed the total of the maximum possible lengths of the legs of the trip.

Class Activity, p. 102
1. $m\angle x = 40°$; acute
3. $m\angle x = 60°$; equiangular
5. $m\angle X = 75°$; $m\angle Y = 50°$; $m\angle 7 = 55°$; acute
7. Equiangular

Home Activity, p. 103
1. $m\angle A = 60°$; $m\angle C = 60°$; equiangular
3. $m\angle R = 74.2°$; acute
5. 100°   7. Same
9. $m\angle J = 30°$; $m\angle O = 60°$; $m\angle JTO = 90°$; Right
11. $\triangle ABC, \triangle CEF, \triangle ACD, \triangle AEB, \triangle BFD, \triangle BCD, \triangle BEC, \triangle BCF$

Critical Thinking, p. 104
13. True   15. False
17. True   19. False
21. True   23. False
25. False

Class Activity, p. 108
1. 120°   3. 110°
5. $m\angle 1 = 25°$; $m\angle 2 = 35°$

Home Activity, p. 108
1. 6;

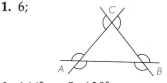

3. 166°   5. 130°
7. $m\angle 1 = 40°$; $m\angle 2 = 140°$; $m\angle 3 = 80°$; $m\angle 4 = 60°$; $m\angle 5 = 120°$
9. 720°
11. Yes; both can be 90°.
13. 20′   15. 1.36 miles
17.–19. Answers may vary. To accomplish the task in Exercises 16–19, the turtle must be turned through the exterior angle.

Critical Thinking, p. 110
21. Bob

Class Activity, p. 113
17. Isosceles, right
19. Scalene, acute
21. A (3, 0) B (0, −9) C (2, 3)
 D (4, 7) E (−6, 1) F (−2, −4)
 G (−5, 0) H (3, 9) I (2, 7)
 J (−7, 5) K (−1, 6) L (−4, −5)

Home Activity, p. 114
1.

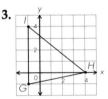

 Isosceles, right

3.

 Scalene, acute

5. A (2, 3) B (0, 0)
 C (7, 9) D (−10, 0)
 E (−3, 2) F (−7, 0)
 G (0, 6) H (0, −4)
 I (−1, 7) J (2, −6)
 K (5, −9) L (10, 0)
 M (0, −9) N (7, −8)
 O (−10, 4) P (8, −5)
7. Right triangle, scalene triangle
9. Scalene triangle, obtuse triangle
11. Ray   **13.** No; the order of the coordinates is different.
15. Parallel
17. Intersecting at (2, −5)
19. A(−8 −2) B(−4, −2) I(4, 4) J(10, 4)
 C(−2, 0) D(−4, 2) K(10, 8) L(6, 8)
 E(−8, 2) F(−12, 4) M(0, 10) N(5, −2)
 G(−2, 4) H(−2, 8) O(0, −2)
 Answers may vary. Possible answers: all have at least one right angle; all have areas of 20; all have at least one side parallel to x-axis; all have at least one side parallel to y-axis.

Critical Thinking, p. 116
21. y = x; line
23. Flexible   **25.** Rigid

Chapter 3 Review, p. 120
 1. 5   **3.** 1, 3, 6, 7
 5. 2, 3   **7.** 6
 9. 93°   **11.** 46°
13. 67°
15. 8; 66
17. Carl damaged the car, Bev the bike.
19. Scalene, obtuse

CHAPTER 4

Class Activity, p. 126
 1.

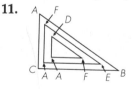

 3. ∠A ↔ ∠X; ∠B ↔ ∠Y; ∠C ↔ ∠Z
 5. Same shape

Home Activity, p. 127
 7. A ↔ D, B ↔ E, C ↔ F, $\overline{AB}$ ↔ $\overline{DE}$, $\overline{BC}$ ↔ $\overline{EF}$, $\overline{AC}$ ↔ $\overline{DF}$, ∠A ↔ ∠D, ∠B ↔ ∠E, ∠C ↔ ∠F
 9. Answers may vary.
11.

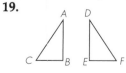

13. $\overline{AC}$ ↔ $\overline{FG}$, $\overline{AC}$ ↔ $\overline{DG}$, $\overline{FG}$ ↔ $\overline{DG}$, $\overline{AB}$ ↔ $\overline{FE}$, $\overline{AB}$ ↔ $\overline{DF}$, $\overline{FE}$ ↔ $\overline{DF}$, $\overline{CB}$ ↔ $\overline{GE}$, $\overline{CB}$ ↔ $\overline{GF}$, $\overline{GE}$ ↔ $\overline{GF}$
15. Flip △FED over. ∠A ↔ ∠E; ∠B ↔ D; ∠C ↔ ∠F
17. Answers may vary.
19.

 C⊿B   E⊿F

21. Flip △ABC onto △DEF, or vice versa.

Class Activity, p. 130
1. All except e.   3. $\frac{1}{4}$
5. $\frac{1}{4}$ 7. $\frac{1}{1}$
9. 2.4   11. 5.489

Class Activity, p. 132
1. $\frac{4}{7}$ 3. $\frac{17}{24}$
5. 5 boys to every 7 girls
7. 41 gallons

Home Activity, p. 132
1. $\frac{1}{2}$ 3. $\frac{5}{8}$
5. $\frac{15}{17}$ 7. $\frac{3}{2}$
9. $\frac{1}{6}$ 11. $\frac{3}{1}$
13. $\frac{1}{2}$ 15. $\frac{1}{2}$
17. $\frac{5}{6}$ 19. 4.5
21. 20   23. $56\frac{2}{3}$
25. $-\frac{2}{3}$

Critical Thinking, p. 133
27. 13-oz box

Class Activity, p. 137
1. By AA
3. $\angle A \leftrightarrow \angle F$; $\angle B \leftrightarrow \angle D$; $\angle C \leftrightarrow \angle E$
5. $EF = 4.4$; $ED = 4.8$
7. Answer may vary.
9. Answer may vary.
11. Answer may vary.

Home Activity, p. 138
1. $\angle A \leftrightarrow \angle D$, $\angle B \leftrightarrow \angle E$, $\angle C \leftrightarrow \angle F$, $\overline{AB} \leftrightarrow \overline{DE}$,
$\overline{BC} \leftrightarrow \overline{EF}$, $\overline{AC} \leftrightarrow \overline{DF}$
3. Yes
5. By SAS
7. Not enough information to tell
9. Yes   11. By AA
13. 3:7; $x = 11\frac{2}{3}$

Class Activity, p. 143
1. 1725 miles
3. Answers may vary.
5. Triangle with sides 3.25 cm, 1.25 cm, and 3 cm

Home Activity, p. 144
1. 1:19.2   3. 8.5 in.
5. 11.3 in.   7. 10.08 m
9. 16 in.   11. 1:2.54
13. 52,081:1   15. 1:0.9
17. 1:3   19. 1:12
21. 1:100
23. Triangle with sides of length 2.5 cm, 1.5 cm, and 2 cm

Class Activity, p. 149
1. 7.5

Home Activity, p. 149
1. Yes; SAS   3. Yes; SAS
5. 100   7. 50
9. Yes; SAS
11. Yes; AA
13. Yes; AA
15. Yes; AA
17. 3, △ABC, △ACD, △BCD
19. 20   21. Yes; SAS
23. 5   25. 20   27. $6\frac{2}{7}$

Critical Thinking, p. 151
$\frac{\$40}{100} = \frac{x}{15}$, or $x = \frac{40 \times 15}{100}$

Class Activity, p. 154
1. $x = 28$, $y = 70$
3. 90°   5. 50°
7. $66\frac{2}{3}$; $26\frac{2}{3}$

Home Activity, p. 155
1. $18\frac{2}{3}''$ and $23\frac{1}{3}''$
3. $x = 15$ cm; $y = 9$ cm
5. $x = 10$; $y = 2\sqrt{3}$
7. 17.5 m   9. 55°
11. 55°   13. Yes; AA
15. Yes; AA   17. Yes; AA
19. 23.75 ft   21. 61.25 ft
23. 54.44 ft

Critical Thinking, p. 158
25. Height to eye; distance from feet to mirror and base of building to mirror
27. Answers will vary.

Chapter 4 Review, p. 162
 1. SSS; $\frac{3}{1}$
 3. AA; $\frac{2}{1}$
 5. Yes; AA   7. 65°
 9. 65°   11. 3
 13. 15

CHAPTER 5

Class Activity, p. 167
 1. True   3. True
 5. Possible answer: Set of one-digit numbers.
 7. {2, 4}, {2, 6}, {2, 8}, {4, 6}, {4, 8}, {6, 8}

Class Activity, p. 169
 1.
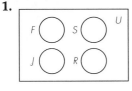

 3.
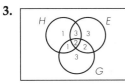

 5. 12; 12; 14
 7. Yes; yes   9. An angle

Home Activity, p. 170
 1. {6, 12, 18, 24}
 3. {a, e, i, o, u}
 5. {101, 102, 103, . . . }
 7. Yes   9. Yes
 11. 30; 10; 18
 13. 317

Critical Thinking, p. 170
 15. 2^n

Class Activity, p. 173
 1.
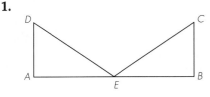

Yes; $\triangle AED \sim \triangle BED$ by SAS; $\frac{DA}{CB} = \frac{AE}{BE} = 1$

Class Activity, p. 174
 1. $NO = 12$; since m$\angle NMP = $ m$\angle OMP$, $\overline{MP}$ bisects $\angle NMO$. Since $MN = MO$ and the included $\angle NMO$ is bisected, by Theorem 5-2-1, $\triangle NMP \cong \triangle OMP$. So $NP = OP$, $OP = 6$, and thus $NO = 12$.

Home Activity, p. 174
 1. Yes   3. Yes
 5. Not enough information given to decide
 7. $\cong$, by Theorem 5-2-1
 9. 41 ft
 11. It is given that $AB = CD$. Then $AC = BD$ by segment addition. $AE = DF$, and m$\angle A = $ m$\angle D$. $AC/BD = AE/DF$ by algebra. Therefore $\triangle AEC \sim \triangle DFB$ by SAS. Since $AC/BD = AE/DF = 1$, $\triangle AEC \cong \triangle DFB$.
 13. $\triangle ABC \sim \triangle DCB$ by AA (rt. angles, alt. int. angles; $\frac{CB}{CB} = 1$
 15. It is given that m$\angle A = $ m$\angle D = 90°$. m$\angle DCE = $ m$\angle ACB$, since they are vertical angles. Then $\triangle ACB \sim \triangle DCE$ by AA. It is given that $AC = DC$, therefore $AC/DC = 1$. Thus $\triangle ACB \cong \triangle DCE$.
 17. Yes; $\triangle BCE \sim \triangle BAE$ by SAS; $\frac{BC}{BA} = \frac{BE}{BE} = 1$
 19. Triangle 1: no
 Triangle 2: yes

Class Activity, p. 179
 1. Yes
 3. Yes
 5. Not enough information
 7. Yes
 9. Sometimes

Home Activity, p. 180
 1. AA, $\frac{6}{6} = \frac{8}{8} = 1$
 3. Yes
 5. Yes; SAS
 7. $\overline{PR}$
 9. Yes; SSS and $\frac{7}{7} = \frac{8}{8} = \frac{5}{5} = 1$
 11. $\angle V$
 13. Always
 15. Sometimes
 17. 1. $AB = BC = CD = DA$ Given
    2. $BD = BD$ Identity
    3. $\frac{AB}{CB} = \frac{AD}{CD} = \frac{BD}{BD}$ Algebra
    4. $\triangle ABD \sim \triangle CBD$ SSS
    5. $\frac{AB}{CB} = 1$ Algebra
    6. $\triangle ABD \cong \triangle CBD$ Definition congruent triangles
    7. m$\angle A = $ m$\angle C$ CPCF

19.

1.	Isosceles $\triangle ABC$ with $AC = BC$	Given
2.	$\overline{CD}$ bisects $\angle ACB$	Given
3.	$m\angle ACD = m\angle BCD$	Definition angle bisector
4.	$CX = CX$	Identity
5.	$\frac{AC}{BC} = \frac{CX}{CX}$	Algebra
6.	$\triangle ACX \sim \triangle BCX$	SAS
7.	$\frac{CX}{CX} = \frac{AC}{BC} = 1$	Algebra
8.	$\triangle ACX \cong \triangle BCX$	Definition congruent triangles
9.	$AX = BX$	CPCF
10.	$\triangle AXB$ is isosceles.	Definition isosceles triangle

Class Activity, p. 184

1. If you are tardy, then you are late to class.

3. If you are informed, then you watch the news on TV.

5. If a triangle is equilateral, then it is isosceles.

Class Activity, p. 185

1. If the temperature is above 75°F then John goes swimming; false.

3. If it is not snowing, then Maria plays tennis; false.

5. If Alex is a teenager, then he is twelve years old; false.

7. If Juan owns a convertible, then he owns an automobile; true.

Class Activity, p. 186.

1.

1) Given
2) Theorem 5-4-1
3) Algebra
4) Substitution

Home Activity, p. 187

1. If it is a Siamese, then it is a cat.

3. If he is honest, then he is a judge.

5. If the triangle has two equal angles, then it is isosceles.

7. If Barb is riding her horse, then she is not walking; true.

9. If two lines meet at right angles, then they are perpendicular; true.

11. If the date is January 31, then it is winter in New York; true.

13. If an animal has two legs, then it is a dog; false.

15. If two angles are right angles, then they are congruent; true.

17. $AC = BC$ is given, and m ACD = m BCD by the definition of an angle bisector. $CD = CD$, so $ADC \sim BDC$ by SAS. m ADC = m BDC because corresponding angle are equal, and m ADC + m BDC = 180° because they form a straight angle. By substitution we find both angle equal 90°, so $CD \perp AB$.

Critical Thinking, p. 188

19. 225

Class Activity, p. 190

1. If you are an NBA star, then you wear these shoes.

3. You can be as good as an NBA star just by wearing a certain brand of basketball shoes.

5. If two angles of a triangle are equal in measure, then the sides opposite the equal angles are equal in length. Yes.

Home Activity, p. 192

1. If you are a champion, then the skis were designed for you.

3. Yes

5. If you wear these clothes, then you will be in style.

7. Yes

9. If you attend the summer camp, then you will have a positive self-image, personal confidence, self-reliance, and academic achievement.

11. Yes

13. M is the midpoint, so $DM = CM$. $m\angle C = m\angle D$ = 90°, $DA = BC$, and $DA/CB = 1$. Thus $\triangle MDA \cong \triangle MCB$ by SAS, and ratio = 1. $MA = MB$ by CPCF. Then $m\angle 1 = m\angle 2$ by Theorem 5-4-1.

15. $AM = BM$; $\angle PMA$ is a right angle; so $\angle PMB$ is a right angle; $PM = PM$ and $AM/BM = 1$. $\triangle PMA \cong \triangle PMB$ by SAS, and ratio = 1. $AP = BP$ by CPCF.

17. $AC = AE$; $AB = AD$; m$\angle CAB$ = m $\angle EAD$. $AC/EA = AB/AD = 1$. $\triangle CAB \cong \triangle EAD$ by SAS and ratio = 1. Then $CB = ED$ by CPCF.

Critical Thinking, p. 194

19. $\frac{n(n-3)}{2}$; 252

Chapter 5 Review

1. It is a combination of the symbols for similarity and equality.

3. True   **5.** True

7. b & g; c & e

9. 23

11. Yes; $\triangle ACD \sim \triangle BCD$, AC = BC, so $\frac{AC}{BC} = 1$

Cumulative Review, Chapters 1–5, p. 201

1. line segment $\overline{BA}$ or $\overline{AB}$

3. angle; $\angle ABC$ or $\angle CBA$ or $\angle B$

5. 80°; acute angle

7. $x° = 65°$; $(2x - 15)° = 115°$

9. -4; -1; 2; 6

11.

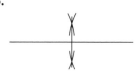

B L C G

13. Answers may vary.

15. $\angle 3$, $\angle 6$; $\angle 4$, $\angle 5$

17. $\angle 1$, $\angle 8$; $\angle 2$, $\angle 7$

19. $\overleftrightarrow{ML}$, $\overleftrightarrow{DH}$

21. $\overleftrightarrow{RS}$, $\overleftrightarrow{DH}$; $\overleftrightarrow{RS}$, $\overleftrightarrow{LH}$

23. M, T, L; R, T, S

25. 140°   **27.** 140°

29. 140°

31. $2\frac{1}{2}$; error: $\frac{1}{4}$ in.

33. rectangular prism

35. b, c, f   **37.** c

39. a, e   **41.** c

43. 28°   **45.** 90°

47. 140°   **49.** 130°

51. 127.5 mi

53. Similar; AA

55. Similar; SSS

57. 14 m **59.** 15

61. True   **63.** a, e; b, f

65. Yes. If a triangle has two equal sides and the included angle is bisected, then the triangle is divided into congruent triangles.

67. 21 Students

69. If you are informed, then you read the newspaper.

71. If the person has been to England, then the person has visited London.

CHAPTER 6

Class Activity, p. 208

1. Answers may vary.

Class Activity, p. 209

1.

CA = CB	Construction
AD = BD	Construction
CD = CD	Identity
$\triangle CDA \cong \triangle CDB$	SSS
m$\angle$ACD = m$\angle$BCD	CPCF
$\overrightarrow{CD}$ bisects $\angle$ACB	Definition of Bisector

3.

5.

Home Activity, p. 210

1. See students' constructions.
3. Draw a point on a line *1*. Label it *M*. From *M*, mark an arc for the length of $\overline{AB}$. Place the compass point on the intersection of this arc. Mark off an arc for the length of $\overline{CD}$. Label the intersection as point *N*.

5.
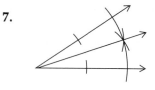

7.

9.
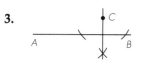

Critical Thinking, p. 210

11. 99

Class Activity, p. 213

1. Answers may vary.
3.
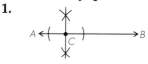

Class Activity, p. 214

1. Answers may vary.
3. See students' constructions.

Home Activity, p. 215

1.

3.
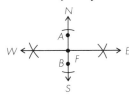

5.
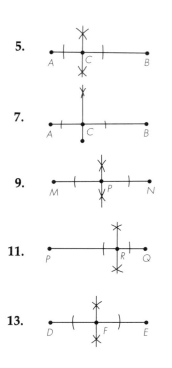

7.

9.

11.

13.

15. Answers may vary.
17.

19. Construct perpendicular lines, then bisect the 90° angle.

Class Activity, p. 220

1. They are parallel; yes.
3. Possible construction:

5. Possible construction:

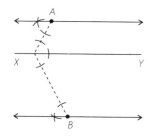

Home Activity, p. 221

1.

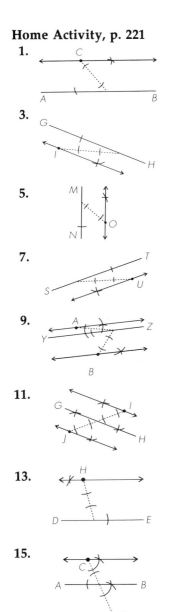

3.

5.

7.

9.

11.

13.

15.

17. Answers may vary.
19. They intersect at one point.

Critical Thinking, p. 222
21. Sara

Class Activity, p. 227
1. See students' constructions.
3.

5. Scalene; acute

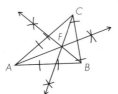

7. No; each segment is drawn from the midpoint of a side to the angle opposite the side.

Home Activity, p. 228
1. See students' constructions.
3.

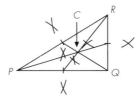

5.

7. They are the same.
9.

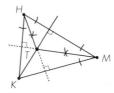

11. On the hypotenuse

Critical Thinking, p. 229
13. Angle bisectors

Class Activity, p. 233
1. Answers may vary.
3.

Home Activity, p. 234
1. Students' constructions
3. Students' constructions
5. Students' constructions
7. Students' constructions
9. Students' constructions
11. Students' constructions
13. Students' constructions

Critical Thinking, p. 234
15. No

Class Activity, p. 238
1. Construct by using ⊥ bisectors of sides.

Class Activity, p. 239
1.

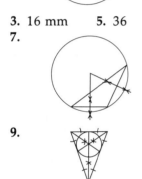

Home Activity, p. 239
1.

3. 16 mm   **5.** 36
7.

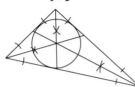

9.

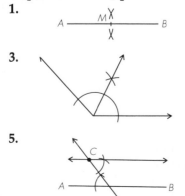

11. They all lie on the median/altitude/angle bisector from the vertex to the base.

Critical Thinking, p. 240
13. Choose a point C, and draw △ABC. Construct the midpoints of $\overline{AC}$ and $\overline{BC}$, labeling the midpoints D an E. Measure $\overline{DE}$. AB = 2(DE).

Chapter 6 Review, p. 244
1.

3.

5.

7.

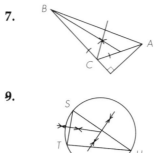

9.

CHAPTER 7

Class Activity, p. 249
1. Yes   **3.** Yes
5. 12.5

Class Activity, p. 251
1. 39   **3.** 22.4
5. 3.9   **7.** 8.9

Home Activity, p. 252
1. Yes   **3.** No
5. 34   **7.** 26.9
9. 17   **11.** 9.2, 36, 49
13. 39.1 ft   **15.** 8.94 ft

Critical Thinking, p. 252
Yes

Class Activity, p. 256
1. 6.6   **3.** 8.2
5. 7.1 m

Home Activity, p. 257
1. 17.0   **3.** 58.0
5. 12.7   **7.** 70.7
9. 18.4   **11.** 70.7
13. 1.0   **15.** 4.2
17. 10.0
19. about 17 ft

Critical Thinking, p. 258
about 35.4 ft

Class Activity, p. 261
1. 31 in.   **3.** 41 yd

Class Activity, p. 262
1. 10.4   **3.** 13.0 cm

Class Activity, p. 263
1. 589 ft   **3.** 95 m

Home Activity, p. 264
1. $x = 17.3$ ft
3. $x = 70.4$ cm; $y = 61.0$ cm
5. $a = 15$ ft
7. Answers may vary.
9. Suppose that $\triangle ABC$ is equilateral. Since an equilateral triangle is also isosceles, $m\angle A = m\angle B$. Similarly, $m\angle B = m\angle C$. Thus, all three angles have the same measure.

Critical Thinking, p. 264
1. 2   3. 8
5. 32   7. 2^n

Class Activity, p. 267
1. $a > b$ 3. $a = b$
5. $a < b$
7. Let $\angle A$ and $\angle B$ be the two non-right angles. Then in $m\angle A + m\angle B + 90° = 180°$. Thus, $m\angle A + m\angle B = 90°$. Since $m\angle A$ and $m\angle B$ are each positive, $m\angle A < 90$ and $m\angle B < 90$. By Theorem 7-4-1, the hypotenuse is the longest side.

Class Activity, p. 269
1. $m\angle A > m\angle B$
3. Insufficient info.
5. $\overline{RT}$, $\overline{ST}$, $\overline{RS}$

Home Activity, p. 269
1. $a > b$
3. Insufficient info.
5. Insufficient info.
7. $m\angle A > m\angle B$
9. $\overline{RQ}$, $\overline{PR}$, $\overline{PQ}$
11. Since $m\angle 2 < \angle m\ 1$, $PM < PO$ by Theorem 7-4-1. But $PN < PM$, and so $PN < PO$. So by Theorem 7-4-2, $m\angle 2 < m\angle 3$.
13. Since $AC > BC$, it follows from Theorem 7-4-2 that $m\angle B > m\angle A$. Thus, $\frac{1}{2} m\angle B > \frac{1}{2} m\angle A$. Therefore, by Theorem 7-4-1, $AD > BD$.

Critical Thinking, p. 270
15. Since $\angle 1$ is an exterior angle of $\triangle ADC$, $m\angle 1 = x + m\angle A$. Thus, $m\angle 1 > x$. Thus, $BC > BD$ by Theorem 7-4-1.

Class Activity, p. 273
1. 5   3. $\sqrt{41}$, or 6.4

Home Activity, p. 273
1. 10   3. 17
5. $AB = \sqrt{45}$; $DC = \sqrt{45}$

Critical Thinking, p. 273
Square both numbers to see which has the larger square. $(\sqrt{10} + \sqrt{17})^2 = 10 + 2\sqrt{170} + 17$ and $(\sqrt{53})^2 = 53$. $10 + 2\sqrt{170} + 17 = 27 + 2\sqrt{170}$. Note that $13^2 = 169$. Therefore, $27 + 2\sqrt{170} > 27 + 2\sqrt{169} = 27 + 2(13) = 53$. Since $\sqrt{10} + \sqrt{17}$ has a larger square than $\sqrt{53}$, $\sqrt{10} + \sqrt{17}$ is the larger number.

Chapter 7 Review
1. No
3. Yes
5. 11.7
7. about 19.8 in.
9. 364
11. x, z, y 13. 8   15. 7

CHAPTER 8

Class Activity, p. 283
1. $\overline{GH}$ and $\overline{KJ}$, $\overline{GK}$ and $\overline{HJ}$; $\angle G$ and $\angle J$, $\angle H$ and $\angle K$
3. GJ and KH
5. No; line segments intersect at point that is not an end point.
7. Yes

Class Activity, p. 285
1. Convex; 360°
3. Convex; 180°

Home Activity, p. 285
1. $\overline{LN}$ and $\overline{MO}$
3. $\angle L$ and $\angle N$; $\angle M$ and $\angle O$
5. $\overline{QR}$ and $\overline{RS}$, $\overline{RS}$ and $\overline{ST}$, $\overline{ST}$ and $\overline{TQ}$, $\overline{TQ}$ and $\overline{QR}$.
7. 2, 3, 4 9. 2, 6
11. 1, 5, 6 13. 360°
15. 360°
17. Answers may vary.

Critical Thinking, p. 286
1080° (6 × 180°)
1440° (8 × 180°)
1800° (10 × 180°)
$(n - 2) × 180°$

Class Activity, p. 288
1. $m\angle B = 70°$; $m\angle C = 110°$; $m\angle D = 110°$
3. $m\angle MPN = 72°$; $m\angle NPO = 43°$; $m\angle PON = 115°$; $m\angle ONP = 22°$
5. Convex. All angles are less than 180°.

Class Activity, p. 290
1. Bases: $\overline{AB}$, $\overline{DC}$; Median: $\overline{EF}$; Legs: $\overline{AD}$, $\overline{BC}$
3. Bases: $\overline{MN}$, $\overline{PO}$; Median: $\overline{QR}$; Legs: $\overline{MP}$, $\overline{NO}$
5. $x = 44$ cm

Home Activity, p. 291
 1. No; triangle
 3. No; concave
 5. No; hexagon
 7. Bases: $\overline{DC}$, $\overline{AB}$; legs: $\overline{DA}$, $\overline{CB}$; median: $\overline{EF}$
 9. m∠1 = 110°; m∠2 = 130°
 11. m∠1 = 125°; m∠2 = 125°; m∠3 = 55°
 13. m∠1 = 59°, m∠2 = 81°, m∠3 = 40°, m∠4 = 19°
 15. x = 17 ft
 17. 16 cm, 24 cm
 19. By AA, △PKO ∼ △MKN

Critical Thinking, p. 292
54 diagonals

5	5
6	9
7	14
8	20
9	27
10	35

Class Activity, p. 294
 1. 15   **3.** 110°
 5. 110°   **7.** 26
 9. 60°   **11.** 60°
 13. 152°; 28°; 152°
 15. 51°; 129°; 51°; 129°

Class Activity, p. 296
 1. 17.5, 14, 14, 30, 10, 17.5
 3. 17, 17, 12, 23
 5. 57°, 123°, 57°
 7. 42°, 138°, 42°, 138°
 9. Yes; opposite angles ≅
 11. No; opposite angles not ≅
 13. Yes; diagonals bisected

Home Activity, p. 297
 1. 100°, 80°, 100°, 80°
 3. 89°, 91°, 89°, 91°
 5. 90°, 90°, 90°
 7. 44, 48, 48, 70, 32, 32
 9. Answers may vary.
 11. 85°
 13. Yes; diagonals bisected

15. No; diagonals not bisected
 17. Yes; opposite sides ≅

Critical Thinking, p. 298
 e. 360°

Class Activity, p. 300
 1. Opposite angles are equal in measure, so m∠A = m∠C. m∠A = 90°, so m∠A + m∠C = 180°. Thus, m∠B + m∠D = 180°, since the sum of the measures of all four angles is 360°. m∠B = m∠D so the measure of each must be 90°.

Class Activity, p. 301
 1. Isosceles; JO = OK
 3. 30   **5.** Not a rectangle
 7. Yes   **9.** No

Home Activity, p. 302
 1. 30   **3.** 30
 5. 15
 7. 3-4-5 triangle
 9. Isosceles
 11. b and d
 13. 360°; m∠2 + m∠3 = 180°; m∠1 + m∠4 = 180°

Critical Thinking, p. 303
Each is about 0.6.

Class Activity, p. 305
 1. Yes; since a rhombus is a parallelogram, the diagonals bisect each other.

Class Activity, p. 307
 1. 5 cm   **3.** 5 cm
 5. 7.07 cm   **7.** 3.54 cm
 9. Parallelogram
 11. Rhombus

Home Activity, p. 308
 1. Parallelogram: X, ___, X, X, ___, X, ___, ___, X, X, X, ___.
   Rectangle: X, ___, X, X, X, X, X, ___, X, X, X, ___.
   Square: X, X, X, X, X, X, X, X, X, X, X, X
   Rhombus: X, X, X, X, ___, X, ___, X, X, X, X, X
   Trapezoid: ___, ___, ___, ___, ___, ___, ___, ___, ___, X, X, ___.
 3. Parallelogram
 5. Rhombus

Chapter 8 Review, p. 312
1. $\overline{SR}$ and $\overline{PQ}$ or $\overline{SP}$ and $\overline{RQ}$
3. 360°
5. Yes; opposite angles are congruent and supplementary.
7. No; both pairs of opposite sides are not congruent.
9. 56°
11. No; can't tell whether *ABCD* is isosceles
13. 32   15. 21
17. 18
19. 45°, 135°, 135°
21. 45°, 45°, 90°, 90°
23. 50 ft

CHAPTER 9

Class Activity, p. 319
1. 96 sq in.   3. 363 m²
5. 12 ft, 4 ft
7. 18 in., 6 in.
9. 33 cm²   11. 208 sq ft

Home Activity, p. 319
1. 32 ft   3. 60 m
5. (Answers may vary.)
7. (Answers may vary.)
9. 16 cm²
11. 35 cm, 21 cm
13. 15 in.
15. 1,492 m²   17. $1,937.50

Critical Thinking, p. 320
19. 300 sq units; 70 units
21. *A* quadruples; *P* doubles

Class Activity, p. 323
1. 800 sq ft   3. about 177 cm²
5. b = 22 m
7. h = 21 cm

Class Activity, p. 324
1. 72 sq in.   3. 42.44 sq ft
5. (Answers may vary.)

Class Activity, p. 325
1. 30 cm   3. 26.6 m
5. 24 cm, 32 cm, 24 cm
7. 45 ft   9. 50 ft

Home Activity, p. 326
1. 96 sq ft   3. 389.7 cm²
5. b = 3.8 m   7. 50 cm²
9. 86.6 cm²   11. 76 m
13. 24   15. Answers may vary.
17. 2,300 sq ft
19. A = 1,350 sq ft; P = 210 ft

Critical Thinking, p. 327
21. 44 ft
23. Possible answers: 10 ft by 12 ft; 8 ft by 14 ft; 9 ft by 13 ft

Class Activity, p. 330
1. 10,000 sq ft
3. 28,210 cm²
5. A = 78 sq in.
7. A = 777 m²
9. A = 182.25 sq in.
11. s = 15 cm 13. 28 in.

Class Activity, p. 332
1. 400 ft   3. 530 yd
5. 360 ft   7. w = 100 cm
9. l = 42 m

Home Activity, p. 333
1. 1,450 cm²   3. 3,500 sq ft
5. 3,120 cm²   7. A = 775 cm²
9. b = 54 m
11. $d_2 = 18$m   13. Answers may vary.
15. A = 54 m²   17. 1,050 cm²
19. 210 sq ft
21. (Answers may vary.)
23. 154 m   25. 74 ft
27. 396 sq ft; 640 sq ft
29. $93.75

Class Activity, p. 338
1. 120 sq ft   3. 90 m²
5. 510 sq in.
7. $500 + 50\sqrt{3}$ sq in.
9. 768 sq ft   11. 43 in.
13. 8 in.
15. $(b_1 + b_2)h$

Class Activity, p. 340
1. P = 92 ft; A = 384 sq ft
3. P = 207.2 cm, A = 1410 cm²
5. P = 100 cm; A = 270 cm²

Home Activity, p. 341
1. $P = 162.6$ ft; $A = 1327.2$ sq ft
3. $P = 310$ ft; $A = 1690$ sq ft
5. $A = 3948.5$ cm²; $h = 53$ cm
7. $A = 874$ m² **9.** $A = 864$ m²
11. 138.75 m²
13. (Answers may vary.)
15. (Answers may vary.)
17. 121   **19.** 785.5 ft
21. 25,745.75 sq ft

Critical Thinking, p. 342
23. Steps
1. $A(ABCD) = A(\text{I}) + A(\text{II}) + A(\text{III})$
2. $A(ABCD) = \frac{1}{2}ah + \frac{1}{2}ch + bh$
3. $A(ABCD) = \frac{1}{2}ah + \frac{1}{2}ch + \frac{1}{2}bh + \frac{1}{2}bh$
$= \frac{1}{2}h\,(a + c + b + b)$
4. $a + b + c = AB$ and $b = CD$
5. $A(ABCD) = \frac{1}{2}h\,(AB + CD)$
Reasons
1. Addition
2. Definition of area for triangle and rectangles
3. Algebra
4. Identity
5. Substitution

Class Activity, p. 344
1. $A = 485.5$ m²; $P = 105$ m
3. $A = 180$ sq ft; $P = 60$ ft

Class Activity, p. 347
1. $\frac{P_1}{P_2} = \frac{1}{3}$; $\frac{A_1}{A_2} = \frac{1}{9}$
3. $\frac{P_1}{P_2} = \frac{1}{6}$; $\frac{A_1}{A_2} = \frac{1}{36}$

5. 1708 sq ft

Home Activity, p. 347
1. $A = 300$ cm²; $P = 98$ cm
3. $A = 117$ sq in., $P = 57$ in.
5. (Answers may vary.)
7. (Answers may vary.)
9. $\overline{ZY} = \frac{1}{2}\,\overline{QR}$, $\overline{ZX} = \frac{1}{2}\,\overline{SR}$, $\overline{YX} = \frac{1}{2}\,\overline{SQ}$; $\overline{ZY}/\overline{QR} = \overline{ZX}/\overline{SR} = \overline{YX}/\overline{SQ} = \frac{1}{2}$, so $\frac{A\,(\triangle XYZ)}{A\,(\triangle QRS)} = (\frac{1}{2})^2 = \frac{1}{4}$.

Critical Thinking, p. 348
10. Square;   **11.** triangle;   **12.** It is a square number.

Chapter 9 Review, p. 352
1. $P = 72$ m; $A = 216$ m²
3. $P = 102$ m; $A = 584.4$ cm²
5. $P = 87.2$ in.; $A = 250$ sq in.
7. $P = 100$ ft; $A = 600$ sq ft
9. $A = 108$ sq ft; $P = 36 + 12\sqrt{2}$ ft
11. $A = 125$ cm²; $h = 5$ cm, $b = 25$ cm
13. 110 sq in.   **15.** 456 cm²
17. $\frac{1}{3} \times \frac{1}{9}$ **19.** $\frac{1}{6} \times \frac{1}{36}$
21. 2,400 tiles

CHAPTER 10

Class Activity, p. 356
Shapes may vary.

Class Activity, p. 357
1. Triangle, convex, nonregular
3. Quadrilateral, concave, nonregular
5. Square/quadrilateral, convex, regular
7. Heptagon, convex, regular

Class Activity, p. 359
1. 720°   **3.** 3240°
5. 128.6°; 51.4°   **7.** 140°; 40°
9. 15

Class Activity, p. 359
1. 40 cm   **3.** 23 in.
5. 195 m

Home Activity, p. 360
1. Triangle, convex, regular
3. Decagon, concave, nonregular
5. 360°   **7.** 180°
9. 1080°
11. Answers may vary.
13. 90°   **15.** 12.9°
17. $n = 10$ **19.** $n = 25$
21. 180 sq ft

Critical Thinking, p. 360
23. 1,296,000-gon

Class Activity, p. 363
1. $A = 120$ cm²   **3.** $A = 220.8$ m²
5. $3750\sqrt{3}$ sq ft   **7.** 261 m²
9. Both 60°; angle was bisected.
11. Yes; AA or SAS and $\frac{\text{base}}{\text{base}} = 1$

Home Activity, p. 364
1. $64\sqrt{3}$ cm²
3. 345.96 sq in.
5. 110 sq in.
7.

Number of Sides	Area
6	2.61 m
7	3.64 m
8	4.84 m
9	6.17 m
10	7.70 m

9. 6 m, 7 m, 8 m, 9 m, 10 m
11. about 1,459,000 sq ft; about 162,000 sq yd
13. about 30
15. about 1600 sq ft
17. Hexagon; no; the sides are not equal in length.
19. $360° \div 9 = 40°$
21. 72°

Class Activity, p. 369
1. 16.2 cm   3. 7.66 cm
5. 7.5 ft   7. 64.2 mm
9. 3.1415927
11. 12.57 ft   13. 314.16 yd

Class Activity, p. 371
1. $A = 176.71$ sq ft
3. $A = 19.63$ sq yd
5. 12.56 sq ft   7. 50.24 m²
9. 7850 m²   11. 1.77 km²
13. $d = 120$ cm   15. $d = 63.24$ m

Home Activity, p. 372
1. 16 in., 50.24 in., 200.96 sq in.
3. 42 yd, 21 yd, 1384.74 sq yd
5. 20 in., 62.8 in., 314 sq in.
7. 4π
9. $25\sqrt{3}$ sq ft or 43.3 sq ft
11. 16.9 sq ft

Critical Thinking, p. 372
13. 18.84 cm   15. 0.6
17. 28.26 cm²
19. 0.36; ratio of radii squared

Class Activity, p. 376
1. Radii: $\overline{OA}$, $\overline{OB}$, $\overline{OC}$, $\overline{OH}$, $\overline{OG}$
Diameters: $\overline{GH}$ and $\overline{BC}$
Chords: $\overline{AF}$, $\overline{GH}$, $\overline{BC}$
Secant: $\overleftrightarrow{DE}$
Tangent: $\overleftrightarrow{BJ}$
Central angles: $\angle COH$, $\angle COA$, $\angle AOG$, $\angle GOB$, $\angle BOH$
Inscribed angles: none

3. 65°   5. 48°

Class Activity, p. 378
1. It is tangent to the circle; no; yes.
3. 80°   5. 80°
7. 90°   9. 45°
11. 135°

Home Activity, p. 378
1. $\overleftrightarrow{QB}$, $\overline{OA}$, $\overline{OC}$ 3. O
5. $\overline{PT}$ 7. $\overline{TC}$, $\overline{TA}$, $\overline{AC}$
9. 70°   11. 20°
13. 20°   15. 50°
17. 110°

Class Activity, p. 382
1. $x = 9$ 3. $x = 9$
5. The products are equal.

Home Activity, p. 383
1. Tangent   3. Secant
5. Center   7. Radius
9. $CX = 10$ or 14
11. $AC = 16$ 13. 20 cm
15. 314 cm²   17. 8
19. 40°   21. 140°

Class Activity, p. 386
1. Rectangle   3. 13.76 sq ft

Class Activity, p. 387
1. 942 sq ft
3. $A = 380.16$ sq in.

Home Activity, p. 388
1. $A = 113.04$ cm²
3. $A = 123.84$ sq in.
5. $\frac{4}{9}$ 7. $x = 60°$
9. $x = 6.9$ 11. 614 ft

Critical Thinking, p. 389
13. Draw a pillar on its side, 3 lines

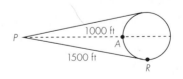

$$PA(PA + \text{diameter of pillar}) = PR^2$$

Chapter 10 Review
1. 120°   3. 162.9°
5. 12
7. $C = 75.36$ m; $A = 452.16$ m^2
9. $C = 31.4$ ft; $A = 78.5$ sq ft
11. 776 revolutions
13. 60°   15. 40°
17. 120°   19. 60°
21. $x = 12$ 23. $x = 18.5$

CHAPTER 11

Class Activity, p. 404
1. Edges: $\overline{PN}$, $\overline{NO}$, $\overline{OP}$, $\overline{PS}$, $\overline{NQ}$, $\overline{OR}$, $\overline{SQ}$, $\overline{QR}$, $\overline{RS}$
   Bases: $\triangle PNO$, $\triangle SQR$
   Lateral Faces: $\square NORQ$, $\square OPSR$, $\square PNQS$
3. Edges: $\overline{PQ}$, $\overline{QR}$, $\overline{RS}$, $\overline{SO}$, $\overline{OP}$, $\overline{PK}$, $\overline{QL}$, $\overline{RM}$, $\overline{SN}$, $\overline{OJ}$, $\overline{KL}$, $\overline{LM}$, $\overline{MN}$, $\overline{NJ}$, $\overline{JK}$
   Bases: Pentagon $OPQRS$,
   Pentagon $JKLMN$
   Lateral Faces: $\square OPJK$, $\square PQLK$, $\square QRML$, $\square RSNM$, $\square SOJN$

Class Activity, p. 406
1. 9.75   3. 162.9
5. $4.84 \pi \approx 15.2$ in^2
7. $22 \pi \approx 69.1$ in^2

Home Activity, p. 406
1. Lateral face $MORP$ is a rectangle, as are the other faces. Right; Edges: $\overline{MN}$, $\overline{NO}$, $\overline{OM}$, $\overline{MP}$, $\overline{NQ}$, $\overline{OR}$, $\overline{PQ}$, $\overline{QR}$, $\overline{RP}$; Bases: $\triangle PQR$, $\triangle MNO$; Lateral faces: $\square MORP$, $\square ONQR$, $\square NMPQ$.

3. No right angle is shown. Oblique; Edges: $\overline{WX}$, $\overline{XY}$, $\overline{YZ}$, $\overline{ZW}$, $\overline{WP}$, $\overline{XQ}$, $\overline{YR}$, $\overline{ZS}$, $\overline{PQ}$, $\overline{QR}$, $\overline{RS}$, $\overline{SP}$; Bases: Quadrilaterals $WXYZ$ and $PQRS$; Lateral faces: Parallelograms $PWZS$, $SZYR$, $RYXQ$, $QXWP$.
5. Lat: 220; Tot: 276
7. Lat: 600; Tot: 660
9. Lat: 4,368 cm^2;
   Tot: 5,688 cm^2
11. Lat: about 305 in^2
    Tot: about 316 in^2

Critical Thinking, p. 406
Same: 16
The cylinder on the left: $48 \pi > 24 \pi$

Class Activity, p. 410
1. 1,071 in.^{3}

Class Activity, p. 411
1. about 396 yd^3

Home Activity, p. 412
1. 36   3. 1,520 mm^3
5. $192 \pi \approx 603$ mm^3
7. $500 \pi \approx 1570$ m^3
9. about 442,740 gal
11. 30 ft

Critical Thinking, p. 412
The second has the greater volume.

Class Activity, p. 416
1. 211.9 in.^{2}

Class Activity, p. 418
1. $176 \pi \approx 553$ cm^2
3. $278.4 \pi \approx 874$

Home Activity, p. 418
1. 480 in.^{2}
3. 1,559 cm^2
5. $96 \pi \approx 301$ ft^2
7. about 3,965 ft^2

Critical Thinking, p. 418

$(\frac{60}{13})^2 \pi$, or about 67

Class Activity, p. 422
1. 220 cm³

Class Activity, p. 424
1. 1,536 $\pi \approx$ 4,823 in.³

Home Activity, p. 424
1. 391 ft³   3. 37.3 in.³
5. 48.3 km³
7. 2,592,100 m³
9. about 2.1 m

Class Activity, p. 426
1. 2,100 ft³

Class Activity, p. 428
1. about 13,572 cm³

Home Activity, p. 429
1. 6.93 yd³
3. 14.4 π, or about 45.2 in.²
5. 5.25 in.²
7. about 232 π, or 728 ornaments
9. 718.8 π, or about 2,257 ft²

Critical Thinking, p. 430

Each domino covers a black square and a red square. Therefore, the total number of covered black squares has to equal the total number of covered red squares. But this cannot happen since the two removed squares are of the same color. The number of black squares in the modified board is no longer equal to the total number of red squares. As a result, the squares cannot be covered by dominoes.

Class Activity, p. 435
1. $\frac{2,048}{3} \pi$, or 2,144 in.³; 804 in.²

Home Activity, p. 435
1. 4,188 yd³; 1,257 yd²
3. 7,240 in.³; 1,810 in.²
5. 11.5 in.³; 24.6 in.
7. 8.38 in.³; 20.0 in.²
9. about 23 people/mi²

Critical Thinking, p. 435

11. $\dfrac{\frac{1}{3}\pi r^3 + \pi r^3}{2} = \frac{2}{3}\pi r^3$
$= \frac{1}{2}(\frac{4}{3}\pi r^3)$

Chapter 11 Review, p. 440
1. 68 m²; 94 m²
3. 36 ft²; 54 ft²
5. 1700 cm³
7. 1,100 m²; 2,310 m³
9. 7235 in³; 1808 in.²
11. cylindrical; 1,843 ft³

CHAPTER 12

Class Activity, p. 446
1. Answers may vary.
3. This is true by Theorem 6-6-2. If a point is on the perpendicular bisector, then it is equidistant from the endpoints of the segment.

Home Activity, p. 448
1. ⊥ bisector
3. Two parallel lines
5. A line parallel to each and equidistant from each
7. Two parallel lines and two semicircles
9. Two rays from (0,0)   11. Circle

Class Activity, p. 451
1. Translation
3. Reflection
5. Answers for the procedure may vary.

Class Activity, p. 452
1. Reflection
3. Rotation
5. Rotation

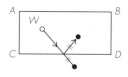

Home Activity, p. 453
1. Translation
3. Reflection
5. Reflection
7. Vertical: A, H, I, M, O, T, U, V, W, X, Y;
   Horizontal: B, C, D, E, H, I, O, X
9.

11. Hexagon; 5
13. Translation 2 units up

Critical Thinking, p. 454
15. Not a mirror image
16. Translation
17. Not a mirror image
18. Rotation about a point

Class Activity, p. 457
1. 4 **3.** 1 **5.** 0
7. 1 **9.** 0

Class Activity, p. 458
1–7. Answers may vary.

Home Activity, p. 459
1. Equilateral triangle, Square, Rectangle, Trapezoid, Parallelogram, Circle
3. Triangle: 120°; Square: 90°; Circle: angle approaches 0°
5. Rotation
7. Rotation
9. All
11. Square 90°; circle: approaches 0°; triangle: 120°; octagon: 45°
13. Pollen and virus

Critical Thinking, p. 460
15. 2

Class Activity, p. 463
5. Parabolic curve made of straight lines

Class Activity, p. 464
1. Draw reflection of points over line, space 3 lines

3. Measure AP and PB and AP and PB' in Ex. 1.
5. $AQ + QB$

7. A straight line is the shortest distance between two points.

Class Activity, p. 465
1. 2 lines parallel to and 4 mi from each side of the highway
3. Circle with radius of 3 mi, center on school

Home Activity, p. 465
1. Circumference of circle
3. Area of 90° sector
5. Circle
7. 2 concentric circles
9. Translation **11.** Rotation
13. Japan, Peru; 180°
15. Rotation or reflection over either axis
17. 2 reflections equal to 1 90° rotation
19. Reflection over $y = x$

Critical Thinking, p. 466
21. 9
23. Yes; 1 parallel to ends, infinitely many perpendicular to end

Class Activity, p. 469
1. Statement is false. Converse: If a triangle is right, then it is equilateral; false. Inverse: If a triangle is not equilateral, then it is not right; false. Contrapositive: If a triangle is not right, then it is not equilateral; false.
3. Statement is true. Converse: If a triangle is isosceles, then it is equilateral; false. Inverse: If a triangle is not equilateral, then it is not isosceles; false. Contrapositive: If a triangle is not isosceles, then it is not equilateral; true.
5. Statement is false. Converse: If a figure is a square, then it has four equal sides; true. Inverse: If a figure does not have four equal sides, then it is not a square; true. Contrapositive: If a figure is not a square, then it does not have four equal sides; false.

Home Activity, p. 470

1. Statement is false. Converse: If it is night, then the sun is not shining; true. Inverse: If the sun is shining, then it is not night; true. Contrapositive: If it is not night, then the sun is shining; false.

3. Statement is true. Converse: If the measure of two angles are equal, then they are right angles; false. Inverse: If two angles are not right angles, then their measures are not equal; false. Contrapositive: If the measures of two angles are not equal, then they are not right angles; true.

5. Statement is true. Converse: If I don't pitch at the game, then I stay at home; false. Inverse: If I do not stay at home, then I pitch at the game; false. Contrapositive: If I pitch at the game, then I do not stay at home; true.

7. Statement is true. Converse: If two lines are coplanar, then they are parallel; false. Inverse: If two lines are not parallel, then they are not coplanar; false. Contrapositive: If two lines are not coplanar, then they are not parallel; true.

9. Statement is false. Converse: If a quadrilateral is a rectangle, then it has two pairs of parallel sides; true. Inverse: If a quadrilateral does not have two pairs of parallel sides, then it is not a rectangle; true. Contrapositive: If a quadrilateral is not a rectangle, then it does not have two pairs of parallel sides; false.

11. See answers for Exercises 1–10; statement and contrapositive, converse and inverse.

Critical Thinking, p. 470

Friday

Chapter 12 Review, p. 474

1. Perpendicular bisector of chord, or diameter
3. 2 concentric circles with radii of 1 cm and 4 cm
5. Reflection and translation   7. No
9. Yes   11. Yes

13.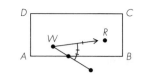

CHAPTER 13

Class Activity, p. 479

1. Point
   (3, 3)
   $(2\frac{1}{2}, 2)$
   (1, −1)
   (0, −3)
   (−2, −7)
   (−4, −11)

3. Point
   (3, 7)
   $(2\frac{1}{2}, 6)$
   (1, 3)
   (0, 1)
   (−2, −3)
   (−4, −7)

5. Point
   (6, −7)
   $(1, −4\frac{1}{2})$
   (0, −4)
   $(−1, −3\frac{1}{2})$
   $(−3, −2\frac{1}{2})$
   (−4, −2)

Class Activity, p. 481

1. $y = \frac{1}{2}x + (-\frac{1}{2})$
3. $x = \frac{5}{2}$
5. $y = 3x + (-2)$
7. $x = 7$ 9. $x = -5$

Home Activity, p. 482

7. Nonvertical   **9.** Vertical

11. $x = 4$

13. $y = -\frac{3}{2}x + \frac{1}{2}$

15. $y = -x + 11$

Critical Thinking, p. 482

Step 5 involves dividing by $(x - y)$. But $x - y = 0$, since $x = y$. Division by zero is not allowed.

Class Activity, p. 485

1. 7   **3.** no slope

5. 1, up   **7.** $-\frac{3}{10}$, down

9. $-\frac{19}{4}$ **11.** $-\frac{9}{5}$

Class Activity, p. 487

1. $-\frac{1}{2}$, 7

3. -1, 4

5. 0, -10

7–9.

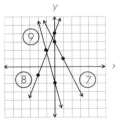

Home Activity, p. 488

1. -3, down   **3.** 0, level

5. -3, $-\frac{1}{2}$

7. 7, -18

9. 0, $3\frac{2}{5}$

11. 3, -8

13–15.

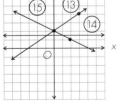

17. $y = -6x + 2\frac{1}{2}$

Critical Thinking, p. 488

19. $\frac{12}{19}$

Class Activity, p. 491

1. No   **3.** Yes

5. No

Class Activity, p. 492

1. No   **3.** Yes

5. $\overline{WZ}$ has slope 0 and $\overline{YZ}$ has slope $\frac{4}{3}$. The product of these slopes is not -1, so these sides are not perpendicular.

7. $\triangle PQR$ is isosceles because $PR = QR = \sqrt{34}$. The point $S(-1, 1)$ is the midpoint of $\overline{PQ}$, because $PS = SQ = \sqrt{2}$ and $PQ = 2\sqrt{2}$. The line $y = -x$ passes through $(-1, 1)$ and $(3, -3)$. Its slope is -1. The slope of $\overline{PQ}$ is 1. The product of the slopes is -1. So $y = -x$ is perpendicular to $\overline{PQ}$ at its midpoint.

Home Activity, p. 493

1. Yes   **3.** Yes

5. No   **7.** Yes

9. $\frac{4}{7}$; $\frac{4}{7}$

11. $ABCD$ is a parallelogram. Opposite sides have equal slopes and so are parallel.

13. No   **15.** Yes

17. Yes, because the product of the slopes of $\overline{AB}$ and $\overline{AC}$ is -1.

19. $KL = KM = 5$, so KL and KM are the legs of an isosceles triangle. Slope of $\overline{KL} = \frac{3}{4}$ and slope of $\overline{KM} = -\frac{4}{3}$, so $\overline{KL} \perp \overline{KM}$ and $\triangle KLM$ is a right triangle.

21. Parallel   **23.** Parallel

25. $\overline{PQ}$ and $\overline{RS}$ each have slopes of $\frac{1}{3}$; $\overline{PS}$ and $\overline{QR}$ have slopes of -3. Since $-3 \times \frac{1}{3} = -1$, all the angles of the quadrilateral are right angles.

Critical Thinking, p. 494
27. 3   **29.** 18
31. 40

Class Activity, p. 497
1. $x^2 + y^2 = 16$
3. Circle 1
5. $x^2 + y^2 = 49$
7. $x^2 + y^2 = 10,000$
9. $(-6, 0)$ $(0, -6)$
   $(-5, 3.32)$ $(1, 5.92)$

Class Activity, p. 498
1. $x^2 + y^2 = 36$
3. $x^2 + y^2 = \frac{9}{16}$
5. $x^2 + y^2 = 26$

Home Activity, p. 499
1. $x^2 + y^2 = 25$
3. $-\frac{3}{4}$
5. The line $y = \frac{4}{3}x$ is tangent to the circle. From Exercise 4, it is known that this line is perpendicular to the line containing the radius to point $(-4, 3)$. The line passes through $(-4, 3)$ because the coordinates of $(-4, 3)$ satisfy the equation $y = \frac{4}{3}x + \frac{25}{3}$.
7. No   **9.** Yes
11. Yes
13. $r = 7$ **15.** $r \approx 4.5$
17. $r = 3\frac{1}{3}$
19. No   **21.** Yes
23. $x^2 + y^2 = 169$
25. $\frac{12}{5}$ **27.** Yes

Critical Thinking, p. 500
29. 3   **31.** $2^n - 1$

Class Activity, p. 502
1. $y = \frac{1}{8}x^2$
3. $y = \frac{1}{16}x^2$
5. $y = 2x^2$

Class Activity, p. 504
1.

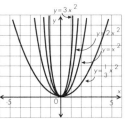

3. Up
5. They will be like the parabolas in Exercise 1 but will open down.
7. Yes   **9.** No
11. No

Home Activity, p. 505
1. $y = \frac{1}{20}x^2$
3. $y = \frac{1}{3}x^2$

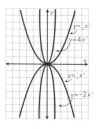

9. Down   **11.** Up
13. Down   **15.** Down
17. $y = 25x^2$ and $y = -25x^2$
19. $(0, 0)$

Critical Thinking, p. 506

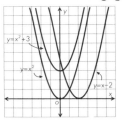

Adding 3 to x^2 moves the graph of $y = x^2$ straight up 3 units. Replacing x with $x - 2$ moves the graph of $y = x^2$ to the right 2 units.

Class Activity, p. 507
1. $MP = \sqrt{(a - [\frac{a+c}{2}])^2 +}$
   $(b - [\frac{b+d}{2}])^2$
   $MQ = \sqrt{([\frac{a+c}{2}] - c)^2 +}$
   $([\frac{b+d}{2}] - d)^2$
   $PQ = \sqrt{(a - c)^2 + (b - d)^2}$
3. $MP + MQ$
   $= 2\sqrt{(\frac{a-c}{2})^2 + (\frac{b-d}{2})^2}$
   $= \sqrt{2^2 (\frac{a-c}{2})^2 + 2^2 (\frac{b-d}{2})^2}$
   $= \sqrt{(2[\frac{a-c}{2}])^2 + (2[\frac{b-d}{2}])^2}$
   $= \sqrt{(a - c)^2 + (b - d)^2} = PQ$

Class Activity, p. 509

1. (17.75, 29.25)

3. (−4.06, 5.63)

5. The figure on the left

7. Students should use the distance formula to show that both $\overline{DF}$ and $\overline{EG}$ have a length of $\sqrt{(b+a)^2 + c^2}$.

Home Activity, p. 510

1. (3, 2)   **3.** (1, 4)

5. $(\frac{3}{2}, -\frac{1}{2})$

7. Slope of $\overline{AC} = \frac{b-0}{b-0} = \frac{b}{b} = 1$.
Slope of $\overline{BD} = \frac{a-0}{0-2}$
$= \frac{a}{-a} = -1$, so $\overline{AC} \perp \overline{BD}$.

9. Use the midpoint theorem to find the coordinates of $P, Q, R,$ and S. Then show that $\overline{PQ}$ and $\overline{SR}$ have the same slope and that $\overline{SP}$ and $\overline{RQ}$ have the same slope.

Chapter 13 Review, p. 514

1. Vertical   **3.** Nonvertical

5. $x = \frac{7}{3}$

7. $x^2 + y^2 = 81$

9. Down   **11.** Up

13. Perpendicular

15. $-\frac{1}{9}$

17.

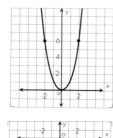

19.

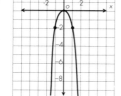

21. Circle; (0, 0), $r = 3\sqrt{5}$

23. Parabola; down

CHAPTER 14

Class Activity, p. 521

1. $\triangle DEF \sim \triangle OMN$
$\triangle DEF \sim \triangle XYZ$
$\triangle ONM \sim \triangle XYZ$

3. 0.725

Class Activity, p. 522

1. $\cos A = 0.8020$
$\cos B = 0.5974$

Home Activity, p. 522

1. $\sin A = 0.3846$ $\sin B = 0.9231$
$\cos A = 0.9231$ $\cos B = 0.3846$

3. $\sin A = 0.7071$ $\sin B = 0.7071$
$\cos A = 0.7071$ $\cos B = 0.7071$

5. 0.5736   **7.** 0.1736

9. 0.4067   **11.** 0.6561

13. 0.8829   **15.** 0.2250

Critical Thinking, p. 522

$28 + 7\sqrt{2}$

Class Activity, p. 525

1. 0.5   **3.** about 28°

5. about 30°

Class Activity, p. 527

1. $\sin A$ or $\cos B$

3. g = 14.2 m

Home Activity, p. 528

1. 0.75   **3.** 0.9325

5. 67°   **7.** 18.5 cm

9. 34.4 mi   **11.** 208 m

Class Activity, p. 531

1. −332°   **3.** −180°

Class Activity, p. 533

1. −0.5172   **3.** −0.7071

Home Activity, p. 534

1. −290°   **3.** −120°

5. 0.80   **7.** 0.00

9. −0.87   **11.** −0.50

13. $r = 10$, $\cos x = 0.60$

15. $r = 10$, $\cos x = -0.60$

17.

$\cos x°$	$\cos x°$	$\cos x°$
1.00	−0.64	−0.34
0.97	−0.77	−0.17
0.87	−0.94	0.00
0.77	−0.98	0.17
0.64	−1.00	0.34
0.50	−0.94	0.71
0.34	−0.82	0.87
0.17	−0.64	0.98
0.00	−0.50	1.00
−0.42		

Class Activity, p. 538

1. 28.3, or about 28 mi

Class Activity, p. 540
1. 16.7

Home Activity, p. 540
1. 13.8   3. 13.6
5. 5.7   7. 39°
9. 40.297, or about 40.3 mi

Class Activity, p. 543
1. about 355 ft

Class Activity, p. 545
1. 8.6 m

Home Activity, p. 546
1. 87.2 ft
3. Station A; 6.3 mi
5. 25.8   7. 131°
9. 82°

Critical Thinking, p. 546
20

Class Activity, p. 551
Methods will vary. Possible answer:
1. If $x = 45°$, then $\tan 45° = 1$, $\sin 45° = 0.7071$. "$1 = 0.7071 + 1$" is false.

Home Activity, p. 551
Methods will vary. Possible answers:
1. If $x = 0$, $\sin 0° + \cos 0° = 0 + 1 \neq 0$.
3. If $x = 0$, $\tan 0° \times \sin 0° = 0 \times 0 = 0$; $\cos 0° = 1$; $0 \neq 1$.
5. If $x = 90$, $\sin 90° = 1 \neq 90$.
7. $\dfrac{(\sin^2 x + \cos^2 x) - \cos^2 x}{\cos^2 x}$
   $\dfrac{\sin^2 x + 0}{\cos^2 x}$
9. $\dfrac{\frac{\sin^2 x + \cos^2 x}{\cos x}}{\frac{1}{\sin x}} = \dfrac{\frac{1}{\cos x}}{\frac{\tan x}{\sin x}} \times \dfrac{\sin x}{\sin x} = \dfrac{\sin x}{\cos x} \times \dfrac{1}{\sin x} = \tan x \times$

Critical Thinking, p. 551
$\cos 2x = \left(\dfrac{\cos^2}{x} = \dfrac{\sin^2}{x}\right) \times 1$
$= \left(\dfrac{\cos^2}{x} - \dfrac{\sin^2}{x}\right) \times$
$\left(\dfrac{\cos^2}{x} + \dfrac{\sin^2}{x}\right)$
$= \dfrac{\cos^4}{x} - \dfrac{\sin^4}{x}$

Chapter 14 Review, p. 556
1. 0.4706; 0.8824; 0.5333
3. $\sin x = \dfrac{3}{9} = 0.600$
   $\cos x = -\dfrac{4}{5} = -0.8000$
5. 9.8 cm   7. 199.2 m
9. If $x = 90°$, then $\sin x = 1$ and $\cos x = 0$. So, $\sin^2 x + 1 = 2$ and $\cos x = 0$. Hence for $x = 90°$, $\sin^2 x + 1 \neq \cos x$.

GLOSSARY

A

Acute Angle: If the number assigned to an angle is between 0° and 90°, then the angle is called an acute angle. [p. 10]

Acute Triangle: An acute triangle is a triangle with the measures of all three angles less than 90°. [p. 100]

Alternate Exterior Angles: Alternate exterior angles are two angles formed by two parallel lines and a transversal such that they have different vertices, are outside of the parallel lines, and are on opposite sides of the transversal. [p. 66]

Alternate Interior Angles: Alternate interior angles are two angles formed by two parallel lines and a transversal such that they have different vertices, are between the parallel lines, and are on opposite sides of the transversal. [p. 57]

Altitude: An altitude of a triangle is a perpendicular line segment from a vertex of the triangle to the opposite side or to the line determined by the opposite side. In △ ABC, AD is the altitude to BC. [p. 227]

Angle: An angle is a figure formed by two rays that have a common endpoint. [p. 5]

Angle Bisector: An angle bisector of a given angle is a ray with the same vertex that separates the given angle into two angles of equal measure. [p. 15]

Angle Bisector of a Triangle: An angle bisector of a triangle is a line segment from a vertex of the triangle to the point where the angle bisector of that angle intersects the opposite side. [p. 226]

Arc: The arc of the circle has the same measure as the central angle that intersects the arc. The symbol for arc is $\overset{\frown}{AB}$ or $\overset{\frown}{AOB}$. [p. 347]

Area: The area of a plane figure is the number of square units contained in the interior. [p. 318]

B

Betweenness: On a number line, point C is between points A and B if the coordinates of A, B, and C (a, b, and c respectively) meet the condition that $a < c < b$, or $a > c > b$. [p. 16]

C

Central Angle: A central angle is an angle whose vertex is at the center of a circle. [p. 374]

Centroid: The point where the medians of a triangle intersect is the centroid or the center of gravity of the triangle. [p. 225]

Chord: A chord of a circle is a line segment whose endpoints are two points on the circle. [p. 380]

Circle: A circle is the set of all points on a plane equidistant from a point called the center. [p. 237]

Collinear Points: Collinear points are points on the same line. [p. 42]

Complementary: Two angles are complementary if their measures add up to 90°. [p. 21]

Concurrent Lines: Concurrent lines are lines which intersect at the same point. [p. 224]

Cone: A cone is the figure formed by a region with a closed, curved boundary and all the line segments joining points on the boundary to a point not in the plane of the region. This point is called the vertex of the cone. [p. 417]

Congruent: If two triangles are similar and the ratio of corresponding sides is one, then the triangles are congruent. [p. 172]

Converse: The converse of an "If A, then B" statement is "If B, then A." [p. 184]

Convex Quadrilateral: A convex quadrilateral is a quadrilateral with the measure of each interior angle less than 180°. [p. 284]

Coplanar Points: Coplanar points are points on the same plane. [p. 42]

Corresponding: Corresponding parts of congruent figures are equal. This may be abbreviated as CPCF. [p. 178]

Corresponding Angles: Corresponding angles are angles that are in the same relative position with respect to the two lines cut by a transversal and the transversal. [p. 49]

Cosine: The cosine of an acute angle of a right triangle is the ratio of the length of the leg adjacent to the angle to the length of the hypotenuse. In symbols, this is written as:

$$\cos A = \frac{\text{length of leg adjacent to angle A}}{\text{length of hypotenuse}}$$ [p. 521]

Cylinder: A cylinder is a three-dimensional figure consisting of two congruent curved regions in parallel planes and the line segments joining corresponding points on the curves that determine the regions. Segments joining corresponding points of the curves are parallel. [p. 404]

D

Diagonal: A diagonal is a line segment determined by two nonadjacent vertices. [p. 283]

E

Equiangular Triangle: An equiangular triangle is a triangle with the measure of each angle equal to 60°. [p. 100]

Equilateral Triangle: An equilateral triangle is a triangle with all three sides having the same measure or length. [p. 94]

Exterior Angle: An exterior angle of a triangle is the angle less than 180° in measure, formed by extending one side of the triangle. [p. 106]

G

Geometric Locus: A geometric locus is the set of all points that satisfy the given conditions. [p. 444]

H

Hypotenuse: The hypotenuse of a right triangle is the side opposite the right angle. [p. 248]

I

Identity: An equation is an identity if it is true for all values of the variables. [p. 547]

Inscribed Angle: An inscribed angle is an angle whose vertex is on the circle and whose rays intersect the circle. [p. 374]

Intersection: The intersection of n sets is the set consisting of the members common to the n sets. [p. 168]

Isosceles Trapezoid: An isosceles trapezoid is a trapezoid with two non-parallel equal sides. [p. 287]

Isosceles Triangle: An isosceles triangle is a triangle with two sides equal in measure or length. [p. 95]

L

Length of a Horizontal Line Segment: The length of a horizontal line segment with endpoints (x_1, y) and (x_2, y) is $|x_2 - x_1|$. For a vertical line segment with end points (x, y_1) and (x, y_2), the length $|y_2 - y_1|$. [p. 272]

Length of a Tangent: The length of a tangent is the length of a segment from a point outside a circle to the point of intersection of the tangent with the circle. [p. 381]

Line: A line is the set of all ordered pairs (points) in the coordinate plane that satisfy the equation of the form $y = mx + b$ or of the form $x = a$, where m, b, and a are real numbers. [p. 481]

Line Segment: A line segment is a subset of a line, consisting of two points A and B and all points between A and B. [p. 4]

Line Symmetry: Line symmetry is a reflection transformation that maps the figure onto itself. [p. 456]

M

Median: The median of a triangle is a line segment joining the vertex of an angle and the midpoint of the opposite side. [p. 224]

Median: The median of a trapezoid is the line segment joining the midponts of the two non-parallel sides. [p. 289]

Midpoint of a Line Segment: The midpoint of a line segment is the point that divides the segment into two equal segments. [p. 13]

N

Noncollinear: Points that are not on the same line are noncollinear. [p. 42]

Noncoplanar: Points that are not on the same plane are called noncoplanar. [p. 42]

O

Obtuse Angle: If the number assigned to the angle is between 90° and 180°, then the angle is called an obtuse angle. [p.11]

Ordered Pair: A point on the x-y plane is an ordered pair of numbers, (x,y). [p. 112]

P

Parabola: A parabola is the set of points on a plane equidistant from a line and a point not on the line. [p. 462]

Parallel: Two lines are parallel if they are coplanar and do not intersect. The symbol ∥ means "is parallel to." [p. 44]

Parallelogram: A parallelogram is a quadrilateral with both pairs of opposite sides parallel. [p. 293]

Perimeter: The perimeter of a figure is the distance around that figure or the sum of the lengths of all the sides. [p. 317]

Perpendicular Lines: Perpendicular lines are lines that intersect to form right angles. Symbol is ⊥. [p. 67]

Point: A point is an ordered pair (x, y) of real numbers. [p. 478]

Point Symmetry: Point symmetry or rotation symmetry is a rotation transformation that maps a figure onto itself. [p. 458]

Polygon: A polygon is a closed plane figure consisting of line segments. [p. 357]

Polygonal Region: A polygonal region is a polygon together with all the points inside the polygon. [p. 403]

Prism: Suppose two congruent polygons are situated in parallel planes so that all line segments joining corresponding vertices are parallel. The union of the two congruent polygonal regions and all the line segments joining corresponding points of the polygons is a prism. [p. 403]

Proportion: A proportion is an equation consisting of two equal ratios. [p. 130]

Pyramid: A pyramid is a polyhedron formed by a polygonal region and all the line segments connecting a point P not in the plane of the region to the points of the polygon that determine the region. [p. 414]

Pyramid
Vertex
Lateral edge
Lateral face
Base

Q

Quadrilateral: A quadrilateral is a closed, plane, four-sided figure. [p. 283]

R

Radius: The radius of a circle is a line segment whose endpoints are the center of a circle and a point on the circle. [p. 367]

Ratio: A ratio is one number divided by another number, or a fraction $\frac{n}{d}$ where d is not zero. [p. 129]

Ray: A ray is a subset of a line consisting of a point A on the line and all points of the line that lie to one side of point A. [p. 4]

Rectangle: A rectangle is a parallelogram with one right angle. [p. 300]

Reflection: A reflection is a transformation such that a line (mirror) is the perpendicular bisector of the segments joining corresponding points. [p. 450]

Regular Polygon: A regular polygon is a polygon that is equilateral and equiangular. [p. 357]

Rhombus: A rhombus is a parallelogram with sides of equal length. [p. 305]

Right Angle: If the number assigned to an angle is exactly 90°, then the angle is called a right angle. [p. 111]

Right Triangle: A right triangle is a triangle with a 90° angle. [p. 100]

Rotation: A rotation is a transformation that maps each point in a figure A to the corresponding points in A' by revolving figure A about a point. [p. 452]

S

Scalene Triangle: A scalene triangle is a triangle with no equal sides. [p. 95]

Secant: A secant is a line segment that intersects a circle in two points and has one endpoint outside and the other endpoint on the circle. [p. 381]

Set: A set is a well-defined collection. [p. 166]

Similar Figures: Two figures are similar if the corresponding angles are equal in measure and the corresponding sides are in equal ratio. [p. 141]

Similar Triangle: Two triangles are similar if:
a. The corresponding sides have the same ratio, *or*
b. two angles of one triangle are equal in measure to two angles of another triangle, *or*
c. one angle of a triangle is equal to the measure of the other triangle and the corresponding sides that include the equal angles are in the same ratio. [p. 135]

Sine: The sine of an acute angle A in a right triangle is the ratio of the length of the leg opposite the angle to the length of the hypotenuse. In symbol, this is written as
$$\sin A = \frac{\text{length of leg opposite angle A}}{\text{length of hypotenuse}}$$ [p. 520]

Skew: Skew lines are lines in three dimensions that do not intersect and are not parallel. [p. 45]

Slope: The number m in $y = mx + b$ is the slope of the line graph. [p. 483]

Sphere: A sphere is the set of all points equidistant from a point called the center. The distance from any point on the sphere to the center is called the radius of the sphere. [p. 431]

Square: A square is a rectangle with sides that are equal in length. [p. 306]

Straight Angle: If an angle measures 180°, then the angle is a straight angle. [p. 11]

Subset: A subset is any set contained in the given set. [p. 167]

Supplementary: Two angles are supplementary if their measures add to 180°. [p. 19]

Symmetry: Symmetry is a transformation that maps a figure onto itself. [p. 455]

T

Tangent: If a line intersects a circle in one point, then the line is tangent to the circle. Tangent $\overleftrightarrow{GI}$ intersects circle O at H. [p. 377]

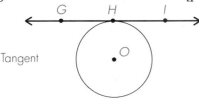

Tangent

Tangent: The tangent of an acute angle A of a right triangle is the ratio of the length of the leg opposite the angle to the length of the leg adjacent to the angle. In symbols, this is written
$$\tan A = \frac{\text{length of leg opposite angle A}}{\text{length of leg adjacent to angle } A}$$ [p. 524]

Transformation: A transformation is a mapping of a geometric figure, preserving shape but not necessarily size. [p. 449]

Translation: A translation is a transformation that slides the figure from one position to another, preserving congruency. [p. 449]

Transversal: A transversal is a line that intersects two or more coplanar lines at different points. [p. 48]

Trapezoid: A trapezoid is a quadrilateral with one and only one pair of parallel sides. [p. 287]

Triangle: A triangle is a figure consisting of three noncollinear points and their connecting line segments. [p. 83]

U

Union: The union of n sets is the set consisting of all the members in the n sets. [p. 168]

Universal Set: The universal set or universe is all the possible members in the well-defined set. [p. 167]

V

Vertex: Suppose that an angle with measure $x°$ has its vertex at the origin with one ray on the positive horizontal axis and a point $P\,(a,\,b)$ on the other ray r units from the origin. Then the sine and cosine of the angle are
$\sin x° = \frac{b}{r} \qquad \cos x° = \frac{a}{r}$ [p. 534]

Vertical Angles: Vertical angles are the opposite angles formed when two lines intersect. [p. 55]

Volume: The volume of a figure is the number of cubic units that the figure contains. [p. 408]

Y

Y-intercept: The y-intercept of a nonvertical line in the coordinate plane is the y-coordinate of the point where the line intersects the y-axis. [p. 486]